C0-ALP-124

Results and Problems in Cell Differentiation

A Series of Topical Volumes in Developmental Biology

8

Biochemical Differentiation in Insect Glands

Edited by W. Beermann

With Contributions by

B. K. Baker W. Baudisch H. M. Blau A. Efstratiadis R. E. Gelinas M. R. Goldsmith U. Grossbach J. R. Hunsley F. C. Kafatos M. Koehler G. D. Mazur P. B. Moore M. R. Nadel J. Nardi M. Paul W. H. Petri J. C. Regier Y. Suzuki J. N. Vournakis A. R. Wyman

With 110 Figures

Springer-Verlag Berlin Heidelberg New York 1977

Professor Dr. WOLFGANG BEERMANN
Max-Planck-Institut für Biologie
Spemann-Straße 34, D-7400 Tübingen

ISBN 3-540-08286-7 Springer-Verlag Berlin · Heidelberg · New York
ISBN 0-387-08286-7 Springer-Verlag New York · Heidelberg · Berlin

Library of Congress Cataloging in Publication Data. Main entry under title: Biochemical differentiation in insect glands. (Results and problems in cell differentiation; 8) Includes bibliographical references. 1. Insects-Development. 2. Insects-Cytology. 3. Glands. 4. Cell differentiation. 5. Biological chemistry. I. Beermann, W. II. Baker, B. III. Series. QH607.R4 vol. 8 [QL495.5] 574.8'761s 77-23423 [595.7'08'761]

Typesetting, printing, and binding: Brühlsche Universitätsdruckerei, Lahn-Gießen
2131/3130-543210

Preface

The majority of studies devoted to animal development traditionally start out from questions of morphogenesis. Of course, visible differentiation, as well as the events leading to it, should ultimately become describable in molecular terms. Nevertheless, even "simple" morphogenetic processes may have a complex biochemical basis which makes it difficult to recognize the key functions involved. This difficulty obviously does not exist in the case of glands, i.e., organs and tissues primarily concerned with, and characterized by, the nature of their products, i.e., one, or a few secretory proteins synthesized in huge quantities. In these systems, when we observe differences between different portions of a gland, or when switches of the synthetic activity occur during development, there is no question as to what we have to look for: we are directly faced with the fact of differential protein synthesis and the problem of its control. Insect glands, in addition, share other significant properties, i.e., the absence of cell division during growth and, concomitantly, the formation of giant cells with polyploid or polytene nuclei. This unique set of peculiarities can be fully exploited only if one compares various representative systems, every one of which, when considered by itself, might appear too exotic to invite generalization.

In the present volume, the editors have endeavored to bring together contributions covering selected insect glands from various points of view, including the developmental, molecular genetic and cytogenetic aspects. Each contribution independently analyzes the particular system under study as comprehensively as possible. The multitude of data and problems thus brought to light should keep developmental biologists busy for some time and prevent early obsolescence.

Summer 1977

W. Beermann
W. Gehring
J. B. Gurdon
F. C. Kafatos
J. Reinert

Contents

The Salivary Gland of *Chironomus* (Diptera): A Model System for the Study of Cell Differentiation

By U. Grossbach. With 30 Figures

Balbiani Ring Pattern and Biochemical Activities in the Salivary Gland of *Acricotopus lucidus* (Chironomidae)

By W. Baudisch. With 13 Figures

Contributors

BAKER, B. K., 50 Melville Avenue, Dorchester, MA 02124, USA

BAUDISCH, WALTER, Zentralinstitut für Genetik und Kulturpflanzenforschung der Akademie der Wissenschaften der Deutschen Demokratischen Republik, 4325 Gatersleben, GDR

BLAU, H.M., Dept. of Pediatrics, Biochemistry and Biophysics, School of Medicine, University of California, San Francisco, CA 94143, USA

EFSTRATIADIS, A., The Biological Laboratories, Harvard University, Cambridge, MA 02138, USA

GELINAS, R. E., Cold Spring Harbor Laboratories, Cold Spring Harbor, NY 11724, USA

GOLDSMITH, M. R., School of Biological Sciences, Dept. of Cellular and Developmental Biology, University of California, Irvine, CA 92664, USA

GROSSBACH, ULRICH, I. Zoologisches Institut, Lehrstuhl für Entwicklungsphysiologie, 3400 Göttingen, FRG

HUNSLEY, J. R., Dept. of Biochemistry, St. Louis University School of Medicine, St. Louis, MO 63104, USA

KAFATOS, F. C., The Biological Laboratories, Harvard University, Cambridge, MA 02138, USA

KOEHLER, M., The Biological Laboratories, Harvard University, Cambridge, MA 02138, USA

MAZUR, G. D., The Biological Laboratories, Harvard University, Cambridge, MA 02138, USA

MOORE, P. B., Dept. of Biochemistry, Oklahoma State University, Stillwater, OK 74074, USA

NADEL, M. R., Dept. of Viral Oncology, Tower Building, Rockefeller University, 1230 York Avenue, New York, NY 10021, USA

NARDI, J., Dept. of Genetics and Development, University of Illinois, Urbana, IL 61801, USA

PAUL, M., Dept. of Biology, University of Victoria, Victoria, British Columbia, Canada

PETRI, W. H., Dept. of Biology, Boston College, Chestnut Hill, MA 02167, USA

REGIER, J. C., The Biological Laboratories, Harvard University, Cambridge, MA 02138, USA

Suzuki, Yoshiaki, Carnegie Inst., Dept. of Embryology, Baltimore, MD 21210, USA

Vournakis, J.N., Dept. of Biology, Syracuse University, Syracuse, NY 13210, USA

Wyman, A.R., Dept. of Biology, Bennington College, Bennington, VT 05201, USA

Differentiation of the Silk Gland
A Model System for the Study of Differential Gene Action

Y. SUZUKI

Department of Embryology,
Carnegie Institution of Washington, Baltimore, MD 21210, USA

I. Introduction

Differential gene transcription is one mechanism which controls development and differentiation in animal cells. One way to study this mechanism is to isolate genes of known function and reconstruct their normal controls (see Brown et al., 1970). Toward this goal Brown and his colleagues have undertaken a long-range study on the genes for ribosomal RNA (rDNA), 5S RNA, and transfer RNA from *Xenopus laevis* (Brown et al., 1970; Dawid et al., 1970; Reeder and Brown, 1970; Roeder et al., 1970; Wensink and Brown, 1971; Brown et al., 1971; Reeder and Roeder, 1972; Brown and Sugimoto, 1973; Reeder, 1973). Meanwhile, it might be possible to search for substances which control their transcription during development. We recognize, however, that even a complete elucidation of the factors regulating the rDNA, 5S DNA, and transfer RNA genes may not necessarily explain how tissue-specific genes are controlled in embryogenesis. The genes for these structural RNAs are not characteristic of any tissue; they are genes controlling "housekeeping" products found in all cells and should be regulated differently from genes which are expressed in a single cell-type such as globin, immunoglobulin and silk fibroin genes.

For this reason we wanted to study the genes which function in a specialized tissue (Suzuki and Brown, 1969; Brown et al., 1970). In order to study any gene, its RNA product must be available in pure form for molecular hybridization. Several groups have been studying isolation methods and characterization of tissue-specific messenger RNAs, e.g. mRNAs for hemoglobin (Lockard and Lingrel, 1969, 1971; Housman et al., 1971; Lane et al., 1971; Mathews et al., 1971; Nienhuis et al., 1971; Aviv and Leder, 1972; Pemberton et al., 1972; Sampson et al., 1972), immunoglobulin (Stavnezer and Huang, 1971; Delovitch et al., 1972; Swan et al., 1972; Brownlee et al., 1973), myosin (Heywood, 1969; Rourke and Heywood, 1972), ovalbumin (Rhoads et al., 1971; Comstock et al., 1972a, b; Means et al., 1972; Palmiter et al., 1972; Rosenfeld et al., 1972a, b; Schutz et al., 1972), lens crystallins

(Berns et al., 1972a, b; Mathews et al., 1972; Williamson et al., 1972), and fibroin (Tanaka and Shimura, 1965; Shimura et al., 1967; Greene et al., 1973). The assay methods for these different mRNAs have always been their ability to direct specific polypeptide synthesis in a cell-free system or recently in frog oocytes (Lane et al., 1971; Berns et al., 1972a). These assay methods document the presence of a mRNA in a functional form, but do not assess its purity. Because of this inability to measure mRNA purity and quantity, we sought a mRNA which could be identified by physical and chemical criteria (Suzuki and Brown, 1969).

With this in mind, we selected the messenger RNA for silk fibroin synthesized by the posterior silk gland of the silkworm *Bombyx mori* (Suzuki and Brown, 1969, 1972). One reason for this selection was the simplicity of the protein, fibroin. Using the same idea Yčas and Vincent in 1959 purified the RNA of the silkworm *Epanaphe (Anaphe) moloneyi* but failed to find any difference between the total RNA from the posterior gland and that from other tissues of the same animal (Yčas and Vincent, 1960). Since their study preceded the messenger RNA concept and since only a few percent of the posterior gland RNA is fibroin mRNA (Suzuki and Brown, 1972; Suzuki and Suzuki, 1974), they measured rRNA and tRNA. The silkworm *Epanaphe* inhabits Nigeria, Africa and its silk fibroin consists essentially of two amino acids, glycine (42 mol %) and alanine (53 mol %), which must thereby qualify it to be the simplest of naturally occurring proteins (see Table 2; Lucas et al., 1958). The posterior silk gland of *B. mori* produces a silk fibroin only slightly more complex than that of *E. moloneyi*. In addition, it has the following advantages for the isolation of fibroin mRNA.

1. *B. mori* eggs are available commercially and the larvae can be raised in any laboratory throughout the year by feeding mulberry leaves or an artificial diet which is now commercially available (see Suzuki and Brown, 1972; Brown and Suzuki, 1973; Suzuki and Suzuki, 1974).

2. The posterior silk glands from a mature *B. mori* larva contain as much as 4 to 5 mg of RNA and 0.2 mg of DNA (Tashiro et al., 1968). These nucleic acids can be labeled in vivo with $^{32}PO_4^{3-}$ or ^{3}H-nucleoside precursors to a high specific activity.

3. The posterior gland synthesizes exclusively a single protein, silk fibroin at the end of larval life (Yamanouchi, 1921; Machida, 1926; Shimura et al., 1958; Shibukawa, 1959; Tashiro et al., 1968).

4. The fibroin has a very simple amino acid composition and sequence (Kirimura, 1962; Kirimura and Suzuki, 1962a, b; Lucas et al., 1957, 1958, 1960, 1962; Lucas and Rudall, 1968). About 90 % of the amino acid residues of fibroin are glycine, alanine, serine and tyrosine, with glycine comprising 45 % of the total and alternating with other amino acids throughout most of the molecule (Lucas and Rudall, 1968). Assignment of codons for these amino acids (Nirenberg et al., 1966) predicts that the fibroin mRNA should be at least 57 % G + C (Suzuki and Brown, 1972), a value distinctly higher than that of bulk DNA or rRNA of *B. mori* (39 % and 48 %, respectively: Neulat, 1967; Suzuki and Brown, 1972). This high G + C content should consist of an unusually high G content of about 40 % or more of the total base residues. In addition the repetitious amino acid sequence of fibroin predicts a simple repeating nucleotide sequence for its mRNA (see Table 7).

Making use of these circumstances the mRNA for fibroin has been isolated from *B. mori* and *Bombyx mandarina* in a pure form and identified by partial sequence analysis (Suzuki and Brown, 1972; Suzuki and Suzuki, 1974). Furthermore, using heighly radioactive mRNA of known purity it has been possible to quantitate its genes in *B. mori* DNA. There is only a single fibroin gene per haploid complement of DNAs from the posterior gland as well as from the animal's carcass and the middle gland (Suzuki et al., 1972; Lizardi and Brown, 1973; Gage and Manning, 1976). This information clearly tells us that specific amplification of fibroin genes is not occurring in the posterior silk gland, and that differential gene transcription and/or post-transcriptional control must account for the specialized synthesis of silk fibroin.

The present review refers to the description of the system including a historical summary, the isolation and identification of the fibroin mRNA, and the quantification of the mRNA and its genes in *B. mori*. Other kinds of silk are described as well as a discussion of the kinds of problems for which the system seems suited to answer in the future.

II. Description of the Experimental System

A. Historical Domestication of Bombyx mori

The main silkworm of interest for this study is the larva of the moth *B. mori* L. which, according to Chinese literature, was domesticated in China for the purpose of silk production more than 4000 years ago. The closest relatives of the domestic silkworm *B. mori* are *Bombyx (Theophila) mandarina* Moore which inhabits China, Korea and Japan, and *Theophila religiosae* Helf. *(T. huttoni)* which inhabits the Himalayan mountainsides (see Tazima, 1964). It seems unlikely that *B. mori* could have originated from *T. religiosae* since the haploid chromosome number of *B. mori* is 28, while that of *T. religiosae* is 31 (Tazima and Inagaki, 1959). From morphological studies of *B. mandarina* Sasaki (1898) proposed that it is the possible ancestor of *B. mori*.

As in other Lepidoptera, the chromosomes in *Bombyx* are almost round in shape, which makes the identification of individual chromosomes including the sex chromosomes very difficult. Linkage mapping has been constructed for more than 100 markers on 21 chromosomes out of the 28 chromosomes in *B. mori* (see Tazima, 1964; Chikushi, 1972; Doira et al., 1972). Kawaguchi undertook an extensive chromosome study of *B. mandarina* harvested in Japan and of F_1 hybrids obtained by the cross of *B. mandarina* with *B. mori* (1923a, b, c, 1928). *B. mandarina* has 27 haploid chromosomes, and the individual hybrids have either 27 or 28 haploid chromosomes in a population ratio of about 1 to 1. He observed that at maturation division of the F_1 hybrids there are only 27 chromosome pairs one of which is composed of one larger *B. mandarina* chromosome and two small *B. mori* chromosomes, and he assumed that one of the *mandarina* chromosomes might have split into two during the history of domestication.

Recently, Astaurov et al. (1959) studied the chromosome number of *B. mandarina* which inhabits the Ussuri district, Manchuria, and found that it has 28

Table 1. Sequences recognized or probable in the silks[a]

Type	Sequence
1.	-Gly-Gly-Gly-Gly-Gly-Gly-
2.	-Ala-Ala-Ala-Ala-Ala-Ala-
3.	-Gly-Ala-Gly-Ala-Gly-Ala-
4.	-Ser-Gly-Ala-Gly-Ala-Gly-
5.	-Ala-Glu(NH_2)-Ala-Glu(NH_2)-Ala-Glu(NH_2)-
6.	- X -Gly-Pro- X -Gly-Pro-
7.	- X -Gly-Ala- X -Gly-Ala- or -Ser-Gly-Ala-Ser-Gly-Ala-

[a] From Lucas and Rudall, 1968.

Table 2. Amino acid compositions of silk fibroins[a]

Type[b]	1	3	4	5	6
X-ray group	Polyglycine	2[a]	1	6	Collagen
	Phymatocera aterrima	*Anaphe moloneyi*	*Bombyx mori*	*Digelansinus diversipes*	*Nematus ribessi*
	Number of amino acid residues per 1000 residues				
Glycine	662	424	445	22	275
Alanine	15	530	293	382	138
Serine	34	3	121	91	110
Aspartic acid	76	5	13	22	58
Glutamic acid	7	3	10	363[c]	98
Arginine	29	1	5	5	40
Tyrosine	57	1	52	21	25
Proline	8	2	3	—	97
Ammonia	31	12	—	350[d]	—

[a] Data were taken from several Tables by Lucas and Rudall (1968), and only major amino acids are listed.
[b] Types shown in Table 1.
[c] and [d] should be taken as glutamine.

chromosomes. It seems, therefore, that there are at least two different varieties of *B. mandarina* inhabiting different regions. Since the most immediate ancestor of the domesticated silkworm is more likely to have had 28 chromosomes, Tazima (1964) emphasized the importance of carrying out an extensive chromosome survey of wild silkworms inhabiting China.

In order to discuss this problem further we should refer to the surprising variety of silk fibroin produced by various organisms. Fibroins are classified into several groups by their drastically different amino acid compositions and by their X-ray diffraction patterns (see Sect. II.C. and Tables 1 and 2). This shows a marked contrast to the conserved homology of amino acid sequence or nucleotide sequence of functionally related proteins or RNAs in different organisms (see Sober, 1970; Sinclair and Brown, 1971). Rudall and Kenchington (1971) speculated how so many different fibroins could have evolved in related organisms of the family

Arthropoda. Fibroins are secreted and have an "extracellular function" such as cocoons, webs and various attachment threads. Once produced they do not have contact with intracellular components of the animal. They proposed that this lack of contact enabled animals to tolerate a variety of these proteins. One important selective influence on the structure of fibroin and the mechanism of its production is the fact that silk does not coagulate within the lumen of the glands (Lucas and Rudall, 1968).

The amino acid compositions of fibroins from *B. mori* and *B. mandarina* are indistinguishable (cf. Table 2 with Kirimura, 1962). Although the amino acid sequence of the latter fibroin has not been analyzed, X-ray diffraction patterns of the two fibroins (Warwicker, 1960) strongly support the idea that they have similar sequences. Recently the fibroin mRNAs from *B. mori* (Suzuki and Brown, 1972) and *B. mandarina* (Suzuki and Suzuki, 1974) have been isolated and partially sequenced. The molecular size of the two RNAs under denaturing conditions as well as their nucleotide sequences are indistinguishable except some possible slight modifications in the third nucleotides of some codons (Suzuki and Suzuki, 1974). These results may be taken as an additional supporting evidence that *B. mandarina* is closely related to *B. mori.*

Comparison of isozymes of haemolymph acid phosphatase and esterase as well as integument esterase from these species have further documented their close evolutionary relatedness (Yoshitake, 1968a, b).

B. Development of the Silk Gland

The silk glands, a pair of thin, long and closed organs (Fig. 1), are derived from the ectoderm, and begin as invaginations of the basal parts of the second maxillae (Nunome, 1937). The invagination takes place in between the longest embryonic stage and the stage of blastokinesis when reversion from dorsal concavity to convexity occurs (Nunome, 1937). The stage has been assigned the number 19 in the developmental table by Takami and Kitazawa (1960). At the beginning of their development the right and left basal regions invaginate separately and then these basal parts unite later at stage 20 (Nunome, 1937). At this stage there are about 30 cells each for the presumptive middle and posterior part of the silk gland, but they are still indistinguishable from each other morphologically (Shigematsu and Takeshita, 1968). The cells continue to divide very rapidly and 24 h after the blastokinesis the characteristic morphological appearance of the silk glands is apparent (stage 23 to 24). Silk-like materials can be observed in the gland at stage 28 at the latest (Nunome, 1937). Cell division in the silk gland is limited to embryonic stage; no cell division takes place during the entire larval period (Ono, 1942) which lasts 4 weeks and includes 5 instars. The number of cells in a single silk gland varies from strain to strain and from gland to gland, being approximately 300 in the anterior, 250 in the middle, and 500 in the posterior part (Nunome, 1937; Ono, 1942; Shimizu and Horiuchi, 1952; Shigematsu and Takeshita, 1968).

When an embryo hatches, the small silk glands which weigh only 10 μg (Ono, 1951) are fully developed morphologically and are capable of spinning a small amount of silk. Larvae at all stages except during moulting periods can spin small

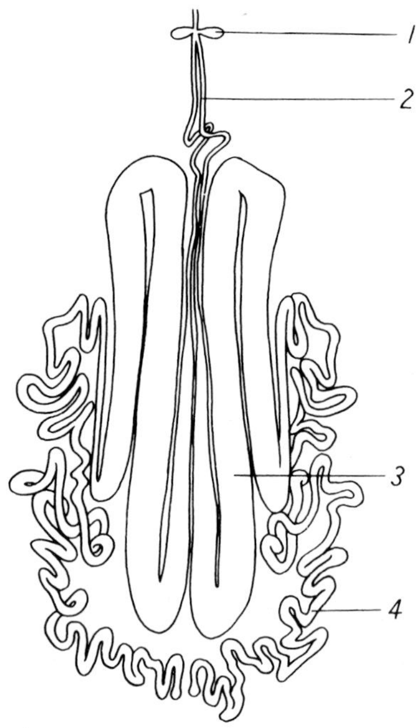

Fig. 1. Half-schematic drawing of silk gland from matured *Bombyx mori* larva: *1*: the Filippi's glands; *2*, *3*, and *4*: anterior, middle and posterior part of silk glands ($\times$ 1.5)

amount of silk which is used to anchor the old skin to the substratum so that the larva can moult easily. This suggests repeated occurrence and cessation of silk synthesis during the feeding stage and moulting stage, respectively. The amounts of liquid silk secreted from the gland cells into the lumen increase during each feeding stage. At the end of the feeding stage, a small amount of the silk is spun out, but the majority is stored in the gland lumen (Akai, 1964). During each moulting stage the majority seems to be converted to components of low molecular weight (Akai, 1965). The rate of fibroin synthesis becomes extremely low during the moulting stage as compared with that during the feeding stage (Akai and Kobayashi, 1966; Prudhomme and Chavancy, 1969). Using another silkworm *Philosamia cynthia ricini* Akai (1971a) reaffirmed that both fibroin and sericin, which were deposited in the lumen of the gland during the fourth instar are dissolved in the early and middle period of the moulting stage. Furthermore, he confirmed that the production of these materials in the gland cell stopped during moulting stage. Recent studies show that fibroin mRNA synthesis occurs as expected during the feeding stages of the third, fourth and fifth instars but cannot be detected during the third and fourth moulting stages (Suzuki and Brown, 1972; Suzuki and Suzuki, 1974, see Sect. III.E.). In contrast, synthesis of rRNA, tRNA and DNA-like RNA continues even in the moulting stages (Suzuki and Suzuki, 1974).

A newly hatched larva which weighed only about 0.45 mg grows into a mature larva of about 4.5 g in four weeks. During this period its silk glands weighing 10 µg at the hatching stage also increase in size about 1.6×10^5 times without cell division

(Ono, 1951). The growth of the larva and its silk glands increase in parallel until the fourth moulting stage.

In the early fourth instar the wet weight of the gland and the amounts of DNA, RNA and protein increase logarithmically for three days until they reach a stationary state (Morimoto et al., 1968). Electron microscopical studies show that in early stages the cytoplasm is filled with free ribosomes and with rough endoplasmic reticulum (ER), first of the lamellar type and then of the vesicular or tubular type. The Golgi apparatus becomes well developed. At the beginning of the moulting stage, most of the ER becomes lamellar in type, and the Golgi vacuoles disappear. Autophagosomes and lysosomes increase markedly (Morimoto et al., 1968). During this moult, the body weight drops slightly because the larvae do not eat. The wet weight of the gland and the amount of RNA increase towards the end of this moulting stage (Morimoto et al., 1968). This increase precedes without interruption the next logarithmic increase in gland weight and RNA content in the fifth instar (Morimoto et al., 1968; Tashiro et al., 1968). During the fifth instar the body weight increases about six times, whereas the posterior silk glands increase their weight about 20-fold (Tashiro et al., 1968).

Tashiro et al. (1968) have undertaken a precise biochemical and ultrastructural analysis of the posterior silk gland during the fifth instar. Under their conditions the fifth instar lasts about 192 h (8 days). In the early stage of the instar (0–96 h) the amounts of DNA, RNA and protein in the posterior glands increase rapidly in accordance with the exponential increase in the wet weight of the gland. However, biosynthesis of fibroin proceeds mainly in the later part of the instar (120–192 h). Electron microscopical observations have shown that a number of free ribosomes exist in the cytoplasm (0–12 h). Rough ER with closely spaced cisternae was also observed. Then the rough ER starts to proliferate rapidly, and at the same time lamellar ER is transformed into vesicular or tubular forms. By the later stage of the instar (120–192 h) the cytoplasm is filled mostly with tubular or vesicular ER. Golgi vacuoles, free vacuoles which are filled with fibroin globules, and mitochondria are also observed. From these observations Tashiro et al. (1968) have concluded that in the early stage of the instar, the cellular structures necessary for the biosynthesis of fibroin are organized rapidly, while in the later stage the biosynthesis of fibroin proceeds at a maximum rate and utilizes these structures effectively. By this time the mRNA for fibroin has been synthesized and accumulated to comprise about 4.4% of the cellular RNA (Suzuki and Brown, 1972; Suzuki and Giza, 1976).

The posterior glands in a matured larva are thin tubes of about 20 cm long with many convolutions, closed at one end and leading to the middle gland at the other end (Fig. 1). Fibroin is excreted as very concentrated (about 30%) aqueous solution from the posterior gland cells into the lumen and is transmitted continuously to the middle gland to be stored there until the larva starts spinning its cocoon (Shimizu et al., 1957; Fukuda and Florkin, 1959; Fukuda et al., 1960).

As shown in Figure 1 the silk glands are divided into three parts, i.e. the anterior, middle and posterior part. The middle and posterior parts are composed of gland cell and tunica propria, while the anterior part is composed of tunica intima, duct cell and tunica propria. The cells in the glands are hexagonal, have huge dendritic nuclei, and each cell extends as much as half the circumference of the gland (Figs. 2 and 3). The largest cells are located in the midsection of the middle silk gland, the site

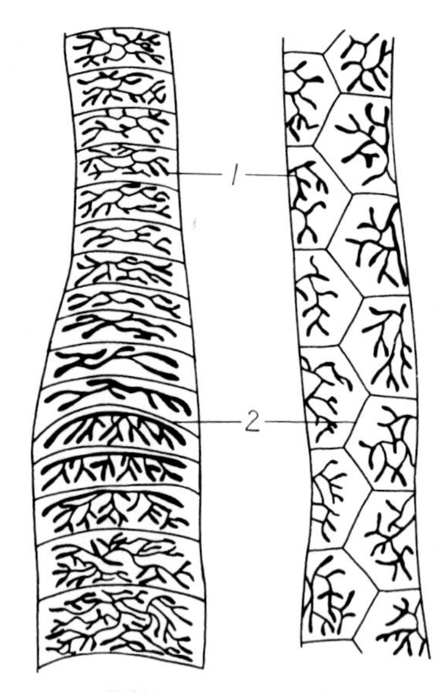

Fig. 2. Half-schematic drawing of the silk glands from a matured *Bombyx mori* larva. *Left:* transitional region from the anterior to the middle part of the silk gland (× 34). *Right:* middle portion of the anterior silk gland seen from the backside (× 34). *1*: ramified nucleus; *2*: protoplasmic membrane. (From Yamanouchi, 1921)

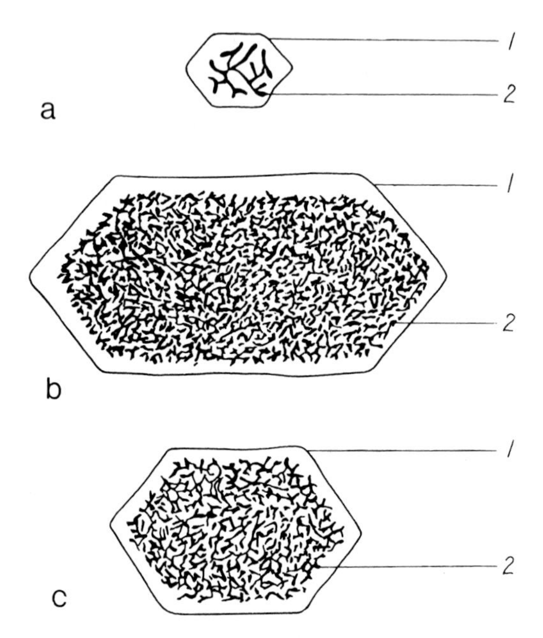

Fig. 3a—c. Half-schematic drawing of the individual silk gland cells of a matured *Bombyx mori* larva. (a) from the anterior region of the anterior silk gland (× 34), (b) from the posterior region of the middle silk gland (× 34), (c) from the posterior silk gland (× 34), *1*: protoplasmic membrane; *2*: ramified nucleus (From Yamanouchi, 1921)

of sericin synthesis (Yamanouchi, 1921; Machida, 1926; Shibukawa, 1959); the individual cells reach dimensions of 4 × 1 mm and can be seen with the naked eye after staining.

The silk proteins, a columnar fibroin and three layer sericins, move anteriorly together, without mixing, into the anterior section of the gland and the spinneret. Just before the spinneret the two glands join together (Fig. 1) and the proteins at the spinneret consist of twin cores of fibroin each sorrounded by cementing layers of sericin.

Cells in the posterior and middle silk glands are the sites of intense protein synthesis during the later stages of larval development and represent one of the most efficient protein-synthesizing systems so far recognized. The rate of fibroin synthesis is more than 60 times that of serum albumin synthesis in chicken liver per wet weight of tissue (Shimura, 1964). The cells are supplied with oxygen by a complex network of fine tracheal tubes and obtain soluble precursors like amino acids through the haemolymph. Fukuda (1960) showed that about 90% of the mulberry leaf protein was incorporated by the insect, and that whereas in the fourth instar only 2 to 3% of the incorporated material was used by the silk gland, in the fifth instar up to 97% of the ingested protein nitrogen was used for silk protein synthesis. The amino acid composition in mulberry leaf protein is quite different from that of fibroin (Katayama, 1915; Fukuda, 1951; Watanabe, 1953). However, glycine, alanine and tyrosine in the haemolymph are absorbed mainly into the posterior silk gland; glutamic acid enters mainly into the middle gland (Fukuda, 1956).

A cocoon spun by a single silkworm weighs about 0.4 g in dry weight, and is mainly composed of fibroin (70–80%) and sericin (20–30%) with trace amount (2–3% in total) of waxy materials, carbohydrates, inorganic substances, and colored materials for which many mutants are known (see Tazima, 1964; Chikushi, 1972).

Thus the main quantitative function of the silkworm and its silk gland after the middle of the fifth instar is the biosynthesis of silk proteins.

Matsuura et al. (1968) have done biochemical and ultrastructural analyses of the cytolytic processes in the silk gland during metamorphosis from mature larva to pupa. Both the middle and posterior parts of the silk glands are degenerated in the pupa. DNA from the middle and posterior gland are fragmented and possibly transferred into the fat body; some of the nucleotides derived from these DNAs are precursors for DNA of adult tissues (Chinzei and Tojo, 1972).

C. Physical and Chemical Characteristics of Fibroin

The production of silk is particularly associated with the phylum Arthropoda. The most notable silk-spinning groups of arthropods are in the classes Insecta and Arachnida. Of the insects the main silk spinners are included in the order Lepidoptera, and it is in this group that we find those species which produce silk of commercial value.

The fibroin synthesized by *B. mori* attracted a good deal of attention from chemists during the late 19th and early 20th centuries, largely because it was one of the few proteins which could be obtained in large quantities and in a comparatively

Table 3. Polypeptide constituents of *B. mori* fibroin[a]

Peptide sequences	Relative abundance (% of total)
Crystalline region	60
Gly-Ala-Gly-Ala-Gly-[Ser-Gly-(Ala-Gly)$_n$]$_8$-Ser-Gly-Ala-Ala-Gly-Tyr	
n is usually 2 and has a mean value of 2	
Amorphous region	40
Gly-Ala-Gly-Ala-Gly-Ala-Gly-Tyr	5.8
Gly-(Gly$_3$, Ala$_2$, Val)-Tyr	5.6
Val-(Gly$_2$, Ala, Asp)-Tyr	4.5
Gly-Ala-Gly-Tyr	—
Gly-Val-Gly-Tyr	—
Ser-Gly-Tyr	—
Gly-Pro-Tyr	—

[a] All data in this Table are taken from Lucas and Rudall (1968).
— Not quantitated.

pure state. The physical and chemical properties of fibroin, however, are not fully elucidated because fibroin is covered with sericin and it is very difficult to obtain pure fibroin without degradation. Furthermore, fibroin once spun is almost insoluble in aqueous solution and even a fresh fibroin solution obtained directly from silk glands occasionally coagulates into an insoluble form.

Extensive analyses on various silk proteins are summarized in several review articles (Shimizu et al., 1957; Lucas et al., 1958; Lucas and Rudall, 1968; Rudall and Kenchington, 1971).

Amino acid composition and partial sequence analysis of fibroins coupled with X-ray diffraction pattern studies have yielded information on the primary structure of various fibroins. Lucas and Rudall (1968) classify fibroins into several groups (see Table 1). Long sequences within silk proteins can be of the types shown in Table 1. Some representative amino acid compositions of various fibroins are listed in Table 2.

The fibroin from *B. mori* is the best characterized. Chymotrypsin digestion of fibroin produces an insoluble or crystalline fraction which comprises about 60% of the protein (Lucas et al., 1957). This polypeptide fraction is composed of only four amino acids arranged in a very simple repeating sequence (Table 3). The main feature of the polypeptide is that glycine residues alternate with three other amino acids almost throughout its entire sequence. The remaining 40% of fibroin, the soluble or amorphous region of the protein contains the minor amino acids. However, even in these polypeptides, glycine residues mostly alternate with other amino acids, and no adjacent glycine residues were found (Lucas et al., 1962; Table 3). This structural information is sufficient to predict most of the nucleotide sequences of fibroin mRNA (see Sect. III.A.).

The exact molecular weight of fibroin has not yet been determined conclusively. Earlier reports gave values ranging from 3.3×10^4 to the order of 10^6 daltons (Coleman and Howitt, 1947; Mercer, 1954; Lucas et al., 1958; Hyde and Wippler, 1962). One reason for these variations was due to drastic treatment (Coleman and

Howitt, 1947; Shimizu et al., 1957; Lucas et al., 1958) of the silk to remove sericin from fibroin.

In order to avoid these harsh procedures recently native fibroin has been extracted directly from the most posterior part of the middle silk gland where fibroin is still not covered with sericin (Tashiro and Ohtsuki, 1970a). Sedimentation analysis of this fibroin have shown that the main sedimenting species present is a 10s component corresponding to 3.7×10^5 daltons in molecular weight (Tashiro and Ohtsuki, 1970a). This value is comparable to 3.5×10^5 (Sprague, 1975), 3.65×10^5 (Sasaki and Noda, 1973a), and 4×10^5 daltons (Lucas, 1966) for fibroin extracted also from the silk gland. Tashiro and Ohtsuki (1970b) then found that the 10s component splits reversibly into smaller polypeptides of 6.8s in the presence of a sulphydryl compound. The 6.8s component was originally estimated to be about 1.7×10^5 daltons, and thus they proposed a two chain model for the structure of fibroin (Tashiro and Ohtsuki, 1970b). However, the situation was found to be more complicated. Addition of dithiothreitol to the fibroin solution resulted in release of minor components leaving a major homogeneous component of 6.8s (Tashiro et al., 1972). The molecular weight of the major homogeneous component determined in the presence of 6M guanidine-HCl and sulphydryl compound was about 3.0×10^5 daltons. Although the value is smaller than the molecular weight of the 10s component, it is definitely larger than one half of the 10s component. From these results they abandoned the two chain model and now suggest that one native fibroin molecule is composed of minor heterogeneous low molecular weight components and one major homogeneous component probably connected together by disulphide bonds (Tashiro et al., 1972). They also suggested that the larger component (6.8s) is either a single polypeptide chain or several polypeptide chains bound by covalent bonds other than disulphide bonds. Fibroin molecules stored in Golgi vacuoles and therefore presumed to be newly synthesized, were observed electromicroscopically (Akai, 1971b). The fibers are helical bundles of about 130 Å in diameter composed of 5 to 7 threads, each 20 to 30 Å thick. He suggested that each thread was a single fibroin subunit but this prediction has not been confirmed.

Sasaki and Noda (1973b) also reduced the native fibroin with 2-mercaptoethanol in a 6M guanidine-HCl or 8M urea solution of pH 7.0, and found that the fibroin molecule was composed of three small molecular weight components (2.6×10^4 daltons) and one large molecular weight component (2.8×10^5 daltons).

However, these complex models (Tashiro et al., 1972; Sasaki and Noda, 1973b) have been confronted with a contradiction reported by Sprague (1975). By sodium dodecyl sulfate polyacrylamide gel electrophoresis she estimated the molecular weight of fibroin as 3.5×10^5 daltons, and neither omission or inclusion of reducing agents (up to 5% 2-mercaptoethanol or 0.1 M dithiothreitol), nor carboxymethylation of the reduced protein altered its relative mobility on gels.

D. Polyploidization in the Silk Gland

As already described in Section II.B., the silk gland grows dramatically during larval development without cell division. There are approximately 250 and 500 cells in a single middle and posterior silk gland, respectively. Gillot and Daillie (1968) measured the increase of DNA in silk gland as it developed from the end of the second instar (0.4 μg per gland) to the end of the fifth instar (150 μg per gland). They also measured the DNA amount separately in the middle and posterior gland in the fifth instar. The middle and posterior glands contain 80 and 70 μg of DNA, respectively, late in the fifth instar when DNA synthesis has terminated. Morimoto et al. (1968) and Tashiro et al. (1968) measured an increase in DNA content in the posterior glands from 1.5 μg per gland at the beginning of the fourth instar to 100 μg per gland at the middle of the fifth instar. The slight difference between the values by Gillot and Daillie (1968) and Tashiro et al. (1968) may be due to differences in strains and rearing conditions of the animals.

This increase has been called "polyploidization", a term which infers that all the DNA is replicated uniformly in the absence of cell division. The gigantic nucleus in the silk gland (Fig. 3) looks like an interphase one and does not show any apparent chromosome structure such as is seen in polytene chromosomes of Diptera. Until the middle stage of the third instar, nuclei are roundish rectangular in shape (Suzuki and Giza, 1976). After this stage each nucleus enlarges by ramifying into all parts of the cell (Fig. 3). Nakanishi et al. (1969) propose a polytene structure for these chromosomes from their electron microscopic observations on the posterior silk glands.

However, recent studies reveal that fibroin genes and ribosomal RNA genes are distributed randomly in the polyploid nucleus (Thomas and Brown, 1976) and that a closer survey of the nucleus after Feulgen staining shows millions of individual chromosome-like granules (Suzuki and Giza, 1975). Furthermore, the approximate number of these granules in a nucleus is a product of the polyploid number and the haploid chromosome number (Suzuki and Giza, 1975). Therefore, I prefer the following model to the polytene structure; multiple endomitotic replications of individual chromosomes are followed by random segregation of the replicated chromosomes, each individual chromosome maintaining its independence.

A preliminary study by Feulgen microspectrophotometry has revealed that a single posterior gland cell nucleus on the 6th day of the fifth instar has 400000 times more DNA than a spermatogonium (Nakanishi et al., 1969). This value suggests that DNA replication has occurred about 17 or 18 times during the growth of the silk gland, about nine of these replications occur between the end of the second instar and the end of the fifth instar (calculated from the results by Gillot and Daillie, 1968). Recent measurement of the DNA content of individual spermatids by Rasch (1974) gave a value of 0.52 pg. Gage (1974a) has derived the same value for genome size by DNA renaturation kinetics method (Britten and Kohne, 1968). From this value and the average DNA content of a single posterior gland cell nucleus (0.2 μg, calculated from Tashiro et al., 1968) a similar value of 400000 ploidy is calculated.

Although polyploidization is assumed to be a uniform replication of the DNA this had never been proven. Examples of differential replication have been demonstrated. A selective amplification of the DNA encoding the large rRNAs takes place during oogenesis in a variety of animals (Brown and Dawid, 1968; Gall et al., 1969; Lima-de-Faria et al., 1969). Another "amplification" is also known in specific bands of polytene chromosomes in Sciaridae (Crouse and Keyl, 1968;

Rudkin, 1972). In *Drosophila*, most of the repetitive, heterochromatic DNA is excluded from replication during polytenization of the salivary gland chromosomes (Dickson et al., 1971; Gall et al., 1971).

The polyploidization process in the silkworm has been studied by Gage (1971, 1974b) by comparison of the DNA of mature silk glands and carcass. It is not known whether carcass cells are diploid, but they are at least three orders of magnitude smaller than posterior or middle silk gland cells in nuclear volume and DNA content. DNAs from the middle and posterior silk glands as well as the carcass were judged to be the same by comparison of their thermal denaturation profiles and buoyant densities in CsCl. The Tm of all three DNAs corresponds to an average G + C content of 39% which agrees with that reported for *B. mori* embryo and pupa DNA (Neulat, 1967).

These tests are not sensitive enough to detect subtle sequence difference. If one class of nucleotide sequences would become a larger fraction of the DNA, other classes must necessarily be diluted. The genes for rRNA, 4s RNA, and fibroin mRNA were tested for their abundance in DNAs from the middle silk gland, posterior silk gland, and carcass, by molecular hybridization, and no difference was found between the DNAs (Gage, 1971, 1974b; Suzuki et al., 1972).

The kinetics of silkworm DNA reassociation indicates that the repetitive fraction constitutes less than 20% of the total DNA (Gage, 1971). Using labeled cRNA transcribed in vitro from DNAs from different tissues a change in as little as 5% of the repetitive fraction would be detected. In other words any DNA sequence which was amplified differentially in a tissue should be detectable if it constitutes more than 1% of total DNA. The results clearly show that little, if any, change has taken place in the sequences within the repetitive fraction during polytenization (Gage, 1971). Thus highly polyploid DNA in the posterior and middle silk gland is not demonstrably different from carcass DNA in the relative abundance of its nucleotide sequences.

Polyploidization cannot itself account for control of fibroin gene transcription since the silk materials are synthesized even in a newly hatched larva which has a low degree of ploidy in its posterior silk gland. Furthermore, the fibroin mRNA synthesis is detected in the third and fourth instar when the degree of ploidy in the posterior silk gland is still low (see Sect. III.E.).

E. Functional Adaptation of the tRNA Population to Fibroin Synthesis in the Posterior Silk Gland

The regulation and economy of cells depend partly on the relationship between the tRNA population and the amino acid composition of the proteins synthesized (Yamane, 1965; Sueoka and Kano-Sueoka, 1970). This relationship could best be studied in some differentiated animal cells which synthesizes mainly one protein. The posterior silk gland would seem ideal for such a study since it synthesizes its own "housekeeping" proteins during the first half of the fifth instar, and in the second half devotes its protein synthesizing apparatus to the production of fibroin (Tashiro et al., 1968) which has an extremely biased amino acid composition (Table 2).

Table 4. Acylation of tRNAs in the posterior silk gland of *B. mori* during the fifth instar[a]

Amino acid	pmol of tRNA on 4th day	pmol of tRNA on 6th and 8th day
Glycine	14.0	120.0
Alanine	12.5	64.3
Serine	23.0	39.6
Tyrosine	14.5	27.7
Aspartic acid	9.0	10.6
Leucine	5.0	5.0
Threonine	8.0	12.6
Glutamic acid	5.0	8.6

[a] Simplified data taken from Garel et al. (1970). Acylations were performed with 20 μg tRNA, at 30° C for 20 min. These reactions were done once with silk gland synthetases followed by two or three further incubations with rat liver synthetases.

Table 5. Quantity in pmol of each isoacceptor tRNA in a posterior silk gland of *B. mori* at the 4th and the 8th day of the fifth instar[a]

Species	iso-tRNA^{Ala}		iso-tRNA^{Gly}	
	4th	8th	4th	8th
I	1[b]	21[b]	40	325
II	42	330	12	432
III	11	180	9	118

[a] Cited from Garel et al. (1971a).
[b] Calculated from the data shown in Table 1 by Garel et al. (1971a).

Matsuzaki (1963, 1966) and Onodera and Komano (1964) observed a remarkable correlation between the amino acid composition of fibroin and the extent of acylation of each tRNA in an in vitro system including tRNAs and aminoacyl-tRNA synthestases obtained from the posterior gland during the late fifth instar. Some of the synthetases, however, were found to be rather unstable. Therefore, Garel et al. (1970) fortified this in vitro acylation system supplying saturation amounts of the enzymes from rat liver and found a stricter correlation between the amino acid composition of fibroin and the extent of tRNA acylation (see Table 4). The correlation was not observed in the first half of the fifth instar (Garel et al., 1970). These results together with recent results by Chavancy et al. (1971) and Garel et al. (1971a, b) strongly suggest that the tRNA population of the posterior gland during the late fifth instar reflects the amino acid composition of fibroin.

Garel et al. (1971a) fractionated the aminoacyl-tRNA on MAK columns and estimated the relative amounts of each isoacceptor tRNA acylated for the four major amino acids in fibroin, that is, glycine, alanine, serine and tyrosine. Some of their results are cited in Table 5. Based on these results they propose the following scheme: the high rate of translation of the fibroin mRNA requires a correspondingly high level of certain isoacceptor tRNAs which recognize synonym codons. These isoacceptor tRNAs might regulate fibroin synthesis by a "modulation" role such as the one proposed by Stent (1964) and others. From this point of view, the species $tRNA_{I}^{Gly}$ and $tRNA_{II}^{Ala}$ (Table 5) can be regarded as the common isoacceptor tRNAs involved in general protein synthesis. The situation is different for the other three isoacceptor tRNAs, $tRNA_{II}^{Gly}$, $tRNA_{III}^{Gly}$ and $tRNA_{III}^{Ala}$. Thus they could exist in the tRNA population at low concentrations during the growth period (the first half of the fifth instar) participating in non-specialized protein synthesis as well as in a constant low level synthesis of fibroin. In the specialized state (the second half of the instar), however, they increase greatly in amounts. They might be specific iso-tRNAs which preferentially decode the fibroin mRNA.

Recently Kawakami and Shimura (1974) purified gly-tRNAs from the posterior silk gland of the late fifth instar into three fractions ($tRNA_{1}^{Gly}$, $tRNA_{2-1}^{Gly}$ and $tRNA_{2-2}^{Gly}$) which were 70 to 90% pure, and ala-tRNAs into 6 fractions, 4 of which were 70 to 90% pure. These three isoaccepting $tRNAs^{Gly}$ were tested for their in vitro binding activity to ribosomes in the presence of triplets. The $tRNA_{1}^{Gly}$ responded to GGA and GGG, and both $tRNA_{2-1}^{Gly}$ and $tRNA_{2-2}^{Gly}$ recognized GGU and GGC (Kawakami and Shimura, 1976). The exact correspondence between each iso-tRNA by Kawakami and Shimura (1974, 1976) and Garel et al. (1971a, b) is not known yet. Kawakami and Shimura (1976) confirmed the previous observations that in a cell-free system from the posterior gland of the late fifth instar individual ^{14}C-amino acids were incorporated into protein from ^{14}C-aminoacyl-tRNAs in the same proportion that they exist in fibroin which is quite different from that of general proteins of the gland. Participation of three species of iso-$tRNAs^{Gly}$ in the fibroin synthesis was analyzed in detail in this cell-free system. The ratio of glycine incorporated into fibroin from gly-$tRNA_{1}^{Gly}$ to that from gly-$tRNA_{2-1}^{Gly}$ and gly-$tRNA_{2-2}^{Gly}$ was 1.0 to 1.4. This observation may be comparable with the 1.0 to 1.4 ratio of GGA to GGU codons found in the fibroin mRNA (Suzuki and Brown, 1972). In a separate experiment (Kawakami and Shimura, 1976) the fibroin molecule which was synthesized in the cell-free system using ^{3}H-gly-$tRNA_{1}^{Gly}$ and ^{14}C-gly-$tRNA_{2-1}^{Gly}$ was digested with chymotrypsin and separated into two fractions, that is, the crystalline part and the amorphous part. The amorphous part was further fractionated into A1 and A2 fractions. The molar ratio of ^{3}H to ^{14}C were 1.0 to 1.0 in the crystalline part, 1.0 to 1.4 in A1 and 2.3 to 1.0 in A2. These results suggest that there is a bias in the distribution of glycine synonym codons in the fibroin mRNA. This problem could be solved by complete sequence analysis of the fibroin mRNA.

Recently Mazima and Shimura (1972) reported that there is no selective amplification of the genes for $tRNAs^{Gly}$ in the posterior silk gland during the fifth instar. The genes for $tRNAs^{Gly}$ and $tRNAs^{Leu}$ comprise about 3.0 to 3.5×10^{-4}% of the genome either at 24 h or at 168 h of the fifth instar.

In the posterior gland during the fifth instar the genes for fibroin and some of the genes for isoaccepting tRNAs are apparently coordinated in their expression. How are they regulated, and do they have a common operator or promotor sequence?

Table 6. Characteristics of *B. mori* fibroin mutants

Mutant name	Genetic natures	Origin	Dominancy over the normal allele	Location of mutation
Rayo (naked pupa) [later called Nd or Nd(1)]		Spontaneous mutation	Dominant	Not mapped
Nd(2)		Spontaneous mutation	Dominant	Not mapped
Nd_1		Spontaneous mutation	Dominant	Not mapped
Nd_2		Spontaneous mutation	Dominant	Not mapped
Nd-s		Spontaneous mutation	Incompletely dominant	14th chromosome 0.0
flc		Spontaneous mutation	Recessive	3rd chromosome 49.0
Rayo B [named Nd b]		Spontaneous mutation	Incompletely dominant	Not mapped

Allelism between mutants	Phenotype	% of normal fibroin spun	defect found	References
Nd(2) (Nd_1, Nd_2)	Failure of cocoon formation or formation of unusual cocoons which are composed mostly of sericin	Not known	Remarkable growth inhibition only in posterior silk glands. Fibroin accumulation in the lumen is not recognizable.	Nakano (1937, 1951), Gamo (1973a, b)
Nd(1) (Nd_1, Nd_2)	Failure of cocoon formation or formation of unusual cocoons which are composed mostly of sericin	Not known	Remarkable growth inhibition only in posterior silk glands. Fibroin accumulation in the lumen is not recognizable.	Gamo (1973a, b)
Nd_2 (Nd(1), Nd(2))	Failure of cocoon formation or formation of unusual cocoons which are composed mostly of sericin	≦0.2	Remarkable growth inhibition only in posterior silk glands. Fibroin accumulation in the lumen is not recognizable.	Hashimoto (pers. comm.), Watanabe (1959), Machida (1970, 1972)
Nd_1 (Nd(1), Nd(2))	Failure of cocoon formation or formation of unusual cocoons which are composed mostly of sericin	Not known	Remarkable growth inhibition only in posterior silk glands. Fibroin accumulation in the lumen is not recognizable.	
None	Failure of cocoon formation or formation of unusual cocoons which are composed mostly of sericin	≦0.7	Remarkable growth inhibition only in posterior silk glands. Fibroin accumulation in the lumen is not recognizable.	Horiuchi et al. (1963, 1964), Horiuchi and Ohi (1966), Horiuchi and Chikushi (1969), Chikushi (1972)
None	Cocoon weight is about one-third of normal and the cocoon looks like flimsy paper.	Not known 20%?	Posterior gland is considerably smaller than that of normal.	Doira (1970), Chikushi (1972), Iijima (1972)
Not known	Failure of cocoon formation	Fluctuate	Constrictions in the middle silk glands. Fibroin is in their silk gland lumen but is disturbed in its spinning.	Nakano (1937, 1951)

F. Mutants for Silk Fibroin Production

A variety of "Rayo (naked pupa) " mutants are known for *B. mori*. They are characterized by low or no fibroin synthesis resulting in an unusual cocoon consisting mainly of sericin or failure of cocoon formation. Thus, phenotypically larvae of these mutants often metamorphose into naked pupae. Table 6 summarizes their genetic natures.

In 1933 Nakano found the first naked pupa mutant, Rayo which was named Nd or Nd(1) later, and he also found another mutant, Rayo B (Ndb) in 1935 (Nakano, 1937, 1951). The Ndb is rather irrelevant here to describe because it carries full of fibroin in its silk gland lumen, but is disturbed in the spinning of this fibroin. It reveals constrictions very often in the middle silk glands (Nakano, 1951). Nd(2) is derived from the Nakano's stock but possibly an independent strain from Nd(1) (Gamo, pers. comm.). Another mutant Nd_1 was found independently by Hashimoto (Gamo, pers. comm.), and during the course of study it was subdivided into two types, Nd_1 and Nd_2 (Machida, 1970, 1972). However, they are almost indistinguishable for these days (Chikushi and Machida, pers. comm.). From all the genetic data listed in Table 6, Nd(1), Nd(2), Nd_1, and Nd_2 are allele and indistinguishable from each other in their characteristics. The F_1 hybrids between these mutants and normal show mutant character in their phenotype indicating dominance of these mutations over the normal allele (Nakano, 1951; Machida, 1970, 1972). Defect is found only in posterior silk glands which are remarkably inhibited in their growth, and fibroin accumulation in the lumen is not recognizable. The cell numbers of posterior glands of the mutants are almost the same as those of normal (Nakano, 1951; Machida, 1970, 1972). The growth inhibition in posterior glands occurs only in the later half of larval life. The posterior gland cells are slightly degenerated in the fifth instar (Machida, 1970; Iijima, 1973). In the cells some unusually large blocks of fibroin-like materials are deposited (Machida, 1970, 1972; Iijima, 1972).

From a Burman strain at Thanghpre another Nd mutant, Nd-s was derived by spontaneous mutation (Horiuchi et al., 1963). Its characteristics are almost the same as the above Nd mutants, but it is not allelic with them (Horiuchi et al., 1964). The gene for Nd-s has been mapped on the 14th chromosome at location 0.0 (Horiuchi and Ohi, 1966; Horiuchi and Chikushi, 1969; Chikushi, 1972) and the Nd-s reveals incomplete dominancy in heterozygote (Horiuchi et al., 1963; Horiuchi and Ohi, 1966). The cell numbers in posterior glands of Nd-s are less than those of normal (Horiuchi et al., 1964).

A recessive mutant has been found by Doira (1970). The mutant, flc named after its cocoon appearance like flimsy paper, is an intermediate one which produces a thin cocoon weighing about one-third of normal cocoon. The cocoon contains both fibroin and sericin. The flc is not allelic with any mutants known, and has been mapped on the 3rd chromosome at location 49.0 (Doira, 1970; see Chikushi, 1972).

Recently quantitative measurements of total cellular RNA and fibroin mRNA in posterior glands in these mutants have been done (Suzuki and Suzuki, 1974) and are described in Section III.E.

III. The Messenger RNA for Fibroin

A. Prediction of the Sequence of the Fibroin mRNA

A summary of the amino acid composition and some of the more common polypeptide sequences of fibroin is shown in Tables 2 and 3. Chymotrypsin digestion of *B. mori* fibroin produces an insoluble (or crystalline) fraction which comprises about 60% of the protein (Lucas et al., 1957). This fraction has been sequenced completely (Lucas et al., 1957) and is composed of repeats of the sequence shown in Table 3. Although some variations are seen in the amorphous region of fibroin which remains in the supernate after the chymotrypsin digestion, the sequences of this region is very similar to that of the crystalline region (Table 3; Lucas et al., 1962).

If we apply the codons, GGX (glycine), GCY (alanine), UCZ or AG^{U}_{C} (serine) and UA^{U}_{C} (tyrosine) (Nirenberg et al., 1966), for the polypeptide of the crystalline region, the sequence of its mRNA can be predicted (Table 7). Assuming that the third nucleotide of these codons (represented here as X, Y, and Z) is neither a G nor C residue, a minimum G + C content of the predicted nucleotide sequence should be about 60%. This value should be 2 or 3% lower for the entire mRNA molecule, due to codons for minor amino acids within the amorphous region. Therefore, the mRNA should have a minimum G + C content of about 57% which is higher than the G + C content of any RNA component of the cell except tRNA. In addition the fibroin mRNA should be distinguished further by an unusually high G content of about 40% or more.

B. Isolation and Purification of Fibroin mRNA

Generally larvae were injected with 0.5 to 2 mCi of $^{32}PO_4^{3-}$ in the middle of the fifth instar. Since a labeling time of 30 to 90 min did not give detectable amounts of a high GC component, a search was carried out for a stable RNA generally 6 to 24 h

Table 7. Messenger RNA sequence which codes for the predominant polypeptide of *B. mori* fibroin[a]

$$\mathrm{GGX\text{-}GCY\text{-}GGX\text{-}GCY\text{-}GGX\text{-}}\left[\begin{array}{l}\mathrm{UCZ}\\ \mathrm{or\text{-}GGX\text{-}(GCY\text{-}GGX)_2}\\ \mathrm{AG^{U}_{C}}\end{array}\right]_8\text{-}\begin{array}{l}\mathrm{UCZ}\\ \mathrm{or\text{-}GGX\text{-}GCY\text{-}GCY\text{-}GGX\text{-}UA^{U}_{C}}\\ \mathrm{AG^{U}_{C}}\end{array}$$

Codon assignments (Nirenberg et al., 1966)

Glycine	GGX
Alanine	GCY
Serine	UCZ or AG^{U}_{C}
Tyrosine	UA^{U}_{C}

X, Y, or Z can be any ribonucleotide

This predicted sequence was deduced by assigning codons to the polypeptide of the crystalline region of fibroin (see Table 3).
[a] (From Suzuki and Brown, 1972).

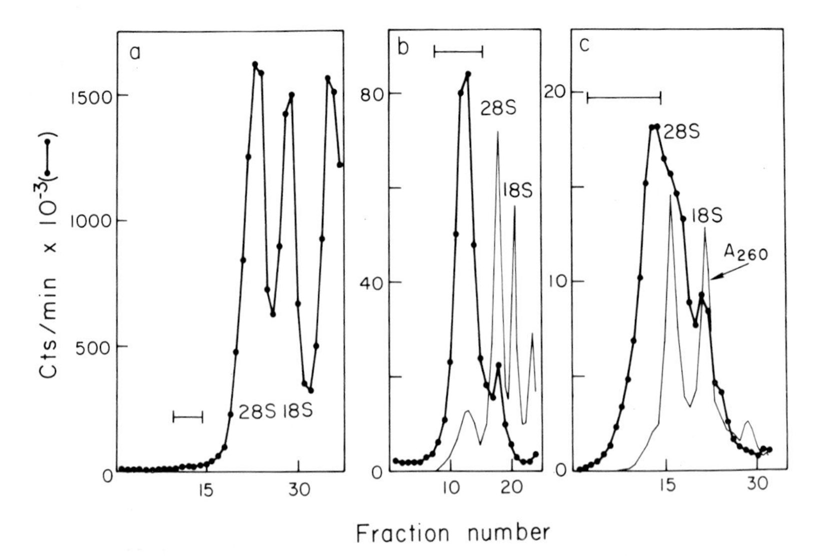

Fig. 4a—c. Purification of fibroin ^{32}P-labeled mRNA from the posterior silk gland of *B. mori* larvae. (—●—●—), Cts/min; (———), optical density at 260 nm. (a) First-round sucrose gradient centrifugation. The RNA designated by the bracket was pooled for a second gradient. (b) Second-round sucrose gradient centrifugation. About 150 µg of *X. laevis* rRNA was added as a marker. The rapidly sedimenting absorbance peak is due to fibroin mRNA. The RNA within the bracket was pooled for the formamide/sucrose gradient step. (c) Fractionation of pooled fibroin mRNA from (b) on a 70% foramide, 4.6 to 22% sucrose gradient. About 150 µg of intact *X. laevis* rRNA was added as the optical density marker. The bracketed ^{32}P-RNA was pooled for hybridization studies (Suzuki et al., 1972)

after the isotope injection. The posterior silk gland, middle silk gland and the animal's carcass were homogenized in the presence of SDS, digested with pronase, the ^{32}P-RNAs were extracted with SDS-phenol and fractionated by sucrose gradient centrifugation (Suzuki and Brown, 1972). A broad radioactive peak ranging from 40 to 65 s was visible in the RNA from the posterior silk gland (Suzuki and Brown, 1972; Fig. 4a). This RNA, as well as the RNAs from equivalent regions of the other two gradients (not shown here), was recovered and applied to a second sucrose gradient (Fig. 4b). The rapidly sedimenting RNA from the posterior silk gland behaved as a more homogeneous RNA of about 48s (Fig. 4b) than that purified from the other two tissues, and it had a G + C content of about 59% with a G content of 40% (Suzuki and Brown, 1972). The equivalent sized but more heterogeneous RNAs from the carcass and the middle silk gland were DNA-like in their base compositions, containing 39% and 43 to 45% G + C, respectively. Thus the rapidly sedimenting ^{32}P-RNA from the posterior gland had the expected base composition, which was different from that of rRNA and DNA-like RNA. Although its GC content is similar to that of 4s RNA, its very high G content and rapid sedimentation constant distinguish the two.

Traces of rRNA are present in a fibroin mRNA preparation from middle or late fifth instar and are removed by a third sucrose gradient centrifugation in the presence of formamide (Fig. 4c). Under the denaturing condition secondary

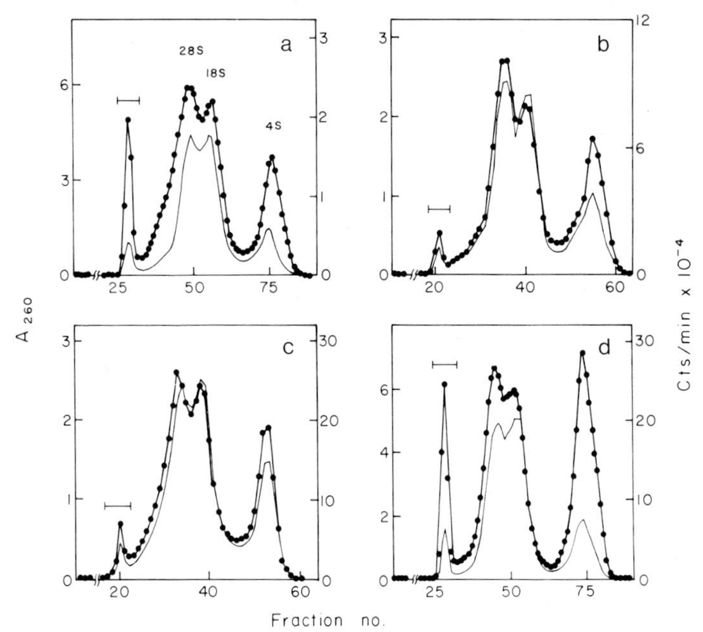

Fig. 5a—d. Large scale purification of fibroin mRNA from posterior silk glands on Bio-Gel A-50m. Mixtures of ^{32}P-labeled and unlabeled RNA from posterior silk glands obtained from *B. mori* larvae of various stages were applied to a Bio-Gel A-50m column (5 × 87 cm), 20 ml fractions (a) and (d) or 28 ml fractions (b) and (c) were collected, and half of each fraction was counted. (a) About 66 mg of RNA obtained from 800 larvae of fourth feeding stage. (b) About 40 mg of RNA from 350 larvae of fourth moulting stage. (c) About 50 mg of RNA from 120 larvae of the early fifth instar (24 h after the fourth ecdysis). (d) About 72 mg of RNA from 30 larvae of late fifth instar (6 days after fourth ecdysis). ———, A_{260}; —●—, ^{32}P cts/min (Suzuki and Suzuki, 1974)

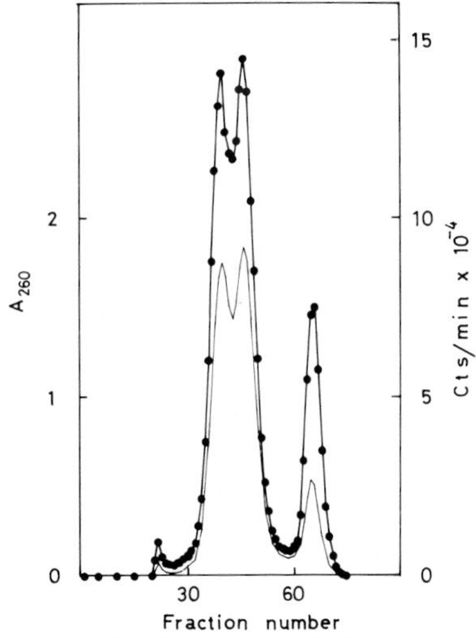

Fig. 6. Purification of fibroin mRNA from *B. mori* fibroin mutant, Nd(1). About 5 mg of RNA from the posterior glands were applied to a 2.6 × 58 cm Bio-Gel A-50m column (Suzuki and Suzuki, unpubl.)

interactions in the mRNA are eliminated and the rRNA contaminant is released (Suzuki et al., 1972).

The method described above for fibroin mRNA isolation involved three cycles of sucrose gradient centrifugation, and was limited to less than 12 mg of total RNA in any single preparation. A much simpler method with the large capacity was the use of an agarose bead column (Bio-Gel A-50 m) which fractionates molecules of 1×10^5 to 50×10^6 daltons. A large Bio-Gel column (5×87 m) can fractionate more than 130 mg of bulk RNA rapidly and yields several mg of a mRNA fraction which is as much as 80 to 90% pure. The fibroin mRNA elutes as a discrete peak at the exclusion/inclusion boundary of the column (Fig. 5a and 5d) even when it comprises only 0.3% of the cellular RNA (Fig. 6).

C. Identification of Fibroin mRNA by Partial Sequence Analysis

The definitive identification of the isolated RNA as fibroin mRNA has been obtained by comparison of oligonucleotides from RNase T_1 digestion with those predicted from the proposed structure of fibroin mRNA (Table 7). The predicted oligonucleotide pattern (Table 8) is based on the crystalline part of fibroin. It was assumed that the third nucleotide (X, Y, or Z) of the three principal codons is never a G residue. This assumption was itself tested by analysis of the oligonucleotides produced by T_1 and pancreatic RNase digestion. Two different oligonucleotide patterns are predicted, depending upon the serine codon (Table 8; Suzuki and Brown, 1972).

The RNA with high G+C content was digested with RNase T_1, and the oligonucleotides were fractionated on a DEAE-Sephadex column (Fig. 7a). The main striking feature is the virtual absence of tetranucleotides, very much in contrast to the situation in a digest of 40s rRNA precursor (Fig. 7b). This fact and the abundant presence of pentanucleotides is prima facie evidence for a pattern predicted for fibroin mRNA with UCZ as the major serine codon (Table 8).

Each size-group of oligonucleotides in Figure 5a was fractionated further on a Dowex 1-X2 column, and the individual components were sequenced (Table 8). The mononucleotide fraction was Gp as expected from the known specificity of RNase T_1. The large amount of Gp attests to the frequency with which G residues are adjacent in this RNA, and is mostly derived from the second nucleotide of the glycine codon. The principal dinucleotide should contain the third nucleotide of the glycine codon (X) and the first nucleotide of the alanine codon (G). The major dinucleotide found was UpGp, and the next frequent one was ApGp (Table 8). The result suggests that the major glycine codon is GGU, GGA being second in frequency.

The principal trinucleotide which comprises more than 60% of the total trinucleotide fraction, is CpUpGp (Table 8). The major trinucleotides should contain the second (C) and third (Y) nucleotides of the alanine codon followed by the first nucleotide of the next glycine codon (G). Therefore, we concluded that the major alanine codon is GCU.

The major pentanucleotide, which comprises about half of the total pentanucleotide fraction, is UpUpCpApGp (Table 8). The predicted major pentanucleotide

Table 8. Oligonucleotides predicted and found after RNase T^{1} digestion of fibroin messenger RNA

Percentage of total nucleotide residues					
Predicted[a]			Found		
Sequence	Serine codon UCZ	Serine codon AG^{U}_{C}	Sequence	each class of oligonucleotide[b]	Each component[c]
Mono-					
Gp	16.4	16.4	Gp	18.8 ± 2.7	18.8
Di-				18.7 ± 3.7	
XpGp	21.5	21.5	UpGp		9.6
			ApGp		6.9
$^{U}_{C}$pGp	0	10.2	CpGp		2.2
Tri-				27.7 ± 1.2	
CpYpGp	33.9	33.9	CpUpGp		17.2
			(Cp, Ap)Gp		4.2
XpApGp	0	15.3	CpCpGp		2.0
			ApApGp		1.6
			(Up, Ap)Gp		1.5
			———		1.2
Tetra-	0	0	———	3.7 ± 0.7	
Penta-				17.2 ± 1.5	
XpUpCpZpGp	25.4	0	UpUpCpApGp		8.3
XpUpAp$^{U}_{C}$pGp	2.8	2.8	(2 Ap, Up, Cp)Gp[d]		3.6
			(2 Cp, Up, Ap)Gp[e]		1.8
			———		3.5
Hexa-	0	0	———	4.5 ± 0.8	
≧Hepta-	0	0	———	9.5 ± 2.5	

(———) Not analyzed further.

[a] Predicted oligonucleotides from the sequences of mRNA which code for the crystalline region only (see Table 6). It is assumed that X, Y and Z are never G residues.

[b] Average of five analyses from DEAE-Sephadex chromatography as in Fig. 5a is shown with standard deviation.

[c] Relative abundance of each oligonucleotide in the total digest. Each class of oligonucleotides was subfractionated on Dowex 1-X2.

[d] Most of this pentanucleotide (>80%) was (ApUp, ApCp)Gp, but it contains some other position isomer(s).

[e] About half of this pentanucleotide was (2 Cp, Up)ApGp, while the remainder was another position isomer(s).

(From Suzuki and Brown, 1972).

consists of the serine codon (UCZ), preceded by the last nucleotide of the glycine codon (X) and followed by the first nucleotide of the next glycin codon (G); therefore, the predominant serine codon is UCA. The analysis also shows that the glycine codon preceding serine codon is more frequently GGU than GGA or GGC.

These results, as well as the analyses of oligonucleotides produced by pancreatic RNase digestion of the mRNA (see Suzuki and Brown, 1972), identified the RNA as fibroin mRNA.

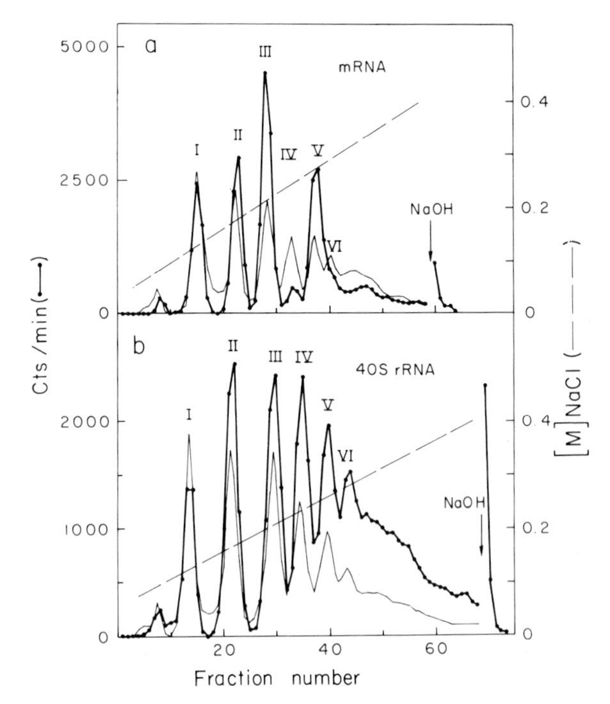

Fig. 7a and b. Fractionation of oligonucleotides produced by RNase T_1 digestion of the mRNA and the 40s rRNA precursor. (a) The mRNA was digested with RNase T_1 together with 0.5 mg of *E. coli* sRNA as carrier, and the oligonucleotides were fractionated on a DEAE-Sephadex A25 column. (b) The 40s rRNA precursor was digested with RNase T_1 together with 0.5 mg of *E. coli* sRNA and fractionated on a DEAE-Sephadex A25 column. *Arrows* position where 1 N NaOH was added to elute residual material. *Roman numerals:* chain length of oligonucleotides. *First unnumbered peak:* cyclic GMP. (———), O.D. at 254 nm as recorded automatically by an ISCO spectrophotometer; (—●—●—), Cts/min; (— —), concentration of NaCl [M] (Suzuki and Brown, 1972)

D. A Purity Calibration of Fibroin mRNA Preparations

Fibroin mRNA fraction obtained in the heavy regions of the sucrose gradient shows about 48, 56, and 60% GC for the samples from the early, middle and late fifth instar, respectively. A pure fibroin mRNA preparation was found to contain about 60% GC (Suzuki and Brown, 1972). Therefore, high molecular weight RNA fractions from the early fifth instar are not pure mRNA; only in the middle or late fifth instar does fibroin mRNA predominate over other DNA-like RNAs. For this reason it was necessary to develop an additional step which would measure the fibroin mRNA content in the excluded fractions from Bio-Gel or the high molecular weight fractions from sucrose gradient.

It is possible to estimate the approximate purity of fibroin mRNA from the oligonucleotide profile after RNase T_1 digestion. The amount of tetranucleotides in the digest is especially important. Preliminary observations indicated that the higher the GC content of a fibroin mRNA preparation, the lower the proportion of

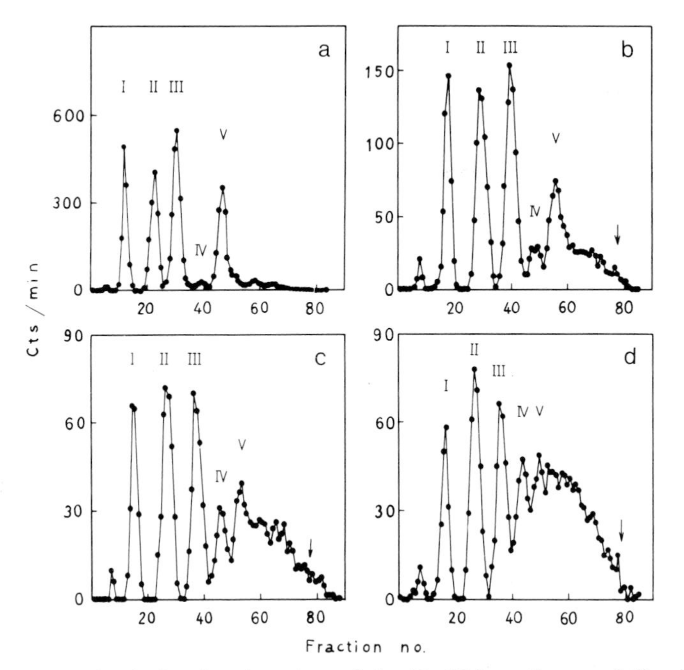

Fig. 8a—d. DEAE-Sephadex fractionation of the T_1 RNase digests of fibroin mRNA, precursor rRNA and mixtures of the two. The T_1 RNase digests of (a) fibroin mRNA obtained from late fifth instar larvae and purified by two cycles of Bio-Gel A-50 m columns and one sucrose gradient centrifugation in presence of 70% formamide, (b) a 65 : 35 mixture of fibroin mRNA and percursor rRNA, (c) a 23 : 77 mixture of fibroin mRNA and precursor rRNA, and (d) precursor rRNA from *Bombyx mori* larva labeled for 50 min with $^{32}PO_4^{3-}$. Nucleotides were eluted with linear NaCl gradient from 0.10 to 0.45 M except in (a) (0.10 to 0.40 M). *Arrows:* position where additional elution buffer containing 0.45 M NaCl was applied. *Roman numerals:* chain length of oligonucleotides. —●—, ^{32}P cts/min (Suzuki and Suzuki, 1974)

tetranucleotides. Therefore, a higher fraction of tetranucleotides in a mRNA digest would be interpreted to represent a lower fraction of mRNA. The purest mRNA preparation so far obtained has only 1.7% of its total digest as tetranucleotides (Fig. 8a), and we have assumed that this preparation is almost 100% pure.

A major contaminant of fibroin mRNA preparations from middle or late fifth instar larvae is rRNA (Suzuki et al., 1972). The RNase T_1 digest of precursor rRNA from *B. mori* larvae contains about 12% tetranucleotides (Fig. 8d). The other possible contaminants, especially in the early fifth instar, is DNA-like RNA which is about 39% GC. A T_1 RNase digest of the Bio-Gel excluded RNA (42% GC) which was isolated from the posterior gland during the fourth moulting stage contained about 9% tetranucleotides (Fig. 10b). A T_1 digest of the Bio-Gel excluded RNA (39% GC) from animal's carcass was found to reveal essentially a similar oligonucleotide profile like Figure 8d or Figure 10b, and the tetranucleotides comprised about 9% of the digest (see Fig. 9).

Mixing experiments were carried out in which known amounts of mRNA and precursor rRNA digests were mixed and fractionated on DEAE-Sephadex A 25

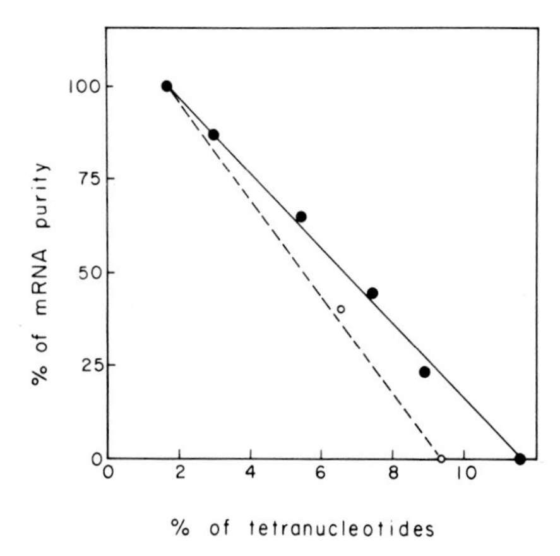

Fig. 9. Purity calibration curves for fibroin mRNA preparations. The purity shown was calculated from the mixing ratio of mRNA digest and precursor rRNA digest or dRNA digest assuming that mRNA preparation shown in Figure 8a was 100% pure. The abscissa is the amount of tetranucleotides found after DEAE-Sephadex column fractionation of the T_1 RNase digest. *Solid line:* calibration curve for rRNA contaminant. *Dashed line:* calibration curve for dRNAs of low GC content (39 to 42% GC) (Suzuki and Suzuki, 1974)

(Figs. 8b and 8c). Messenger RNA purity is plotted against the fraction of tetranucleotides in the digest (Fig. 9; Suzuki and Suzuki, 1974). Since the calibration curve is reasonably linear, it can be used to estimate the purity of any unknown fibroin mRNA preparation from middle or late fifth instar larvae in which the major contamination is known to be rRNA. However, when DNA-like RNA is the major contaminant the calibration curve is slightly different (dashed line, Fig. 9) due to the lower quantity of tetranucleotides (9%). This latter curve was used for mRNA fractions isolated from the early fifth instar as well as any mRNA preparation with a GC content below that of precursor rRNA (about 48% GC). The overall error in estimating the mRNA purity by this method is about ±20% (see Suzuki and Suzuki, 1974).

E. Quantitative Measurements of Fibroin mRNA Synthesis

The induction of a specific protein synthesis may be due to two general mechanisms. The one is the induced increase of the specific mRNA content and the other is the induced translation of a large preexisting population of the mRNA. For the former at least two examples are known; the estrogen-induced synthesis of ovalbumin mRNA in chick oviduct (Comstock et al., 1972b; Rosenfeld et al., 1972a) and the DMSO-induced synthesis of globin mRNA during erythroid differentiation of cultured leukemia cells (Ross et al., 1972). For the latter mechanism there exists the established concept that eggs contain stored and masked mRNAs which are ready to function at or immediately after fertilization (see Davidson, 1968; Brown and Dawid, 1969; Gross et al., 1973; Skoultchi and Gross, 1973). The same question can be studied in the silk gland system by quantitative measurements of fibroin mRNA synthesis at various stages of the silk gland development. For the quantification of fibroin mRNA, (1) we fractionate 40 to 130 mg of RNA with Bio-Gel, obtain mg quantities of fibroin mRNA fraction, and measure the mRNA fraction (both by A_{260} and ^{32}P-radioactivity) to total cellular RNA, (2) we digest this mRNA fraction with RNase T_1, fractionate the resulting oligonucleotides with DEAE-Sephadex column, and measure the amounts of tetranucleotides, and (3) from the tetranucleotide amount we estimate a purity of the mRNA preparation and simply by calculation we know the fibroin mRNA quantity. The same technique has been applied to quantification of fibroin mRNA in *B. mori* fibroin mutants as well as *B. mandarina.*

1. Fibroin mRNA During the Third and Fourth Larval Instars

Studies on fibroin and its mRNA have been restricted mainly to the fifth instar (Tashiro et al., 1968; Suzuki and Brown, 1972). However, a fibroin-like material has been described in silk glands throughout larval life except moulting stages and even before hatching in embryos just after the gland has completed its differentiation (Nunome, 1937; see Sect. II.B.). It has also been known for a long time that removal of the corpora allata of second, third or fourth instar larvae causes diminutive pupation of larvae at the end of the instar. This prematured pupation is often preceded by spinning of fibroin-like material and formation of a small cocoon (Bounhiol, 1936, 1937a, 1937b; Kiguchi, pers. comm.; Ohtaki, pers. comm.). These observations suggest that fibroin mRNA exists throughout larval life. When and how much is the mRNA synthesized? Does it exist continuously? Alternatively, this "fibroin-like" material might be a completely different gene product such as has been described for certain spiders (see Lucas and Rudall, 1968).

For quantitative measurements of fibroin mRNA both labeled and unlabeled RNA from third as well as fourth instar animals were analyzed (Suzuki and Suzuki, 1974). Similar results were obtained with RNAs isolated from both instars. During the feeding stages ^{32}P-mRNA and unlabeled mRNA accumulated to about 5% and 2% of the total posterior gland RNA (see Fig. 5a and Table 9). The latter value corresponds to 0.2 and 2 µg per pair of posterior glands in the third and fourth instar, respectively, indicating that the fibroin mRNA accumulated is increasing as the stage proceeds (Table 9). However, labeled and unlabeled RNA isolated from animals of the third or fourth moulting stage were markedly different in fibroin mRNA content from that in feeding larvae (compare Fig. 5a with 5b, and Fig. 10a with 10b; see also Table 9). In moulting animals the "mRNA" fraction from Bio-Gel (Fig. 5b) had a base composition of 42% GC and 9% of tetranucleotides in RNase T_1 digest (Fig. 10b). Therefore, not only does fibroin mRNA synthesis decrease or stop but pre-existing fibroin mRNA (Fig. 10a) is no longer detectable by either radioactivity or A_{260} criterion (Fig. 10b). The amount of fibroin mRNA in the third and fourth moulting stages is less than 0.5% of cellular RNA by both weight and radioactivity (Table 9). This corresponds to less than 0.05 and 0.6 µg per pair of posterior glands of the third and fourth moulting stages, respectively. In other words, more than 70% of the fibroin mRNA which had accumulated during the feeding stage is degraded during the moulting stages. Control experiments were carried out which excluded the possibility of mRNA degradation during RNA

Table 9. Summary of quantitative data on the synthesis and accumulation of nucleic acids in a pair of posterior silk glands of *B. mori* larve through the third to fifth instar[a]

Stages	Larva weight	Gland weight	DNA	Bulk RNA	rRNA		Fibroin mRNA		
					weight	^{32}P cpm	weight		^{32}P cpm
	grams	mg	µg	µg	µg	% of total	µg	% of total	% of total
3rd instar									
Feeding stage	0.22	1.1	1.3	9	8.2	72	0.21	2.1	5.0
Moulting stage	0.24	1.9	1.6	13	9.6	72	< 0.05	<0.4	< 0.8
4th instar									
Feeding stage	0.95	9.2	12	100	83	72	2.1	2.1	5.1
Moulting stage	1.0	11	13	120	92	71	< 0.6	<0.5	< 0.5
5th instar									
4th ecdysis	0.96	18	19	250	—	—	—	—	—
1 day	1.4	40	70	500	390	74	< 2	<0.5	0.7
2 days	1.9	100	130	2000	1500	77	20	1.0	1.8
4 days	3.4	250	200	4500	3600	72	90	2.0	4.6
6 days	4.8	320	180	4900	3900	56	170	3.5	10

[a] From Suzuki and Suzuki, 1974.

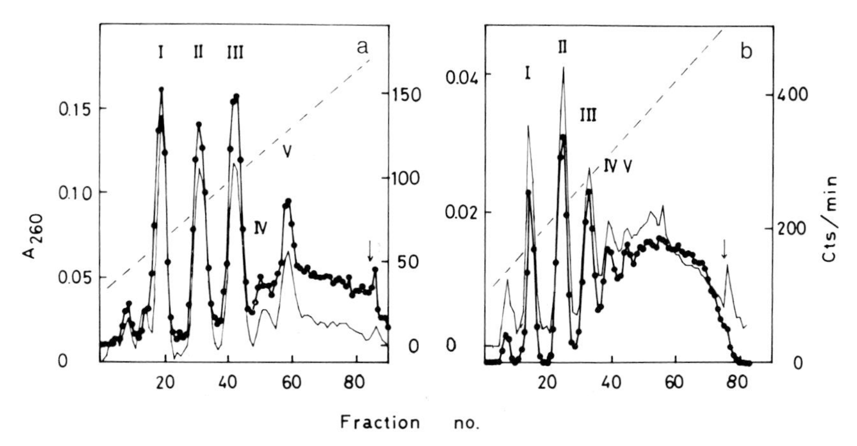

Fig. 10a and b. DEAE-Sephadex fractionation of the T_1 RNase digests of fibroin mRNA from the fourth instar larvae. (a) Fibroin mRNA fraction at the feeding stage. About 390 μg from Figure 5a. (b) RNA fraction at the moulting stage. About 270 μg from Figure 5b. ———, A_{260}; —●—, ^{32}P cts/min (Suzuki and Suzuki, 1974)

extraction (Suzuki and Suzuki, 1974). Therefore, this degradation occurs in vivo. It should be emphasized that between the feeding stage and moulting stage not only fibroin mRNA disappears but also a dramatic qualitative change of RNA syntheses occurs (Figs. 10a and 10b). The syntheses of DNA-like RNAs of large molecular weight which are excluded by a Bio-Gel column predominates over that of fibroin mRNA until 24 h of fifth instar (Fig. 11a). The chronological changes in the pattern of RNA synthesis are displayed in Figures 10a through 11c.

2. *Fibroin mRNA in the Fifth Instar*

At 12 h of the fifth instar $^{32}PO_4^{3-}$ was given to the larvae, and RNA was extracted from the posterior silk glands 12 h later (Fig. 5c). Although the purity of ^{32}P-fibroin mRNA is not high at this stage (Fig. 11a) the ^{32}P-oligonucleotide profile in Figure 11a shows that some mRNA is present. Therefore, fibroin mRNA synthesis can be recognized even during the first day of the fifth instar. At this stage fibroin mRNA comprises only about 0.7% of all ^{32}P-labeled RNA and less than 0.5% by A_{260} of the cellular RNA.

By 48 h of the fifth instar fibroin mRNA predominates over other DNA-like RNAs of similar sizes, i.e., the purity of this fraction becomes higher and higher as the stage proceeds (Figs. 11b and 11c), and the quantity of fibroin mRNA increases (Table 9). The increase continues until fibroin mRNA comprises about 3.5% of the cellular RNA by weight. The fraction of total ^{32}P-RNA synthesis that is fibroin mRNA becomes as high as 10 to 25% during a 24-h period late in the fifth instar (Figs. 5d, 11c and Table 9).

The results on the analysis of fibroin mRNA as well as rRNA are summarized in Table 9. Most rRNA is synthesized during the first half of the fifth instar in contrast to fibroin mRNA (Table 9; Tashiro et al., 1968). These results indicate that transcription of fibroin genes is regulated differently from that of rDNA.

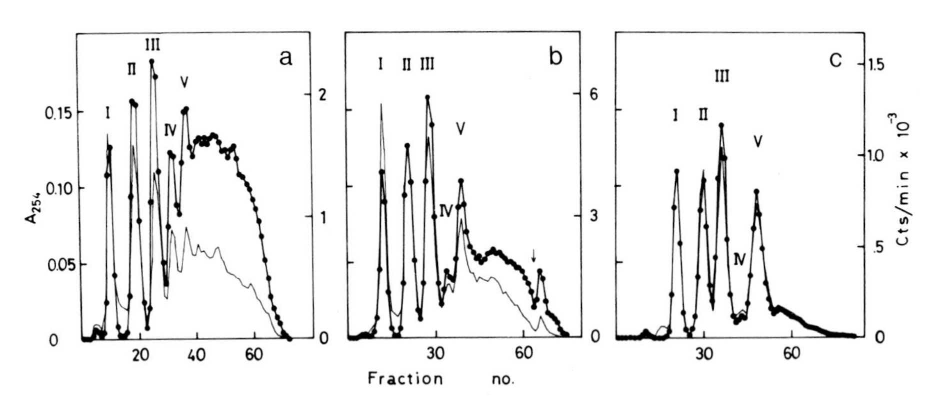

Fig. 11a—c. DEAE-Sephadex fractionation of the T_1 RNase digests of fibroin mRNA from fifth instar larvae. (a) Fibroin mRNA fraction at 24 h of instar. About 500 μg recovered from Figure 5c. (b) Fibroin mRNA fraction at 2 days of instar. About 260 μg obtained by Bio-Gel fractionation. (c) Fibroin mRNA fraction at 6 days of instar. About 240 μg purified by 2 cycles of Bio-Gel fractionation (from Fig. 5d). ———, A_{254}; —●—, ^{32}P cts/min (Suzuki and Suzuki, 1974)

Furthermore, we can conclude that more than 99% of the fibroin mRNA which is detected in the fifth instar for the massive synthesis of cocoon fibroin is synthesized de novo during the fifth instar (Table 9).

3. *Fibroin mRNA in B. mori Mutants*

A variety of *B. mori* fibroin mutants have been known which do not produce appreciable amount of fibroin (Table 6). RNA analysis for these mutants has been undertaken (Suzuki and Suzuki, 1974).

The four homozygous larvae and the four heterozygotes (Table 10) were injected with $^{32}PO_4^{3-}$ on the fourth day of the fifth instar and RNA was extracted from their posterior silk glands 24 h later. Reflecting the growth inhibition in the posterior glands of these mutants (Table 6) the total cellular RNA content in the glands are greatly reduced (Table 10). In Nd_1, Nd(1) and Nd(2) the total RNA is only 6 to 7% of normal at the comparable developmental stage of the larva. In heterozygotes, Nd(1)/+ and Nd(2)/+, the RNA contents are 8 and 13%, respectively, of normal. In the case of Nd-s mutant the RNA content is only 4% of normal in homozygote and 22% in heterozygote (Table 10). These results may suggest that the mutations are simply at the level of growth inhibition in the posterior gland, in other words, these mutants might be a phenocopy of normal younger larvae in their posterior gland development. However, more severe damage has been found in their fibroin mRNA contents (see Fig. 6 and Table 10). The mRNA in homozygotes is about 1 μg per pair of posterior silk glands which is only 1% of that in normal larva (Table 10). It indicates that fibroin mRNA synthesis has been depressed 5 times more severely than the bulk RNA. Therefore, the small posterior silk glands in the mutants are not a phenocopy of those from normal larva of earlier stages of development. Heterozygotes, Nd(1)/+ and Nd(2)/+, contain slightly more fibroin mRNA than homozygotes do, but they are only 3 and 4 μg,

Table 10. Fibroin mRNA in the *B. mori* mutants which do not synthesize appreciable amounts of fibroin[a]

Mutants	Bulk RNA[b]	Fibroin mRNA[b]			Fibroin or fibroin-like
	weight µg	weight µg		^{32}P cpm %	material spun[c] weight, mg
Nd[1]	280	1.4	0.5	0.9	≦ 0.6
Nd (1)	310	0.8	0.3	0.5	—
Nd (1)/+	580	4.1	0.7	1.3	—
Nd (2)	260	1.3	0.5	0.6	—
Nd (2)/+	370	3.0	0.8	1.3	—
Nd-s	180	1.3	0.7	1.6	≦ 2
Nd-s/+	990	15	1.5	1.8	—
+/+	4500	90	2.0	4.6	300

[a] From Suzuki and Suzuki, 1974.
[b] Amount in a pair of the posterior silk glands.
[c] Amounts spun by a larva.

respectively, per pair of posterior glands. In Nd-s/+ an intermediate level of fibroin mRNA (15 µg per pair of posterior glands) is found (Table 10); a fact which agrees with the genetic observation that the defect behaves as an incomplete dominant mutation in the heterozygote, Nd-s/+ (Table 6). From all this we must conclude that the primary biochemical lesion in these mutants is still unknown.

4. Fibroin mRNA in the Wild Silkworm B. mandarina

The mRNA of *B. mandarina*, the presumed ancestor of *B. mori*, was isolated and compared with that of *B. mori* (Suzuki and Suzuki, 1974). Although initial studies suggested that its ^{32}P-mRNA sedimented more rapidly in an SDS-sucrose gradient than that of *B. mori* mRNA, subsequent centrifugation in 70% formamide revealed the same sedimentation constant for both mRNAs. The mRNA fraction from *B. mandarina* was excluded from the Bio-Gel column at the same position as *B. mori* mRNA. The former was digested with T_1 RNase, and the digest fractionated by DEAE-Sephadex (Figure not shown). The oligonucleotide profile was indistinguishable from a *B. mori* digest indicating that the sequence and average codon frequency for major amino acids in fibroin are quite similar, if not identical, between the mRNAs. The fibroin mRNA from *B. mandarina* comprises about 2% by weight and 3 to 10% by ^{32}P-labeling of the cellular RNA from the posterior gland in the middle fifth instar.

Fibroin is one of the extreme proteins which have diverged dramatically in evolution (Lucas and Rudall, 1968). The fact that the two fibroin mRNAs from *B. mandarina* and *B. mori* are indistinguishable with the present criteria gives additional supporting evidence for the ancestor-descendant relationship between the two silkworms. However, fine comparison of the two mRNAs should be taken with reservation. None of the methods employed here are precise enough to tell minor differences in their molecular sizes and nucleotide sequences.

IV. The Genes for Fibroin

A. A Test System for Specific Amplification of Specialized Genes

One mechanism which has been proposed for the regulation of specific genes is their selective replication in those cells in which they are expressed (Schulz, 1965; Brown and Dawid, 1968). Specific amplification of rDNA is known to occur in oocytes of amphibians and insects (Brown and Dawid, 1968; Gall et al., 1969) and polytenization of DNA in *Drosophila* is restricted to replication of the euchromatin portion of the genome (Gall et al., 1971; Dickson et al., 1971). The important question remains, however, whether genes coding for specialized proteins are amplified specifically in the cells which express these genes.

Reassociation kinetics of denatured DNA mixed with in vivo labeled globin mRNA have been studied under conditions of DNA excess (Bishop et al., 1972). The results showed about three to five globin genes per haploid equivalent. Specific gene amplification has been ruled out in this system since the reassociation kinetics were the same with the DNA from all tissues (Bishop et al., 1972; see also Packman et al., 1972; Harrison et al., 1972).

The posterior silk gland represents itself as an ideal tissue with which to answer this question. The cell contains about 0.2 μg of DNA as a result of "polyploidization" (Sect. II.D.). In the fifth instar the last replications of DNA occur during the first 3 days (Gillot and Daillie, 1968; Tashiro et al., 1968). Soon after DNA synthesis has stopped the posterior gland cells change from synthesizing a variety of protein to the synthesis of a single protein fibroin. This sequence of cessation of DNA synthesis followed by synthesis of specific proteins is characteristic of many differentiated cells. One explanation for this sequence of events could be that the specific genes to be expressed are amplified just before their expression.

The fibroin system besides being an exaggerated example of cell specialization has other features which make possible the analysis of the fibroin genes. The mRNA for fibroin can be labeled to high specific activity with $^{32}PO_4^{3-}$, ^{3}H-nucleoside precursors or ^{125}I, and it has been purified and fully characterized by partial sequence analysis (Sect. III). The mRNA, and, therefore, the fibroin genes, have an unusually high G+C content (about 60%). This fact made it possible to separate the fibroin genes from the bulk of *B. mori* DNA (39% G+C) by buoyant density centrifugation. This partial purification of the fibroin genes enhances the specificity with which they can be detected by hybridization. Furthermore, the RNA recovered from the hybrid could be identified as fibroin mRNA by its sequence analysis (Suzuki et al., 1972).

B. Identification of the Fibroin mRNA-Fibroin Gene Hybrid

Fibroin ^{32}P-mRNA purified by two sucrose gradient fractionations has been estimated to be greater than 80% pure (Suzuki and Brown, 1972). In an initial hybridization experiment, such a preparation was found insufficiently pure to detect fibroin genes (Suzuki et al., 1972). The major hybrid obtained with the DNA

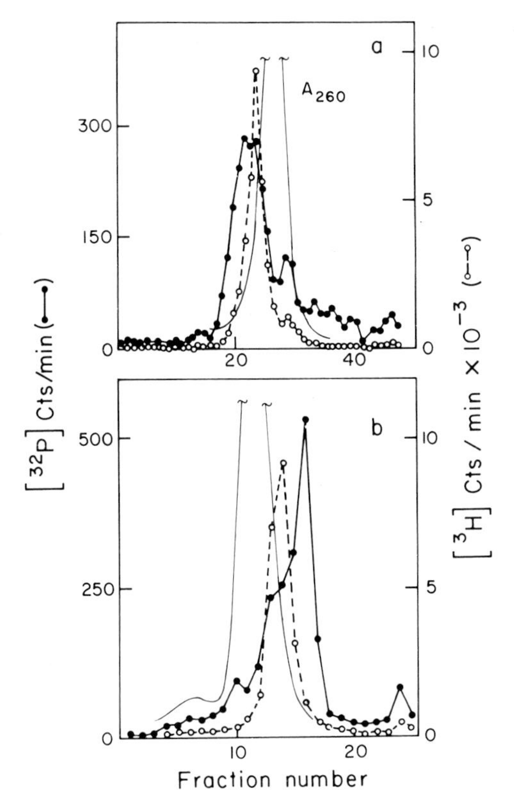

Fig. 12a and b. Hybridization of formamide-purified fibroin mRNA with *B. mori* DNA which had been fractionated in (a) CsCl or (b) Ag^{+}/Cs_2SO_4. In each case, 2 mg of sheared carcass DNA was centrifuged to equilibrium in 50.1 Spinco rotor. Immobilized DNA fractions were hybridized together with 0.6 μg ^{32}P-labeled mRNA/ml (3.2×10^4 cts/min/μg) and 0.6 μg *X. laevis* ^{3}H-labeled rRNA/ml. (—●—●—), ^{32}P (cts/min); (—○—○—), ^{3}H (cts/min); (———), O.D. at 260 nm (Suzuki et al., 1972)

fractionated in a CsCl gradient was identified as ^{32}P-rRNA-rDNA hybrid indicating some contamination of ^{32}P-mRNA preparation with ^{32}P-rRNA. The domination of the hybridization reaction by a small amount of ^{32}P-rRNA contaminant indicated that the genes for rRNA are at least an order of magnitude more abundant in *B. mori* DNA than are fibroin genes. Actually *B. mori* rRNA hybridizes with 0.17% of *B. mori* DNA (Gage, Suzuki and Brown, unpubl.) while fibroin mRNA hybridizes to only 0.0022% as described below.

Further purification of the ^{32}P-mRNA was done by centrifugation through a sucrose gradient containing 70% formamide (Suzuki et al., 1972) which eliminated most of the contaminating ^{32}P-rRNA. The purified ^{32}P-mRNA was mixed with ^{3}H-rRNA from *X. laevis* and hybridized to CsCl fractionated DNA (Fig. 12a). Hybrids of ^{32}P-RNA with DNA of higher buoyant density than rDNA were revealed which we tentatively identified as hybrids of fibroin mRNA with its genes (Fig. 12a).

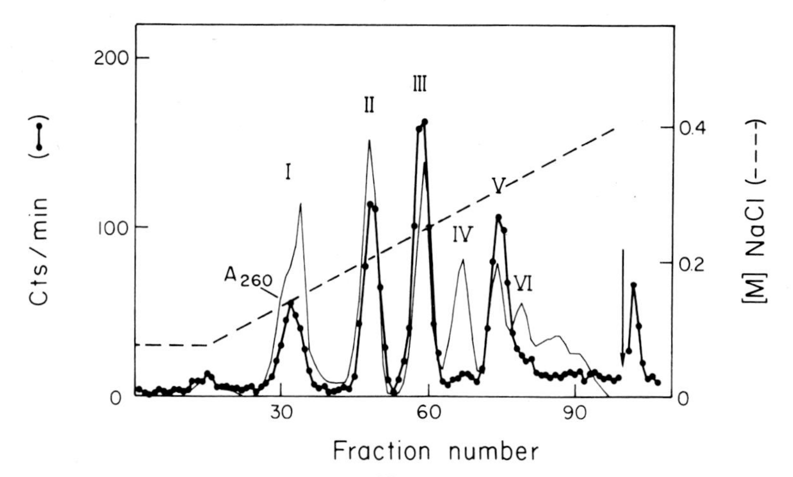

Fig. 13. DEAE-Sephadex fractionation of ^{32}P-labeled oligonucleotides from a T_1 RNase digest of ^{32}P-labeled mRNA which had hybridized to *B. mori* DNA. Pooled ^{32}P-labeled mRNA from hybrids containing 2500 cts/min was digested with RNase T_1 plus 200 µg of carrier *E. coli* sRNA, and fractionated on a DEAE-Sephadex A25 column. (—●—●—), ^{32}P (cts/min); (———), O.D. at 260 nm; (— — —), NaCl concentration [M] (Suzuki et al., 1972)

In order to identify unequivocally and to quantitate the very low level hybridization of fibroin mRNA, we sought a technique by which larger amounts of DNA could be fractionated with a greater density separation of fibroin genes from rDNA and bulk DNA. Both criteria were met by complexing the DNA with Ag^+ followed by centrifugation in Cs_2SO_4 (Fig. 12b). Fibroin genes bind less Ag^+ than does rDNA which binds less Ag^+ than bulk DNA. Consequently, their relative buoyant densities are inverted so that the fibroin genes are the least dense of the three kinds of DNA. Besides reducing the overlap of rRNA and mRNA hybridization shown in Figure 12a, the Ag^+/Cs_2SO_4 centrifugation has the further advantage that 2 mg of DNA can be centrifuged in one gradient without affecting the separation of the fibroin genes (Fig. 12b); this made it possible to identify the ^{32}P-mRNA-fibroin gene hybrid.

Only a small fraction of the input fibroin mRNA actually hybridized with the GC-rich DNA (Fig. 12). Generally it is a level of 0.1% of the input mRNA. Consequently, it was necessary to show that the RNA in the hybrids was identical to the input mRNA. For this purpose, about 8.5 mg of DNA was complexed with Ag^+ and banded in Cs_2SO_4. The DNA pooled from the light side of the gradient was centrifuged again in Cs_2SO_4, and hybridized with ^{32}P-mRNA, ^{3}H-rRNA and an excess of unlabeled carcass RNA to dilute ^{32}P-rRNA contaminants. Essentially all of the ^{32}P-RNA which hybridized did so with a DNA which banded to the light side of the rDNA. The ^{32}P-RNA in hybrids was recovered and characterized by its base composition and oligonucleotide profile (Suzuki et al., 1972). The recovered RNA contained 58% G + C and 42% G. The oligonucleotide profile shown in Figure 13 is the characteristic one of the fibroin mRNA as described in Section III.C. Consequently, the ^{32}P-RNA which hybridized to the DNA of low buoyant density in Ag^+/Cs_2SO_4 is fibroin mRNA.

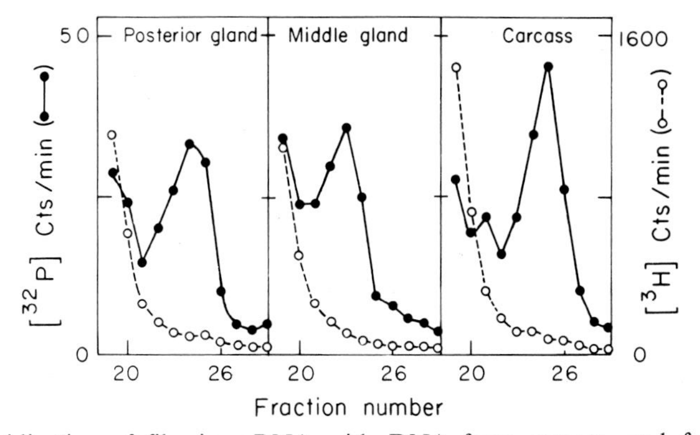

Fig. 14. Hybridization of fibroin mRNA with DNA from carcass, and from posterior and middle silk glands. DNA which had been sheared to about 2.5×10^6 daltons was complexed with silver, banded in Cs_2SO_4 and immobilized on filters. Filters containing DNA from the fibroin gene regions of 3 gradients were hybridized together for 48 h with 0.9 µg formamide-purified fibroin ^{32}P-labeled mRNA/ml (1.4×10^4 cts/min/µg), 6.6 µg *B. mori* ^3H-labeled rRNA/ml and 130 µg unlabeled carcass RNA/ml. In this experiment the fibroin genes from 550 µg of posterior gland DNA, 620 µg of middle gland DNA and 690 µg of carcass DNA were hybridized. (—●—●—), ^{32}P (cts/min); (—○—○—), ^3H (cts/min) (Suzuki et al., 1972)

C. Gene Dosage in Various Tissues

Using the technique mentioned above (Fig. 12b) it was possible to test if there is any specific amplification of fibroin genes in the posterior gland cells. DNA was isolated from the highly ployploid cells of the posterior gland late in the fifth instar when fibroin synthesis was going on, and for comparison, also from the polyploid middle silk gland and from the animal's carcass. Figure 14 shows that a similar quantity of ^{32}P-mRNA hybridized with each DNA. The ^{32}P-mRNA which bound to each of the three DNAs were summed. The amount of ^3H-rRNA which hybridized to each filter assessed the extent to which fibroin genes and rDNA overlapped and permitted a correction for ^{32}P-rRNA hybridization. The amount of DNA which had actually been immobilized on filters was measured. After correction it was known that all DNA samples hybridized closely similar amounts of fibroin mRNA. We conclude, therefore, that there is no selective amplification of the DNA homologous to fibroin mRNA in posterior silk gland cells (Suzuki et al., 1972).

An accurate determination of the number of fibroin genes per genome depends upon three numbers; (1) the genome size (i.e. the amount of DNA per haploid chromosome set), (2) the molecular weight of the mRNA, and (3) the fraction of the genome homologous with fibroin mRNA.

The fraction of DNA homologous to the mRNA was determined by a hybridization saturation experiment. It should be emphasized that the specific activity of the mRNA was unequivocally determined based on the result shown in Section III.E. After suitable corrections in hybridization experiments about 0.0022% of the DNA was found to hybridize with fibroin mRNA under saturating conditions (Suzuki et al., 1972).

Recent measurement of the DNA content of individual *B. mori* spermatids by E. Rasch (1974) has given a value of about 0.52 pg. Gage (1974a) also has estimated a genome of about the same size by DNA renaturation kinetics.

The molecular weight of fibroin mRNA was first estimated by its rate of sedimentation relative to *X. laevis* 28S and 18S rRNA in a formamide-sucrose gradient (Fig. 4c) at roughly 2×10^6 daltons (Suzuki et al., 1972). However, recently it has been found that the formamide-sucrose gradient is not linear giving a steeper gradient toward the bottom. Molecular weight estimation of the mRNA from contour length measurement with electron microscope and electrophoretic mobility in polyacrylamide gel in the presence of formaldehyde or formamide gives a value of 5.7×10^6 daltons as average (Lizardi et al., 1975).

The hybridization saturation value of 0.0022% is subject to several sources of error. Three separate experiments in which hybridization was performed at concentrations of mRNA greater than saturation concentration gave values within 15% of 0.0022%. Hybridizations were performed at 12° C below the Tm of bulk DNA and about 20° C below the Tm predicted for the fibroin genes. These stringent conditions were designed to reduce the hybridization of any contaminating "DNA-like RNA" and to increase the probability of accurate base pairing of fibroin mRNA with its genes. We deduce from the structure of fibroin itself and its mRNA that the genes must be internally repetitive so that an RNA molecule might hybridize in a number of poorly base paired alignments with a single gene sequence causing mismatching as well as free RNA ends not duplexed with DNA. The high criterion of hybridization was designed to reduce mismatching in the hybrid, and RNase treatment of the hybrid should have eliminated regions of the mRNA which were not duplexed. The amount of fibroin genes may have been underestimated by only scoring RNA which hybridized at the low buoyant density in Ag^+/Cs_2SO_4. To minimize this error the DNA was sheared to a weight average molecular weight of approximately 2.5×10^6 daltons. As can be seen in Figure 12b the ^{32}P-mRNA hybridized only very slightly with main band DNA, the main problem being ^{32}P-rRNA contamination for which a correction was made (Suzuki et al., 1972).

Although none of the numbers listed above are highly accurate, we calculated that there are no more than 3, and possibly 1, fibroin genes per haploid complement of DNA (Suzuki et al., 1972; Lizardi and Brown, 1973). One strand of 1 fibroin gene (estimated to be 6×10^6 daltons of DNA; Lizardi and Brown, 1973) would constitute about 0.002% of a haploid genome which contained 0.52 pg of DNA. More recently Gage and Manning (1976) determined the multiplicity of the gene and concluded that there is only one gene per haploid genome.

V. Transcriptional and Translational Control in the Posterior Silk Gland

About 1000 cells in a pair of posterior silk glands from *B. mori* (Ono, 1942) synthesizes about 300 mg of fibroin in about 5 days (Tashiro et al., 1968). Each giant polyploid cell contains as average 0.2 μg of DNA (Tashiro et al., 1968) and the *B.*

mori genome size is estimated at 0.52 pg. Therefore, each cell contains about 4×10^5 fibroin genes whose expression result in the production of 5×10^{14} fibroin molecules. During this developmental period fibroin mRNA accumulates in the fifth instar until it comprises about 4.4% of the cellular RNA or about 2.3×10^{10} molecules per cell (Suzuki and Giza, 1976), most of which are synthesized in the fifth instar (Suzuki and Suzuki, 1974). Since the mRNA molecules are almost as stable as rRNA (Suzuki and Brown, 1972; Suzuki and Giza, 1976) each gene produces roughly 5.8×10^4 mRNA molecules in seven days, each of which in turn can translate on the average of 2×10^4 fibroin molecules in seven days. The approximate transcription rate of fibroin gene is 0.1 mRNA molecules/gene/sec or approximately 1700 nucleotides polymerized/gene/sec; the translation rate is about 0.03 fibroin molecules/mRNA/sec.

In order to prepare for this remarkably efficient transcriptional and translational specialization, each cell need transcribe its fibroin genes and stabilize the resultant mRNA with a higher efficiency than that it transcribes its ribosomal genes and stabilizes rRNA. This can be concluded from the fact that the posterior gland accumulates fibroin mRNA and rRNA to about 4.4% and 80%, respectively, of its total RNA (Suzuki and Giza, 1976), and the fact that rDNA is about 240 times more abundant than fibroin genes (Gage, 1974b).

The most impressive feat which the posterior gland cells can carry out is the stabilization of fibroin mRNA and the efficient programming of the tRNA population and of the majority of ribosomes with the mRNA produced by only 0.002% of its total DNA.

The posterior silk gland is an ideal system in which we can study the mechanism of differential gene action. The specific transcripts of the fibroin genes are unequivocally detected and assayed by its chemical properties. Furthermore, the pure mRNA can be transcribed into complementary DNA in vitro by the reverse transcriptase (Lizardi and Brown, 1972). Using the complementary DNA now we can hope to answer directly questions dealing with the mechanism of differential gene activation as well as processing and transport of mRNA from nucleus to cytoplasm and the mechanism of mRNA stabilization.

Further transcriptional aspects of this system have been described and discussed elsewhere (Suzuki, 1976; Suzuki and Giza, 1976).

Acknowledgements. The manuscript was greatly improved by the suggestions of Drs. D. D. Brown, I. B. Dawid, H. Ikeda, Y. Tashiro, and H. Akai. I am also thankful for Drs. D. D. Brown, L. P. Gage, M. Kawakami, P. Lizardi and E. Rasch who generously permitted me to cite their results before publication. I am indebted to Mr. N. Tsuchida, P. Giza and Mrs. E. Suzuki for their excellent technical assistance, and to Miss A. Sakuma for her expert drawing of Figures 1 to 3. The work has been aided partly by a grant from The Jane Coffin Childs Memorial Fund for Medical Research, grant no. 291 for Y.S. at N.I.H. of Japan.

Note added in proof: Since this review was submitted for publication considerable progress has been made in the structural analysis of the fibroin locus: Using the gene cloning technique we have recently isolated the fibroin locus with its flanking segments in which we presume the existence of regulating sequences (Suzuki and Oshima, Cold Spring Harbor Symp. on Quant. Biol. Vol. 42, in press; Oshima and Suzuki, in preparation). These genes are now available in mg quantity, a circumstance which will facilitate further analysis of the situation in structural as well as functional terms.

References

Akai, H.: Micromorphological changes of the grandular cell in the posterior division of the silk gland during the 5th larval instar of the silkworm, *Bombyx mori*. Bull. Sericult. Exp. Sta. **18**, 475–511 (1964)

Akai, H.: Studies on the ultrastructure of the silk gland in the silkworm, *Bombyx mori* L. Bull. Sericult. Exp. Sta. (Tokyo) **19**, 375–484 (1965), in Japanese with English summary

Akai, H.: An ultrastructural study of changes during molting in the silk gland of larvae of the Eri-silkworm, *Philosamia cynthia ricini* Boisduval (Lepidoptera: Saturnidae). Appl. Ent. Zool. **6**, 27–39 (1971a)

Akai, H.: Ultrastructure of fibroin in the silk gland of larval *Bombyx mori*. Exptl. Cell Res. **69**, 219–223 (1971b)

Akai, H., Kobayashi, M.: Cytological studies of silkprotein and nuclei acid syntheses in the silk gland of the silkworm, *Bombyx mori*. Symposia Cell Chem. **17**, 131–138 (1966), in Japanese with English summary

Astaurov, B.L., Garisheba, M.D., Radinskaya, I.S.: Chromosome complex of Ussuri geographical race of *Bombyx mandarina* M. with special reference to the problem of the origin of the domesticated silkworm, *Bombyx mori* L. Cytology **1**, 327–332 (1959), in Russian

Aviv, H., Leder, P.: Purification of biologically active globin messenger RNA by chromatography on oligothymidylic acid-cellulose. Proc. Natl. Acad. Sci. **69**, 1408–1412 (1972)

Berns, A.J.M., van Kraaikamp, M., Bloemendal, H., Lane, C.D.: Calf crystalline synthesis in frog cells: The translation of lens-cell 14s RNA in oocytes. Proc. Natl. Acad. Sci. **69**, 1606–1609 (1972a)

Berns, A.J.M., Strous, G.J.A.M., Bloemendal, H.: Heterologous in vitro synthesis of lens α-crystalline polypeptide. Nature New Biol. **236**, 7–9 (1972b)

Bishop, J.O., Pemberton, R., Baglioni, C.: Reiteration frequency of haemoglobin genes in the duck. Nature New Biol. **235**, 231–234 (1972)

Bounhiol, J.-J.: Métamorphose après ablation des corpora allata chez le Ver à soie (*Bombyx mori* L.). C. R. Acad. Sci. (Paris) **203**, 388–389 (1936)

Bounhiol, J.-J.: Métamorphose prématurée par ablation des corpora allata chez le jeune ver à soie. C. R. Acad. Sci. (Paris) **205**, 175–177 (1937a)

Bounhiol, J.-J.: La métamorphose des insectes sereit inhibée dans leur jeune age par les corpora allata. C. R. Soc. Biol. **126**, 1189–1191 (1937b)

Britten, R.J., Kohne, D.E.: Repeated sequences in DNA. Science **161**, 529–540 (1968)

Brown, D.D., Dawid, I.B.: Specific gene amplification in oocytes. Science **160**, 272–280 (1968)

Brown, D.D., Dawid, I.B.: Developmental genetics. Ann. Rev. Genet. **3**, 127–154 (1969)

Brown, D.D., Reeder, R.H., Roeder, R.G., Suzuki, Y., Wenski, P.: Isolation, characterization, and control of genes in development. Carnegie Inst. Year Book **69**, 565–574 (1970)

Brown, D.D., Sugimoto, K.: 5S DNAs of *Xenopus laevis* and *Xenopus mulleri*: Evolution of a gene family. J. Mol. Biol. **78**, 397–415 (1973)

Brown, D.D., Suzuki, Y.: The purification of the messenger RNA for silk fibroin. In: Methods in Enzymology, Vol. XXX, Part F, Moldave, K., Grossman, L. (eds.), pp. 648–654. New York: Academic Press 1974

Brown, D.D., Wensink, P.C., Jordan, E.: Purification and some characteristics of 5s DNA from *Xenopus laevis*. Proc. Natl. Acad. Sci. **68**, 3175–3179 (1971)

Brownlee, G.G., Cartwright, E.M., Cowan, N.J., Jarvis, J.M., Milstein, C.: Purification and sequence of messenger RNA for immunoglobulin light chains. Nature New Biol. **244**, 236–240 (1973)

Chavancy, G., Daillie, J., Garel, J.-P.: Adaptation fonctionalle des tRNA à la biosynthèse protéque dans un système cellulaire hautement différencie. IV. Evolution des t-RNA dans la glande séricigène de *Bombyx mori* L. au cours du dernier âge larvaire. Biochimie **53**, 1187–1194 (1971)

Chikushi, H. (ed.): Genes and Genetical Stocks of the Silkworm. Tokyo: Keigaku 1972

Chinzei, Y., Tojo, S.: Nucleic acid changes in the whole body and several organs of the silkworm, *Bombyx mori*, during metamorphosis. J. Insect Physiol. **18**, 1683—1698 (1972)

Coleman, D., Howitt, F.O.: Studies on silk proteins. I. The properties and constitution of fibroin. The conversion of fibroin into a water-soluble form and its bearing on the phenomenon of denaturation. Proc. Roy. Soc. (London) Ser. A **190**, 145—169 (1947)

Comstock, J.P., O'Malley, B.W., Means, A.R.: Stimulation of cell-free polypeptide synthesis by a protein fraction extracted from chick oviduct polyribosomes. Biochemistry **11**, 646—652 (1972a)

Comstock, J.P., Rosenfeld, G.C., O'Malley, B.W., Means, A.R.: Estrogen-induced changes in translation, and specific messenger RNA levels during oviduct differentiation. Proc. Natl. Acad. Sci. **69**, 2377—2380 (1972b)

Crouse, H.V., Keyl, H.G.: Extra replications in the "DNA-puffs" of *Sciara corprophila*. Chromosoma **25**, 357—364 (1968)

Davidson, E.H.: Gene Activity in Early Development. New York: Academic Press 1968

Dawid, I.B., Brown, D.D., Reeder, R.H.: Composition and structure of chromosomal and amplified ribosomal DNAs of *Xenopus laevis*. J. Mol. Biol. **51**, 341—360 (1970)

Delovitch, T.L., Davis, B.K., Holme, G., Sehon, A.H.: Isolation of messenger-like RNA from immunochemically separated polyribosomes. J. Mol. Biol. **69**, 373—386 (1972)

Dickson, E., Boyd, J.B., Laird, C.D.: Sequence diversity of polytene chromosome DNA from *Drosophila hydei*. J. Mol. Biol. **61**, 615—627 (1971)

Doira, H.: Genetics of the flimsy cocoon mutant. Proc. Sericult. Sci. Kyushu **1**, 5 (1970), in Japanese

Doira, H., Chikushi, H., Kihara, H.: The twenty-first genetic linkage group in *Bombyx mori*. Abstracts of the 42nd Ann. Meet. Jap. Soc. Sericult. Sci., p. 29 (1972), in Japanese

Fukuda, T.: Detection by paper chromatography of amino acids in the mulberry leaves and the silkworm, *Bombyx mori*. Bull. Sericult. Exptl. Sta. (Tokyo) **13**, 481—503 (1951), in Japanese with English summary

Fukuda, T.: Studies on the biosynthesis of silk protein by the use of ^{14}C-compound. Proc. 1st Jap. Congr. Radioisotopes, pp. 528—531 (1956), in Japanese

Fukuda, T.: The correlation between the mulberry leaves taken by the silkworm, the silkprotein in the silkgland and the silk filament. Bull. Sericult. Exptl. Sta. (Tokyo) **15**, 595—610 (1960), in Japanese with English summary

Fukuda, T., Florkin, M.: Contributions to silkworm biochemistry. VII. Ordered progression of fibroinogen in the reservor of the silkgland during the 5th instar. Arch. Intern. Physiol. Biochem. **67**, 214—221 (1959)

Fukuda, T., Suto, M., Nakagawa, Y.: Biochemical studies on the formation of the silk protein. X. Ordered progression of fibroin in the inside of the silk gland during the fifth instar. Bull. Agr. Chem. Soc. Japan **24**, 501—505 (1960)

Gage, L.P.: Polyploidization of the silk gland in *Bombyx mori*. Carnegie Inst. Year Book **70**, 39—42 (1971)

Gage, L.P.: The *Bombyx mori* genome: Analysis by DNA reassociation kinetics. Chromosoma **45**, 27—42 (1974a)

Gage, L.P.: Polyploidization of the silk gland of *Bombyx mori*. J. Mol. Biol. **86**, 97—108 (1974b)

Gage, L.P., Manning, R.F.: Determination of the multiplicity of the silk fibroin gene and detection of fibroin gene-related DNA in the genome of *Bombyx mori*. J. Mol. Biol. **101**, 327—348 (1976)

Gall, J.G., Cohen, E.H., Polan, M.L.: Repetitive DNA sequences in Drosophila. Chromosoma **33**, 319—344 (1971)

Gall, J.G., MacGregor, H.C., Kidston, M.E.: Gene amplification in the oocytes of Dytiscid water beetles. Chromosoma **26**, 169—187 (1969)

Gamo, T.: Electrophoretic analyses of the protein extracted with disulphide cleavage from cocoons of the silkworm *Bombyx mori* L. J. Sericult. Sci. Japan **42**, 17—23 (1973a), in Japanese with English summary

Gamo, T.: Genetically different components of fibroin and sericin in the mutants, Nd and Nd-s, of the silkworm *Bombyx mori*. Jap. J. Genetics **48**, 99—104 (1973b)

Garel, J.-P., Mandel, P., Chavancy, G., Daillie, J.: Functional adaptation of tRNAs to fibroin biosynthesis in the silk gland of *Bombyx mori* L. FEBS Lett. **7**, 327—329 (1970)

Garel, J.-P., Mandel, P., Chavancy, G., Daillie, J.: Functional adaptation of tRNAs to protein biosynthesis in a highly differentiated cell system. III. Induction of isoacceptor tRNAs during the secretion of fibroin in the silk gland of *Bombyx mori* L. FEBS Lett. **12**, 249—252 (1971a)

Garel, J.-P., Mandel, P., Chavancy, G., Daillie, J.: Adaptation fonctionnelle des tRNA à la biosynthèse protéique dans un système cellulaire hautement différencié. V. Fractionnement par distribution á contre-courant des t-RNA de la glande séricigène de *Bombyx mori* L. Biochimie **53**, 1195—1200 (1971b)

Gillot, S., Daillie, J.: Rapport entre la mue et la synthese d'ADN dans la glande séricigène du Ver à soie. C. R. Acad. Sci. (Paris) D **266**, 2295—2298 (1968)

Greene, R. A., Morgan, M., Shatkin, A. J., Gage, L. P.: Stimulation of polypeptide synthesis by silk fibroin mRNA in ascites cell extracts. Federation Proc. **32**, 455 Abs (1973)

Gross, K. R., Jacobs-Lorena, M., Baglioni, C., Gross, P. R.: Cell-free translation of maternal messenger RNA from sea urchin eggs. Proc. Natl. Acad. Sci. **70**, 2614—2618 (1973)

Harrison, P. R., Hell, A., Birnie, G. D., Paul, J.: Evidence for single copies of globin genes in the mouse genome. Nature (London) **239**, 219—221 (1972)

Heywood, S. M.: Synthesis of myosin on heterologous ribosomes. Cold Spring Harbor Symp. Quant. Biol. **34**, 799—803 (1969)

Horiuchi, Y., Chikushi, H.: Studies on the genetic linkage group of the gene which is related to the Thanghpre-S sericin cocoon mutant. Abstracts of the 39th Ann. Meeting of Jap. Soc. Sericult. Sci. J. Sericult. Sci. Jap. **37**, 29 (1969), in Japanese

Horiuchi, Y., Namishima, C., Nakamura, K.: Genetics and morphology of the silk glands from the Thanghpre mutant. Abstracts of the 34th Ann. Meeting of Jap. Soc. Sericult. Sci. J. Sericult. Sci. Jap. **33**, 251 (1964), in Japanese

Horiuchi, Y., Namishima, C., Nakamura, K., Yasue, N.: Sericin cocoon mutant isolated from Burman strain of *Bombyx mori*. Abstracts of the 33rd Ann. Meeting of Jap. Soc. Sericult Sci. J. Sericult. Sci. Jap. **32**, 195—196 (1963), in Japanese

Horiuchi, Y., Ohi, H.: Genetics of the sericin cocoon mutant Thanghpre-S. Abstracts of the 36th Ann. Meeting of Jap. Soc. Sericult. Sci. J. Sericult. Sci. Jap. **35**, 228 (1966), in Japanese

Housman, D., Pemberton, R., Tabor, R.: Synthesis of α and β chains of rabbit haemoglobin in a cell-free extract from Krebs II ascites cells. Proc. Natl. Acad. Sci. **68**, 2716—2719 (1971)

Hyde, A. J., Wippler, C.: Molecular weight of silk fibroin. J. Polymer Sci. **58**, 1083—1088 (1962)

Iijima, T.: Ultrastructure of the posterior silk gland of the 'naked pupa' silkworm, *Bombyx mori*. J. Insect Physiol. **18**, 2055—2063 (1972)

Katayama, K.: Studies on the nitrogen compounds from the mulberry leaves. Bull. Sericult. Exptl. Sta. (Tokyo) **1**, 1—40 (1915), in Japanese

Kawaguchi, E.: Cytological studies on the possible transition of *Bombyx mandarina* to *Bombyx mori*. I. New J. Sericult. **31**, 46—52 (1923a), in Japanese

Kawaguchi, E.: Cytological studies on the possible transition of *Bombyx mandarina* to *Bombyx mori*. II. New J. Sericult. **31**, 159—165 (1923b)

Kawaguchi, E.: Cytological studies on the possible transition of *Bombyx mandarina* to *Bombyx mori*. III. New J. Sericult. **31**, 428—434 (1923c)

Kawaguchi, E.: Zytologische Untersuchungen am Seidenspinner und seinen Verwandten: I. Gametogenese von *Bombyx mori* L. und *Bombyx mandarina* M. und ihrer Bastarde. Z. Zellforsch. Anat. **7**, 519—552 (1928)

Kawakami, M., Shimura, K.: Fractionation of glycine-, alanine- and serine-transfer RNA's from the silkglands of silkworms. J. Biochem. (Tokyo) **76**, 187—190 (1974)

Kawakami, M., Shimura, K.: Properties of isoaccepting transfer RNA^{Gly} from silkglands of silkworms. Manuscript in preparation (1976).

Kirimura, J.: Studies on amino acid composition and chemical structure of silk protein by microbiological determination. Bull. Sericult. Exptl. Sta. (Tokyo) **17**, 447—522 (1962), in Japanese with English summary

Kirimura, J., Suzuki, T.: Studies on amino acid composition of silk protein by microbiological determination. III. Amino acid composition of silk protein and silk substance. Bull. Agr. Chem. Soc. Japan **36**, 265—268 (1962a), in Japanese

Kirimura,J., Suzuki,T.: Studies on amino acid composition of silk protein by microbiological determination. IV. Classification of silk fibroins by their amino acid distribution. Bull. Agr. Chem. Soc. Jap. **36**, 336—340 (1962b), in Japanese

Lane,C.D., Marbaix,G., Gurdon,J.B.: Rabbit haemoglobin synthesis in frog cells: the translation of reticulocyte 9s RNA in frog oocytes. J. Mol. Biol. **61**, 73—91 (1971)

Lima-de-Faria,A., Birnstiel,M., Jaworska,H.: Amplification of ribosomal cistrons in the heterochromatin of *Acheta*. Genetics **61**, Suppl. **1**, 145—159 (1969)

Lizardi,P., Brown,D.D.: Enzymatic synthesis of DNA complementary to silk fibroin mRNA. Carnegie Inst. Year Book **71**, 21—22 (1972)

Lizardi,P.M., Brown,D.D.: The length of fibroin structural gene sequences in *Bombyx mori* DNA. Cold Spring Harbor Symp. Quant. Biol. **38**, 701—706 (1973)

Lizardi,P.M., Williamson,R., Brown,D.D.: The size of fibroin messenger RNA and its polyadenylic acid content. Cell **4**, 199—205 (1975)

Lockard,R.E., Lingrel,J.B.: The synthesis of mouse hemoglobin β-chains in a rabbit reticulocyte cell-free system programmed with mouse reticulocyte 9s RNA. Biochem. Biophys. Res. Commun. **37**, 204—212 (1969)

Lockard,R.E., Lingrel,J.B.: Identification of mouse haemoglobin messenger RNA. Nature New Biol. **233**, 204—206 (1971)

Lucas,F.: Cystine content of silk fibroin *(Bombyx mori)*. Nature (London) **210**, 952—953 (1966)

Lucas,F., Rudall,K.M.: Extracellular fibrous proteins: The silks. In: Comprehensive Biochemistry, Vol. 26, Part B, Florkin,M., Stotz,E.H. (eds.), pp. 475—558. Amsterdam-London-New York: Elsevier 1968

Lucas,F., Shaw,J.T.B., Smith,S.G.: The amino acid sequence in a fraction of the fibroin of *Bombyx mori*. Biochem. J. **66**, 468—479 (1957)

Lucas,F., Shaw,J.T.B., Smith,S.G.: The silk proteins. Advan. Protein Chem. **13**, 107—242 (1958)

Lucas,F., Shaw,J.T.B., Smith,S.G.: Comparative studies of fibroins. I. The amino acid composition of various fibroins and its significance in relation to their crystal structure and taxonomy. J. Mol. Biol. **2**, 339—349 (1960)

Lucas,F., Shaw,J.T.B., Smith,S.G.: Some amino acid sequences in the amorphous fraction of the fibroin of *Bombyx mori*. Biochem. J. **83**, 164—171 (1962)

Machida,J.: Studies on the silk substances secreted by *Bombyx mori*. Bull. Sericult. Exptl. Sta. (Tokyo) **7**, 241—262 (1926), in Japanese

Machida,Y.: Studies on the silkglands of silkworms, *Bombyx mori* L. II. The singularity of the silkglands in hereditary trait, Naked pupae (Nd), in the silkworm...(1). Fukuoka Women's Junior College · Studies **3**, 1—21 (1970), in Japanese with English summary

Machida,Y.: Studies on the silkglands in the silkworm, *Bombyx mori* L. III. The singularity of the silkglands in hereditary trait, Naked pupae (Nd), in the silkworm...(2). Fukuoka Women's Junior College · Studies **10**, 39—49 (1972), in Japanese with English summary

Mathews,M.B., Osborn,M., Berns,A.J.M., Bloemendal,H.: Translation of two messenger RNAs from lens in a cell-free system from Krebs II ascites cells. Nature New Biol. **236**, 5—7 (1972)

Mathews,M., Osborn,M., Lingrel,J.B.: Translation of globin messenger RNA in a heterologous cell-free system. Nature New Biol. **233**, 206—209 (1971)

Matsuura,S., Morimoto,T., Nagata,S., Tashiro,Y.: Studies on the posterior silk gland of the silkworm, *Bombyx mori*. II. Cytolytic processes in posterior silk gland cells during metamorphosis from larva to pupa. J. Cell Biol. **38**, 589—603 (1968)

Matsuzaki,K.: The incorporation of ^{14}C-glycine into the soluble RNA of the posterior silkgland. J. Biochem. (Tokyo) **53**, 326—327 (1963)

Matsuzaki,K.: Fractionation of amino acid-specific s-RNA from silkgland by methylated albumin column chromatography. Biochim. Biophys. Acta **114**, 222—226 (1966)

Mazima,R., Shimura,K.: Studies on the biosynthesis of tRNAs. II. Mechanism of $tRNA^{Gly}$ increase in the posterior silk gland of *Bombyx mori*. Abstracts, the 45th Ann. Meet. Jap. Soc. Biochem. Japan J. Biochem. **44**, 488 (1972), in Japanese

Means,A.R., Comstock,J.P., Rosenfeld,G.C., O'Malley,B.W.: Ovalbumin messenger RNA of chick oviduct: Partial characterization, estrogen dependence, and translation in vitro. Proc. Natl. Acad. Sci. **69**, 1146—1150 (1972)

Mercer, E. H.: Studies on the soluble proteins of the silk gland of the silkworm, *Bombyx mori*. Textile Res. J. **24**, 135—145 (1954)

Morimoto, T., Matuura, S., Nagata, S., Tashiro, Y.: Studies on the posterior silk gland of the silkworm, *Bombyx mori*. III. Ultrastructural changes of posterior silk gland cells in the fourth larval instar. J. Cell Biol. **38**, 604—614 (1968)

Nakanishi, Y. H., Kato, H., Utsumi, S.: Polytene chromosomes in silk gland cells of the silkworm, *Bombyx mori*. Experientia **25**, 384—385 (1969)

Nakano, Y.: Studies on the naked pupa mutant. III. J. Sericult. Sci. Jap. **8**, 168—169 (1937), in Japanese

Nakano, Y.: Physiological, antomical and genetical studies on the "Naked" silkworm pupa. J. Sericult. Sci. Jap. **20**, 232—248 (1951), in Japanese

Neulat, M.-M.: Comparison de différents DNA animaux par l'analyse de leurs séquences pyrimidiques. Biochim. Biophys. Acta **149**, 422—434 (1967)

Nienhuis, A. W., Laycock, D. G., Anderson, W. F.: Translation of rabbit haemoglobin messenger RNA by thalassaemic and non-thalassaemic ribosomes. Nature New Biol. **231**, 205—208 (1971)

Nirenberg, M., Caskey, T., Marshall, R., Brimacombe, R., Kellog, D., Doctor, B., Hatfield, D., Levin, J., Rottman, F., Pestka, S., Wilcox, M., Anderson, F.: The RNA code and protein synthesis. Cold Spring Harbor Symp. Quant. Biol. **31**, 11—24 (1966)

Nunome, J.: Development of the silk gland in *Bombyx mori*. Bull. Appl. Zool. Jap. **9**, 68—92 (1937)

Ono, M.: The cell number of the silk gland in *Bombyx mori*. Bull. Kagoshima Agr. Coll. **14**, 123—156 (1942), in Japanese with German summary

Ono, M.: Studies on the growth of the silkgland cell in silkworm larvae. Bull. Sericult. Exptl. Sta. (Tokyo) **13**, 247—303 (1951), in Japanese with English summary

Onodera, K., Komano, T.: Glycine-specific transfer ribonucleic acid in the posterior silkgland of *Bombyx mori*. Biochim. Biophys. Acta **87**, 338—340 (1964)

Packman, S., Aviv, H., Ross, J., Leder, P.: A comparison of globin genes in duck reticulocytes and liver cells. Biochem. Biophys. Res. Commun. **49**, 813—819 (1972)

Palmiter, R. D., Palcios, R., Schimke, R. T.: Identification and isolation of ovalbumin-synthesizing polyribosomes. II. Quantification and immunoprecipitation of polysomes. J. Biol. Chem. **247**, 3296—3304 (1972)

Pemberton, R. E., Housman, D., Lodish, H. F., Baglioni, C.: Isolation of duck haemoglobin messenger RNA and its translation by rabbit reticulocyte cell-free system. Nature New Biol. **235**, 99—102 (1972)

Prudhomme, J.-C., Chavancy, G.: L'incorporation de la glycine dans la glande séricigène chez le Ver à soie au dernier stade larvaire. C. R. Acad. Sci. (Paris) D **268**, 1098—1101 (1969)

Rasch, E.: The DNA content of sperm and hemocyte nuclei of the silkworm, *Bombyx mori* L. Chromosoma **45**, 1—26 (1974)

Reeder, R. H.: Transcription of chromatin by bacterial RNA polymerase. J. Mol. Biol. **80**, 229—241 (1973)

Reeder, R. H., Brown, D. D.: Transcription of the ribosomal RNA genes of an amphibian by the RNA polymerase of a bacterium. J. Mol. Biol. **51**, 361—377 (1970)

Reeder, R. H., Roeder, R. G.: Ribosomal RNA synthesis in isolated nuclei. J. Mol. Biol. **67**, 433—441 (1972)

Rhoads, R. E., McKnight, G. S., Schimke, R. T.: Synthesis of ovalbumin in a rabbit reticulocyte cell-free system programmed with hen-oviduct RNA. J. Biol. Chem. **246**, 7407—7410 (1971)

Roeder, R. G., Reeder, R. H., Brown, D. D.: Multiple forms of RNA polymerase in *Xenopus laevis*: Their relationship to RNA synthesis in vivo and their fidelity of transcription in vitro. Cold Spring Harbor Symp. Quant. Biol. **35**, 727—735 (1970)

Rosenfeld, G. C., Comstock, J. P., Means, A. R., O'Malley, B. W.: Estrogen-induced synthesis of ovalbumin messenger RNA and its translation in a cell-free system. Biochem. Biophys. Res. Commun. **46**, 1695—1903 (1972a)

Rosenfeld, G. C., Comstock, J. P., Means, A. R., O'Malley, B. W.: A rapid method for the isolation and partial purification of specific eukaryotic messenger RNAs. Biochem. Biophys. Res. Commun. **47**, 387—392 (1972b)

Ross,J., Ikawa,Y., Leder,P.: Globin messenger-RNA induction during erythroid differentiation of cultured leukemia cells. Proc. Natl. Acad. Sci. **69**, 3620—3623 (1972)
Rourke,A.W., Heywood,S.M.: Myosin synthesis and specificity of eukaryotic initiation factors. Biochemistry **11**, 2061—2066 (1972)
Rudall,K.M., Kenchington,W.: Arthropod silks: The problem of fibrous proteins in animal tissues. Ann. Rev. Entomol. **16**, 73—96 (1971)
Rudkin,G.T.: Replication in polytene chromosomes. In: Developmental Studies on Giant Chromosomes. Results and Problems in Cell Differentiation, Vol. IV, Beermann,W. (ed.), pp. 58—85. Berlin-Heidelberg-New York: Springer 1972
Sampson,J., Mathews,M.B., Osborn,M., Borghetti,A.F.: Hemoglobin messenger ribonucleic acid translation in cell-free systems from rat and mouse liver and Landcutz ascites cells. Biochemistry **11**, 3636—3640 (1972)
Sasaki,C.: On the affinity of our wild and domesticated silkworms. Anno. Zool. Japan **2**, 33—41 (1898)
Sasaki,T., Noda,H.: Studies on silk fibroin of *Bombyx mori* directly extracted from the silk gland. I. Molecular weight determination in guanidine hydrochloride or urea solutions. Biochim. Biophys. Acta **310**, 76—90 (1973a)
Sasaki,T., Noda,H.: Studies on silk fibroin of *Bombyx mori* directly extracted from the silk gland. II. Effect of reduction of disulfide bonds and subunit structure. Biochim. Biophys. Acta **310**, 91—103 (1973b)
Schultz,J.: Genes, differentiation, and animal development. Brookhaven Symp. Biol. **18**, 116—147 (1965)
Schutz,G., Beato,M., Feigelson,P.: Isolation of eukaryotic messenger RNA on cellulose and its translation in vitro. Biochem. Biophys. Res. Commun. **49**, 680—689 (1972)
Shibukawa,A.: Studies on the silk-substance within the silk gland in the silkworm, *Bombyx mori* L. Bull. Sericult. Exptl. Sta. (Tokyo) **15**, 399—401 (1959), in Japanese with English summary
Shigematsu,H., Takeshita,H.: Effects of gamma-irradiations on development of silk glands and silk formation of the silkworm, *Bombyx mori* L. Bull. Sericult. Exptl. Sta. (Tokyo) **23**, 121—148 (1968), in Japanese with English summary
Shimizu,M., Fukuda,T., Kirimura,J.: The silkprotein. In: The Protein Chemistry. Vol. V, pp. 317—377. Tokyo: Kyoritsu Shuppan 1957, in Japanese
Shimizu,S., Horiuchi,Y.: Relation between the number of silk gland cells and the amount of silk secretion in the silkworm, *Bombyx mori* L. J. Sericult. Sci. Japan **21**, 37—38 (1952), in Japanese
Shimura,K.: Biosynthesis of silk fibroin. Japan J. Biochem. (Tokyo) **36**, 301—316 (1964), in Japanese
Shimura,K., Ejiri,S., Kobayashi,T.: Enzymatic aspects of protein synthesis on the polysomes of silkglands. 7th Intern. Cong. Biochem. (Tokyo), Abstract I, pp. 121—122 (1967)
Shimura,K., Fukai,H., Suto,S., Hoshi,R.: Studies on the biosynthesis of silk fibroin. I. Incorporation of glycine-^{14}C into fibroin in vivo. J. Biochem. **45**, 481—488 (1958)
Sinclair,J.H., Brown,D.D.: Retention of common nucleotide sequence in the ribosomal deoxyribonucleic acid of eukaryotes and some of their physical characteristics. Biochemistry **10**, 2761—2769 (1971)
Skoultchi,A., Gross,P.R.: Maternal histone messenger RNA: Detection by molecular hybridization. Proc. Natl. Acad. Sci. **70**, 2840—2844 (1973)
Sober,H.A. (ed.): Handbook of Biochemistry; Selected Data for Molecular Biology. Cleveland: Chemical Rubber Co. 1970
Sprague,K.U.: The *Bombyx mori* silk proteins: Characterization of large polypeptides. Biochem. **14**, 925—931 (1975)
Stavnezer,J., Huang,R.C.C.: Synthesis of a mouse immunoglobulin light chain in a rabbit reticulocyte cell-free system. Nature New Biol. **230**, 172—176 (1971)
Stent,G.S.: The operon: On its third anniversary. Science **144**, 816—820 (1964)
Sueoka,N., Kano-Sueoka,T.: Transfer RNA and cell differentiation. In: Progress in Nucleic Acid Research and Molecular Biology, Davidson,J.N., Cohen,W.E. (eds.), Vol. X, 23—55. New York: Academic Press 1970

Suzuki, Y.: Fibroin messenger RNA and its genes. Advanc. Biophys. Vol. **8**, 83—114 (1976)
Suzuki, Y., Brown, D.D.: Differentiation of the silk gland in *Bombyx mori*. Carnegie Inst. Year Book **68**, 509—510 (1969)
Suzuki, Y., Brown, D.D.: Isolation and identification of the messenger RNA for silk fibroin from *Bombyx mori*. J. Mol. Biol. **63**, 409—429 (1972)
Suzuki, Y., Gage, L.P., Brown, D.D.: The genes for fibroin in *Bombyx mori*. J. Mol. Biol. **70**, 637—649 (1972)
Suzuki, Y., Giza, P.E.: The regulations of fibroin genes. Carnegie Inst. Year Book **74**, 36—42 (1975)
Suzuki, Y., Giza, P.E.: Accentuated expression of silk fibroin genes in vivo and in vitro. J. Mol. Biol. **107**, 183—206 (1976)
Suzuki, Y., Suzuki, E.: Quantitative measurements of fibroin messenger RNA synthesis in the posterior silk gland of normal and mutant *Bombyx mori*. J. Mol. Biol., **88**, 393—407 (1974)
Swan, D., Aviv, H., Leder, P.: Purification and properties of biologically active messenger RNA for a myeloma light chain. Proc. Natl. Acad. Sci. **69**, 1967—1971 (1972)
Takami, T., Kitazawa, T.: Developmental table of *Bombyx mori* embryo. Tech. Bull. Sericult. Exptl. Sta. (Tokyo) **75**, 1—31 (1960), in Japanese
Tanaka, S., Shimura, K.: Stimulation of amino acid incorporation in an *Escherichia coli* cell-free system by silk gland RNA. J. Biochem. **58**, 145—152 (1965)
Tashiro, Y., Morimoto, T., Matsuura, S., Nagata, S.: Studies on the posterior silk gland of the silkworm, *Bombyx mori*. I. Growth of posterior silk gland cells and biosynthesis of fibroin during the fifth larval instar. J. Cell Biol. **38**, 574—588 (1968)
Tashiro, Y., Ohtsuki, E.: Studies on the posterior silk gland of the silkworm *Bombyx mori*. IV. Ultracentrifugal analysis of native silk proteins, especially fibroin extracted from the middle silk gland of the mature silkworm. J. Cell Biol. **46**, 1—16 (1970a)
Tashiro, Y., Ohtsuki, E.: Dissociation of native fibroin by sulphydryl compounds. Biochim. Biophys. Acta **214**, 265—271 (1970b)
Tashiro, Y., Ohtsuki, E., Shimadzu, T.: Sedimentation analysis of native silk fibroin in urea and guanidine · HCl. Biochi. Biophys. Acta **257**, 198—209 (1972)
Tazima, Y.: The Genetics of the Silkworm. London: Logos Press and Academic Press 1964
Tazima, Y., Inagaki, E.: The chromosome number of an Indian wild silkworm, *Theophila religiosae* Helf. Ann. Rep. Natl. Inst. Genet. (Mishima, Japan) **9**, 24—25 (1959)
Thomas, C., Brown, D.D.: Localization of the genes for silk fibroin in silk gland cells of *Bombyx mori*. Develop. Biol. **49**, 89—100 (1976)
Warwicker, J.O.: Comparative studies of fibroins. II. The crystal structures of various fibroins. J. Mol. Biol. **2**, 350—362 (1960)
Watanabe, T.: The amino acid composition of the mulberry leaf protein. J. Sericult. Sci. Jap. **22**, 102—103 (1953), in Japanese
Watanabe, T.: Studies on the sericin cocoon. I. Chemical properties of the domestic silkworm spinning sericin cocoon. J. Sericult. Sci. Jap. **28**, 251—256 (1959), in Japanese with English summary
Wensink, P., Brown, D.D.: Denaturation map of the ribosomal DNA of *Xenopus laevis*. J. Mol. Biol. **60**, 235—247 (1971)
Williamson, R., Clayton, R., Truman, D.E.S.: Isolation and identification of chick lens crystalline messenger RNA. Biochem. Biophys. Res. Commun. **46**, 1936—1942 (1972)
Yamane, T.: Correlation between aminoacyl-sRNA pool and amino acid composition of proteins. J. Mol. Biol. **14**, 616—618 (1965)
Yamanouchi, M.: Morphologische Beobachtung über die Seidensekretion bei der Seidenraupe. J. Coll. Agr. Hokkaido Imperial Univ. (Sapporo, Japan) **10**, Part 4, 1—49 (1921)
Yčas, M., Vincent, W.S.: The ribonucleic acid of *Epanaphe moloneyi* Druce. Exptl. Cell Res. **21**, 513—522 (1960)
Yoshitake, N.: Phylogenetic aspects on the origin of Japanese race of the silkworm, *Bombyx mori* L. J. Sericult. Sci. Jap. **37**, 83—87 (1968a)
Yoshitake, N.: Esterase and phosphatase polymorphysm in natural population of wild silkworm, *Theophila mandarina* L. J. Sericult. Sci. Jap. **37**, 195—200 (1968b)

The Eggshell of Insects: Differentiation-Specific Proteins and the Control of Their Synthesis and Accumulation During Development

F. C. Kafatos, J. C. Regier, G. D. Mazur, M. R. Nadel, H. M. Blau, W. H. Petri, A. R. Wyman, R. E. Gelinas, P. B. Moore, M. Paul, A. Efstratiadis, J. N. Vournakis, M. R. Goldsmith, J. R. Hunsley, B. Baker, J. Nardi, and M. Koehler

Cellular and Developmental Biology
The Biological Laboratories, Harvard University, Cambridge, MA 02138, USA

I. Introduction

A. Some General Comments on Differentiation

Cell differentiation is the consequence of differential gene expression in space and in time. Differential expression in turn depends on cellular differences, initially established through inheritance of non-chromosomal, cytoplasmic determinants from a polarized progenitor, or through interactions with neighboring cells. Determination, the process of setting up and maintaining the control mechanisms underlying differential gene activity, is obviously of ultimate biological interest. Although a molecular description of determination in eukaryotes is beyond our present grasp, a rigorous biochemical description of differential gene expression is certainly possible. This is easiest with highly differentiated cells, i.e., with cells which, at some point in their life, become specialized for the large-scale production of one or a few cell-specific protein products. Their methodological advantages have been enumerated elsewhere (Kafatos, 1972a). Highly differentiated cells are conceptually relatively simple systems—at least, in terms of quantification of synthesis and accumulation of the characteristic protein products and their mRNA templates. We can be confident that such quantifications will soon enable us to infer at what levels of information flow differential gene expression is controlled in particular systems. We suspect that highly differentiated cells may be simple in another, even more fundamental sense: production of "luxury" proteins (those not required for maintenance) may be devoid of a whole intricate network of homeostatic regulation.

In this introduction, we will summarize our current ideas on developmental regulation of specific protein synthesis, drawing heavily on studies with highly differentiated cells. We will emphasize the temporal dimension of differentiating systems and its implications for regulatory strategies. We can then use this perspective in discussing the insect follicular cells, as a model for stable programs of sequential gene expression.

B. Highly Differentiated Cells with a Long Time-Constant

A general pattern of multiphasic development has emerged from studies of highly differentiated cells (Rutter et al., 1968; Kafatos, 1972a; Rutter et al., 1973). Initially, these cells are preoccupied with growth and multiplication, and synthesize their characteristic products at only very low rates (Phase I). Later, the cell-specific protein synthesis becomes more obvious, and continues to accelerate over a significant period of time, often days (Phase II). Tissue growth usually subsides during this stage, but it may persist long enough to affect the apparent kinetics of cell-specific protein synthesis. Finally, the synthesis is modulated, often decreasing rapidly (Phase III). In many systems, repeated cycles of Phases II and III are possible (e.g., the silk gland during the molt cycles, or the chicken oviduct under estrogen administration and withdrawal). Figure 1 diagrams this developmental pattern, as exemplified in the cocoonase-producing gland of silkmoths, the galea (Kafatos, 1972a).

The dramatic changes of protein synthesis during Phase II of highly differentiated tissues may be explained by a simple regulatory model (Kafatos, 1972a, b). According to this "message accumulation model," the rate of protein synthesis is dictated solely by the amount of mRNA; at the beginning of Phase II specific transcription is reset to a higher rate, bringing about the continued accumulation of the mRNA up to a new steady-state level. If message decay is exponential, message

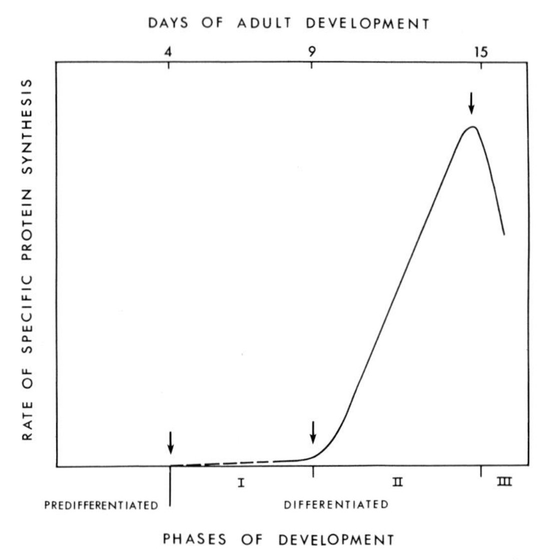

Fig. 1. The phases of differentiation, as exemplified by the changing rate (ordinate) of prococoonase synthesis in the silkmoth galea. (From Kafatos, 1972a)

accumulation will proceed with the following kinetics (under certain assumptions; Kafatos, 1972a, b),

$$M_t = M_{max} - (M_{max} - M_0)e^{-\frac{t \ln 2}{T}} \tag{1}$$

where M_0, M_{max}, and M_t are the message levels at time zero, at the new steady state attained after an increase in transcription, and at an intermediate time, t, respectively; T is the mRNA half-life during this period. It follows that the time, $t_{1/2}$, needed for accumulation of half the maximal mRNA will be nearly equal to the mRNA half-life:

$$t_{1/2} = T \frac{\ln\left(2 - \frac{M_0}{M_{max}}\right)}{\ln 2} \cong T. \tag{2}$$

In other words, in this model the longer the half-life of the message, the longer the duration of Phase II and the higher the steady state level of mRNA. The half-life of mRNAs coding for the cell-specific products of highly differentiated cells is often estimated in days. The "message accumulation model" will hold for these systems only if Phase II is comparably long. We call highly differentiated cells with such a long Phase II "long time-constant systems." Under this model, no specific gene amplification, redundancy or translational control is necessary, since the needed transcriptional activity does not exceed the maximum which can be expected for any single gene (approximately 10–20 mRNA molecules per min at 25°; Kafatos, 1972a, b).

Not all predictions of the model have been tested in any one system. Undoubtedly, minor variations will occur in specific cases. For example, continued genome replication early in Phase II will prolong the kinetics and make them sigmoid, rather than monotonic as predicted by Equation (1); if transcription is readjusted upward slowly rather than as a step function, it will again lead to sigmoidal kinetics; the end of Phase II will come earlier than expected if transcription is shut off before the new steady-state is attained. Nevertheless, information from a variety of systems tends to support the model in its broad features. Gene amplification or redundancy appear not to occur in the translationally very active systems examined: silk glands (Gage, 1974), erythroblasts (Bishop and Freeman, 1974), chick oviduct (Sullivan et al., 1973; Harris et al., 1973). Transcription of stable fibroin mRNA has been documented throughout Phase II in the silk gland (Suzuki, this vol. p. 29). Transcription of a stable, giant mRNA appears to proceed on the Balbiani ring 2 of chromosome IV in *Chironomus tentans* at a rate near but still under the theoretical maximum (at 18°, 2 to 3 mRNA molecules/gene/min, with the RNA polymerases spaced every 450–2100 nucleotides, depending on the size estimate of the RNA, and the RNA chain elongation rate equal to 1.5×10^3 nucleotides/min; estimated from Daneholt, 1973 and 1975; cf. Kafatos, 1972a). The rate of ovalbumin production in the oviduct under secondary estrogen stimulation parallels exactly the accumulation of assayable ovalbumin mRNA; the rate of synthesis of the mRNA has been estimated indirectly and found to be nearly constant from beginning to end of Phase II, at a value approaching the maximum possible with a single gene copy per genome (22 to 34 ovalbumin mRNA molecules/cell/min; Palmiter, 1973).

The shutoff of specific protein synthesis in Phase III may well involve controls more intricate than readjustment of transcription. Estrogen withdrawal from the oviduct leads to a precipitous decline of ovalbumin synthesis, which is explained by a combination of mRNA disappearance and decline of translational productivity (Palmiter and Schimke, 1973). In rabbit reticulocytes, α-globin mRNA may accumulate in the post-ribosomal supernatant (Gianni et al., 1972; Jacobs-Lorena and Baglioni, 1972; Spohr et al., 1972). In the silk gland, fibroin mRNA is rapidly destroyed at the end of each instar, only to begin accumulation again after the molt (Suzuki, this vol., p. 27). The decline of specific protein synthesis during Phase III merits further attention, as does the increase during Phase II.

C. Short Time-Constant Systems

With slight modifications, the "message accumulation model" of Phase II can also describe developing systems characterized by rapid changes in protein synthesis. An example is HTC cells, which synthesize tyrosine aminotransferase (Steinberg et al., 1975). Addition of steroids induces the synthesis of the enzyme, and their removal represses this synthesis. In either case, a ten-fold change in the rate of specific protein synthesis occurs within a few hours, as a result of changes in the amount of active mRNA (rather than its utilization). The data fit well a model in which (1) induction is the result of a nearly ten-fold increase in TAT mRNA production after steroid addition (an increase which is somewhat gradual, and shows a significant lag), and (2) deinduction is the result of a corresponding decrease in TAT mRNA production after steroid removal (as a step function, but again marked by a lag). The rapid kinetics of TAT regulation are consistent with an mRNA half-life of less than 1.5 h during both induction and deinduction. The fully induced TAT production rate is less than 0.5% of total protein synthesis, even though the rate of mRNA production is estimated as 5 molecules/gene/min (assuming one TAT gene per genome), comparable to that estimated for highly differentiated cells (see Sect. I.B., and Kafatos, 1972a, b). These results suggest that the Phase II strategy of highly differentiated long time-constant cells may be applicable to some short time-constant cells, with due allowance for rapid mRNA turnover.

However, in some systems contemporaneous transcription may be inadequate to support exceedingly rapid changes in the absolute rate of protein synthesis. This may be true for systems such as some eggs, seeds and spores which are initially relatively or totally inactive. Upon activation, such systems show a substantial stimulation of protein synthesis within minutes, by recruiting into the translational machinery stored, previously "masked" messengers. This mechanism has been documented particularly well in the sea urchin egg (Gross et al., 1973), in which activation of protein synthesis occurs even in the absence of a nucleus (Denny and Tyler, 1964) and of transcribing mitochondria (Craig and Piatigorsky, 1971).

Gene multiplicity is another possible solution, for synthetic needs beyond those which contemporaneous transcription can support (Kafatos and Gelinas, 1974). This is apparently the explanation for the relatively high reiteration frequency of histone genes (Weinberg et al., 1972). Consider a ca. 500-cell embryo of the sea urchin, *Paracentrotus*. From the rate of cleavage, the amount and size distribution of new histones needed per new genome, and the estimated translational productivity (output of polypeptides per active mRNA per min), it can be estimated that the embryo must have a minimum of 50 pg histone mRNA. Only a small proportion of that may be derived from the stored maternal message. Hybridization data presented by Gross et al. (1973) indicate that the concentration of histone message in the total RNA of unfertilized eggs is 7.5% of the concentration in the histone-synthesizing small polysomes; since the total egg RNA is 1.5×10^{-9} g and certainly less than 2% of the RNA in small polysomes is histone mRNA (Humphreys, 1971), the maternal histone message is less than 3 pg. In less than half an hour, within a single cleavage cycle, the 500-cell embryo must generate another 50 pg of histone mRNA; that entails production of approximately 3×10^4 histone mRNAs per min per genome, and necessitates ca. 2×10^3 total histone genes (at a nearly maximal transcriptional productivity of 15 mRNA molecules/min/gene; Kafatos, 1972a). Similar considerations of embryonic histone synthesis lead to the prediction that histone genes must also be repetitive in Drosophila, as they are (Birnstiel et al., 1974). Histone synthesis during the S-phase of somatic cells is lower than embryonic histone synthesis and may require in itself only a low degree of gene reiteration (Kafatos and Gelinas, 1974).

In conclusion, short time-constant systems face a variety of challenges in terms of kinetics of specific protein synthesis: rapid changes of a small absolute magnitude (TAT), rapid but nonrepeating changes of substantial magnitude (fertilization), rapid and sustained changes of substantial magnitude (embryonic histone synthesis). The strategies employed may vary accordingly: short mRNA half-life, accumulated masked messengers, repetitious genes. We may predict that if specific amplification of translatable genes ever occurs, it will be in highly differentiated somatic cells, faced with rapid sustained increases in protein synthesis beyond what can be supported by contemporaneous transcription.

D. Autonomous Programs of Cell-Specific Protein Synthesis

Although highly differentiated, long time-constant cells are convenient to study, they may not show some important regulatory strategies employed by short time-constant systems—and they may focus attention away from two important aspects of differentiation: its sequential and stable nature.

Much of differentiation, even in long time-constant systems, involves sequential production of a number of proteins, rather than changes in the synthesis of a single characteristic protein (e.g., fibroin or cocoonase). Various secretory enzymes appear in the embryonic pancreas with distinct kinetics (Rutter et al., 1973), and the same holds true for several secretory proteins of the chick oviduct under estrogen induction (Palmiter, 1972). Differentiation in the vertebrate eye lens (Clayton, 1970) and feathers (Kemp and Rogers, 1972) entails sequential production of many structural proteins. Sequential differentiation is even more apparent in somewhat less specialized systems. In slime mold development, specific enzymes appear and disappear in a highly ordered sequence (Newell and Sussman, 1970). Early embryogenesis undoubtedly entails sequential changes in the synthesis of a large number of proteins. It may be methodologically convenient to study sequential differentiation in cells which are highly differentiated but of short time-constant. An example is the salivary gland of Diptera, where ecdysone-triggered differentiation is manifested as a rapid program of sequential puff activation and regression (e.g., Ashburner et al., 1974).

Another key aspect of differentiation is its stability—in the sense of a stably programmed sequence of changes. A program of differentiation may require an initial trigger, but then it usually becomes independent of the stimulus, i.e., autonomous. Good examples are embryogenesis, the late puffing response, and slime mold development. For autonomous development, regulatory circuits must exist which ensure adherence to the normal synthetic timetable. Elucidation of these circuits is one of the major tasks facing developmental biology today. We believe that the circuits can be revealed by studying short time-constant highly differentiated cells which are programmed for sequential synthesis of several distinct products—especially if the biochemical studies can be complemented with genetic analysis.

E. The Follicular Epithelium of Insect Eggs

The follicular epithelium of insects is an exocrine gland which is programmed to produce a series of extracellular products, the last and most prominent of which is the eggshell or chorion. Chorion production entails sequential synthesis of a number of structural proteins, according to an autonomous and apparently short time-constant program. At the time of chorion formation, the follicular cells are almost totally specialized for this synthetic function. The chorion proteins can be purified in substantial amounts, and are generally easy to quantify. Chorion mRNAs have proven relatively easy to identify and assay as a class. Mutants affecting chorion production are available. For these and other reasons, we have chosen to study choriogenesis in insects, with the long-term goal of understanding how autonomous programs of cell-specific protein synthesis are regulated.

Much of our work on this system is as yet unpublished. In this chapter we use our unpublished results without special attribution. In the first several sections, we describe in some detail, both morphologically and biochemically, the chorion and its morphogenesis. Assembly of the newly synthesized chorion proteins into the eggshell is a final expression of the chorion genes. In order to have a full understanding of chorion gene expression, we must understand not only the regulation of protein synthesis but also the posttranslational fate of the proteins, including their secretion and subsequent assembly into an ultrastructurally complex unit. This presupposes a detailed knowledge of chorion morphology and of chorion protein chemistry.

The follicle system has several advantages for morphogenetic as well as protein synthetic studies. The ultrastructure of the chorion suggests that its extracellular morphogenesis is more complex than virus assembly. The proteins are mostly small and structurally related, facilitating the determination of primary sequences which, along with structural studies of higher order, are necessary for understanding assembly. Finally, the structural and chemical variability of the chorion among different groups of insects should be helpful for studies of the relationship between structure and function at many levels.

The chorion is clearly a rich source of problems in cell biology, protein structure and function, molecular evolution, as well as developmental biology. For us, these studies are individually exciting and mutually beneficial.

II. The Developmental Career of Follicular Epithelial Cells

A. Formation of the Follicle

The insect ovary consists of a number of follicle-containing tubes or ovarioles. The organization of the ovariole in different insects is variable in detail (Telfer, 1975). Each follicle consists of a single oocyte and associated accessory cells. Follicles can be classified as meroistic or panoistic. They are *meroistic* if the oocyte is accompanied by a cluster of nurse cells, mitotic siblings of the oocyte which are trophic in function. Meroistic follicles are *polytrophic* if each has a distinct cluster of nurse cells, and *telotrophic* if all follicles of the ovariole share a single terminal nurse cell cluster. If the nurse cells are totally absent, the follicle is *panoistic*.

A common feature of all insect follicles, whether meroistic or panoistic, is the presence of follicular epithelial cells[1]. Unlike the oocyte and the nurse cells, which are derived from the primordial germ cells, the follicular cells are apparently mesodermal in origin (Telfer, 1975). They form a unicellular layer surrounding the oocyte, and play two major roles in oogenesis: participation in the process of yolk uptake (vitellogenesis), and formation of the egg membranes (vitelline membrane plus chorion). The latter function is the main focus of this article.

We shall be concerned chiefly with the follicles of two groups of insects, the silkmoths and *Drosophila*. Both are polytrophic and similar in many respects. In this section we summarize the main features of follicle development in the silkmoths (families Saturniidae and Bombycidae), postponing consideration of some peculairities of the Drosophila[2] follicle until Section XII.

Like most Lepidoptera, the silkmoths have four ovarioles in each of two ovaries. The ovariole is an egg tube which generates new follicles at the anterior tip, the germarium, matures them more posteriorly in the vitellarium, and releases mature, ovulated eggs through its opening to the oviduct. At the apex of the germarium, the stem line oogonia divide mitotically to form one stem line oogonium which remains at the apex, and one which becomes a "cystoblast" (Koch and King, 1966). In the silkmoths, the cystoblast undergoes three more mitoses to form a complex of one oocyte and seven nurse cells (King and Aggarwal, 1965). The cystoblast divisions are incomplete. As a result, intercellular bridges or cytoplasmic strands link

---

[1] For brevity, the terms follicle cell or follicular cell will be used to refer to the cells of the follicular epithelium.

[2] Throughout this paper, we use the following trivial names: cecropia for *Hyalophora cecropia*, polyphemus for *Antheraea polyphemus*, pernyi for *A. pernyi*, Bombyx for *Bombyx mori* and Drosophila for *Drosophila melanogaster*.

Fig. 2. Ovariole dissected from a developing polyphemus moth. Note the abrupt increase in size and change in hue of the first follicle in terminal growth *(arrow)*

all eight cells in an oocyte-nurse cell syncytium (see Telfer, 1975, for a thorough review).

When the oocyte-nurse cell complexes grow to a size approximating the diameter of the ovariole, the confining wall of the ovariole forces them into single file. As each oocyte-nurse cell complex moves from the germarium into the vitellarium, it is surrounded by mesodermal prefollicular tissue derived from the epithelium of the ovariole. At this time the oocyte is in the first meiotic prophase. The follicle cells ahead of and behind each complex develop into interfollicular stalks. Each follicle now moves down the ovariole as new follicles emerge into the vitellarium. As a result, each ovariole contains a graded series of connected follicles which are older and larger the further they are from the apex of the germarium (Fig. 2). The nurse cells are always anterior to the oocyte, thus defining the anterior-posterior axis of the follicle.

As the oocyte grows, the follicular cells divide to form a columnar monolayer. Follicular cells at the nurse cell-oocyte junction proliferate, advance between these two cell types, and seal off the oocyte, except for the area of the cytoplasmic bridges. Follicular cell mitoses soon end, and the epithelium grows by cell enlargement and polyploidization, rather than by cell division.

B. Functions of the Follicular Cells During Vitellogenesis

In the silkmoths, the adults are non-feeding and the eggs are formed during the metamorphosis of the pupa into an adult, using the nutrients accumulated by the caterpillar. Follicles are formed in the germarium late in larval life (King, 1970), and they begin to enter the vitellarium after the initiation of adult development. As adult development proceeds and the earliest follicles continue to mature, additional follicles enter the vitellarium.

During vitellogenesis, the oocyte grows mainly through accumulation of yolk from the blood (Telfer, 1965). At this time, the follicle cells detach from each other to form channels for penetration of hemolymph to the vicinity of the oocyte surface (King and Aggarwal, 1965; Telfer and Smith, 1970). Once there, yolk proteins from the hemolymph are taken up into the oocyte by pinocytosis. In addition, the follicular cells synthesize and release into their intercellular channels a histidine- and glucosamine-rich secretion, probably a glycoprotein, which finds its way into the yolk spheres of the oocyte (Anderson and Telfer, 1969). From studies of Trypan Blue uptake, it has been suggested that the glycoprotein facilitates specific uptake of the yolk protein, vitellogenin, by binding to it selectively. Thus, during vitellogenesis, the follicle cells appear to regulate the access of blood proteins to the oocyte, by physical and chemical means. In addition, the follicle cells are growing and, therefore, must be synthesizing cellular components. They apparently also secrete a thin extracellular layer between themselves and the oocyte, the permeable first layer of the vitelline membrane.

The nurse cells donate ribosomes, centrioles, 4 S RNA, and possibly other substances to the oocyte (Telfer, 1975). Toward the end of vitellogenesis they inject most of their cytoplasmic contents through the intercellular cytoplasmic bridges into the oocyte (King and Aggarwal, 1965; Telfer, 1975); the intercellular bridges are then severed and the residues of the degenerate nurse cells are left embedded in the follicular epithelium.

C. Terminal Growth Phase

In cecropia, when the oocyte has reached two thirds of its final volume the apical regions of neighboring follicle cells begin to adhere more tightly, closing up the yolk transport channels. Hemolymph proteins at this point no longer have access to the oocyte and yolk deposition is terminated. Now a 20–24 h "terminal growth phase" begins (Telfer and Anderson, 1968). During this period the oocyte absorbs water and thus increases by one third to its final volume. The shape changes from a prolate ellipsoid to a more flattened shape characteristic of the mature egg. This rather abrupt increase in volume and a concomitant decrease in coloration are very noticeable in polyphemus and pernyi (Fig. 2). As a result, it is easy to identify macroscopically the last vitellogenic follicle and the first terminal growth follicle in each ovariole; these are defined as being at position 0 and $+1$, respectively, while progressively younger and older follicles are staged as -1, -2, $-3, \ldots$ and $+2$, $+3, \ldots$, respectively.

The follicular epithelial cells continue to enlarge as the follicle enters the terminal growth phase. By the end of terminal growth, the epithelium of a single cecropia follicle has accumulated 27 μg RNA (Pollack and Telfer, 1969). Growth then ends, and the follicular epithelium maintains a nearly constant total dry weight and RNA content (polyphemus; Paul et al., 1972a).

During terminal growth, the oocyte cortex undergoes a series of changes which transform it into the periplasm (where the embryonic blastoderm forms after fertilization). The definitive egg axes also become clear at this stage. As previously mentioned, the anterior-posterior axis is defined by the anterior location of the nurse cells. The flattened sides are bilaterally symmetrical, and correspond to the right and left sides of the embryo. In some cases, at least, the dorsoventral axis is defined by an asymmetry of the two "edges" connecting the flattened sides. In Bombyx, a section along the plane of symmetry of the egg reveals that one of the two edges connecting the anterior and posterior poles is more curved than the other. The curved edge is defined as ventral, and corresponds to the site where the embryonic blastoderm will form (the rest of the oocyte cortex becoming extra-embryonic membranes; Tazima, 1964; Kuwana and Takami, 1968).

In cecropia (Telfer and Smith, 1970), the vitelline membrane (which during vitellogenesis is merely a pale-staining zone 0.2 μm thick) grows to 1.5 μm, by the addition of a dark-staining layer secreted by the follicle cells. When the vitelline membrane approaches its final thickness, a layer of overlapping plates appears at its outer surface (toward the follicular cells). These plates are present in the youngest follicles which resist osmotic shrinkage in hypertonic sucrose solution, and are thought to be a protection against dehydration. After the water barrier is established and the oocyte cortex is transformed into periplasm, the follicle cells stop interacting directly with the oocyte.

The next structure secreted by the follicle cells is an osmiophilic sieve layer (Smith et al., 1971). The follicle cell microvilli remain in contact with this layer, and all subsequent follicle cell secretions must pass through it. Although the sieve layer is extracellular, it remains associated with the follicle cell microvilli even after ovulation (when the epithelium is stripped off the chorion-covered oocyte).

D. Chorion Secretion and Ovulation

The last function of the follicular epithelium is production of a massive proteinaceous eggshell or chorion, which accounts for nearly a third of the dry weight of the egg in some species (Pollack and Telfer, 1969). Chorionating follicles first appear in the saturniids two-thirds of the way through the three-week period of adult development.

In any one follicle, choriogenesis lasts approximately two days (in polyphemus; see Sect. VIII). The follicular epithelium is then shed (ovulation) and the chorion-covered egg is stored in the oviduct. When the adult moth emerges, she contains several dozen (saturniids) to approximately eighty (Bombyx) eggs per ovariole, ready for fertilization.

The synthetic activity of the follicle cells reaches its peak as chorion secretion begins. Many secretory granules appear in the apical cytoplasm and between the microvilli. The sequential deposition of the different layers of the chorion is described in Section III.

Disassembly of the synthetic and secretory apparatus is already evident during the first half of choriogenesis in Bombyx. In cecropia (Smith et al., 1971), disassembly reaches a peak near the time of ovulation. In both cases, membrane-bound isolation bodies containing ribosomes and mitochondria in various stages of degeneration are seen in the cytoplasm.

At ovulation the epithelium is stripped off the chorion-covered oocyte and left within the ovariole, marking the border between chorionating and ovulated eggs. The sloughed epithelium, although atrophied, remains morphologically intact (Telfer and Smith, 1970) throughout the short succeeding life of the adult moth.

III. Structure of the Silkmoth Chorion

A. Macroscopic and Surface View

Silkmoth eggs are large and extremely hard. Their shape is that of a laterally flattened ellipsoid, with a major diameter ranging from 1.5 mm for Bombyx and 3 mm for polyphemus and pernyi, to 3.5 mm in *Antheraea mylitta*. The dry weight of the shell is approximately 40 μg in Bombyx and 460 μg in polyphemus.

A polygonal pattern of ridges marks the surface of the eggshell (Figs. 3 and 4). The appearance and size of the polygons vary not only among species (Figs. 3–11) but also among different strains of Bombyx (Sakaguchi et al., 1973). Each polygon marks the area under one cell, since the ridges are formed at the junction of adjacent follicle cells. In Bombyx the space enclosed by the ridges is not flat, but consists of 4–8 low knobs (Fig. 11), the average number depending on the strain. In polyphemus there are approximately 10000 polygons and therefore an equal number of follicular cells.

The surface of the chorion is dotted with the openings of radial air channels or aeropyles. In many species the aeropyles are found at branching points of the ridges, i.e., at the juncture of three cells (Hinton, 1969; 1970; Figs. 6–9). In other cases (Beament, 1946; Hinton, 1969; Sakaguchi et al., 1973; Fig. 11), they are confined to the space between the ridges. The aeropyles may be simple small holes or elaborate chimney-like structures, reflecting the respiratory physiology of the different species (Sect. III.B.).

In some species, regional differences within the chorion are revealed by the distribution of various types of aeropyles and of other surface structures. In the polyphemus chorion (Figs. 3, 4), the two flat "sides" have simple aeropyles surrounded by a thin rim (Figs. 3, 4, 8). The longest meridian, passing through the anterior, posterior, dorsal and ventral aspects, coincides with a "stripe" of distinctive appearance, approximately 10 cells wide (Figs. 3, 4). Here aeropyles are not apparent and the regular pattern of ridges is disrupted; the imprint of each cell is smaller than at the sides. Between the stripe and the sides, in the area of highest curvature, are bilaterally symmetrical zones of protruding, tall aeropyles (Figs. 3, 4). These zones usually join at two points, near the anterior (Fig. 3) and posterior poles. Each zone shows a gradient in complexity of the tall aeropyles (Figs. 4, 6, 7). In

Fig. 3. Scanning electron micrograph showing the various surface zones of the mature (ovulated) chorion of polyphemus. The relatively flat sides of the eggshell (*f*) (cf. Fig. 8) show polygonal imprints, each left by one follicular epithelial cell. Surrounding a meridional stripe region (*s*) are two bilaterally symmetrical bands of tall aeropyles (*t*) which join (*j*) across the stripe near the anterior and posterior poles. The border (*b*) between the zones of tall and of flat aeropyles is only one cell wide (cf. Figs. 4, 7). At the anterior pole, in the stripe, is the micropyle region (*m*) (cf. Fig. 4, 5). (× 23)

pernyi, these surface regional differences are not apparent (Fig. 10); tall goblet-shaped aeropyles occur throughout (Fig. 9).

At the anterior pole of the chorion (within the stripe region in polyphemus) is the micropyle, the opening through which sperm gain entrance to the egg (Figs. 3–5). It is bordered by a single rosette of petal-shaped cell imprints—8 in Bombyx (Sakaguchi et al., 1973), an average of 15 in polyphemus (Fig. 5) and 20 in *Ephestia kuhniella* (Cummings, 1972). The rosette is surrounded by additional elongate imprints, gradually merging into the surrounding regular pattern of polygons. The opening of the micropyle changes further inward into a species-specific number of fine passageways, which lead to the oocyte (Kuwana and Takami, 1968).

B. Physiology of the Chorion

The insect eggshell must permit absorption of oxygen for the needs of the developing embryo, while avoiding excessive water loss. Since the water molecule is smaller than the oxygen molecule, no membrane has evolved which is permeable to oxygen but not to water (Hinton, 1969). The problem is compounded by the small size of the eggs, which results in a large surface to volume ratio.

Most insect shells contain internal layers of air in interconnected networks. One such network is commonly found in the portion of the chorion closest to the oocyte. The internal air networks presumably facilitate respiration throughout the embryo surface, while minimizing water loss through infrequent openings to the outside (aeropyles); this is analogous to the air-filled space in leaves, which communicates to the outside through stomata. It is likely that the shape of the aeropyles affects gas exchange. In polyphemus, the large aeropyle chimneys meet the necessary conditions for creating ventilatory air currents by a Bernoulli effect (Vogel and Bretz, 1972).

Some eggs find themselves submerged in water, either continuously or occasionally (e.g., after a rain). In such cases, respiration depends on the existence of either breathing tubes or a "plastron" (Hinton, 1969, 1970). The latter is a gas film with extensive air-water interface and constant volume, held in place by the aeropyles or by other hydrophobic surface sculpturing. The plastron takes in oxygen by diffusion from the ambient oxygen-rich water, to replace the oxygen continuously removed by the metabolism of the egg or embryo. In a few cases, such as Drosophila, chorionic horns are elaborated to serve as breathing tubes when they protrude from the water, and as plastrons when they are totally submerged.

Fig. 4. Scanning electron micrograph illustrating at higher magnification some of the surface zones of polyphemus chorion. The zone of tall aeropyles (*t*) shows a gradient in complexity: an intricate network of chimneys with flame-like protuberances (cf. Fig. 6) is present just adjacent to the stripe region (*s*), while farther away the protuberances decrease, and distinct large aeropyles appear. Eventually a transition occurs (*b*) across a single row of cell imprints (cf. Fig. 7) to the flat aeropyles characteristic of the sides (*f*) (cf. Fig. 8). Both the tall and the flat aeropyles occur at corners of the polygonal cell-imprints, points representing the junction of three follicular epithelial cells. At the anterior pole, in the stripe region, a rosette of petal-shaped cell imprints surrounds the micropyle (*m*; cf. Fig. 5), which allows sperm penetration. Cell imprints are oriented in 4 or 5 concentric circles around the micropyle. Farther away in the stripe they become more irregular in outline. (×210)

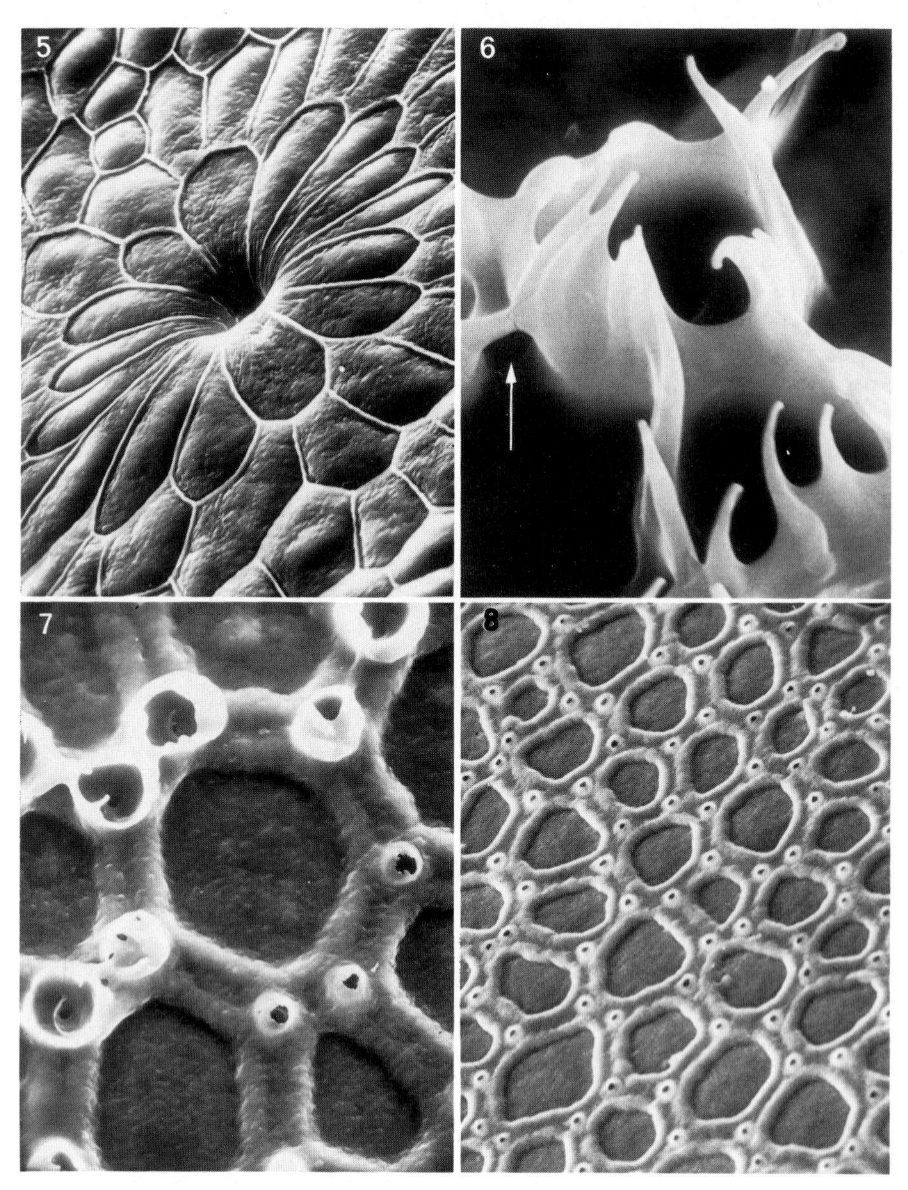

Fig. 5. Scanning electron micrograph of the micropyle in a polyphemus chorion. 15 petal-shaped cell imprints surround the central opening. The ridges formed by junctions of two overlying follicle cells are devoid of aeropyle openings and single in contrast to the ridges in the stripe further from the micropyle (cf. Fig. 4), in the border (Fig. 7) or flat aeropyle (Fig. 8) zones (×875)

Fig. 6. Scanning electron micrograph showing the tall aeropyles from a polyphemus chorion. At this location their sides are coalesced, obscuring the ridge between cell imprints. Tunnels *(arrow)* appear between contiguous aeropyles (×2450)

Fig. 7. Scanning electron micrograph illustrating at higher magnification than Figure 4 the single-cell transition between zones of high and low aeropyles (×875)

Fig. 8. Scanning electron micrograph of the zone of flat aeropyles. The double ridges formed by follicle cell junctions show a flat rimmed aeropyle at each corner. The polygonal cell imprints have an average of 6 sides (×440)

This is particularly efficient, since uptake of underwater oxygen is maximized by the length while evaporation under dry conditions is limited by the cross-sectional area.

For a plastron to be defined as efficient, it must supply, in well aerated water, a significant part of the oxygen need. In most insects with efficient plastrons the water-air interface is about 10^5–10^6 μm^2/mg wet body weight. In pernyi eggshells, the aeropyle openings (Figs. 9, 10) occupy about 12% of the surface, corresponding to a water-air interface of about 4×10^5 μm^2/mg wet body weight (Hinton, 1969). In some moth eggs the aeropyles are too few or too small to form a plastron. The minute size of the Bombyx aeropyles (Fig. 11) is probably related to the occurrence of a long egg diapause, which must impose a requirement for minimal water loss.

C. Ultrastructure of the Cecropia Chorion

The mature chorion of cecropia (Smith et al., 1971) is about 60 μm thick, and consists of two ultrastructurally distinct layers: the trabecular and the lamellate regions. The 1 μm thick trabecular layer forms first, adjacent to the vitelline membrane. It appears as a system of radial columns or trabeculae, spanning a 0.4 μm space between a fibrous base and the first layer of lamellate chorion. The space is filled with air in the mature egg; during choriogenesis it contains thick loose fibers, 150 Å in diameter, occurring either singly or in bundles.

The bulk of the cecropia chorion contains about 120 lamellae, which are made up of thin fibers. Initially these fibers are distributed sparsely, but as choriogenesis progresses they thicken and/or increase in number. In the first half of choriogenesis, any oblique section of lamellate chorion gives a striking pattern of embedded arcs or paraboloid figures (Fig. 12). The arrangement responsible for the parabolic lines has been elucidated by Bouligand, who reviews its occurrence in such widely diverse systems as crab and locust cuticle, ascidian tunica and dinoflagellate chromosomes (Bouligand, 1972). According to his theory, the chorion consists of layers, made up of fibers, more or less straight and roughly parallel to the surface of the oocyte and to each other. In each successive layer, the fiber direction is rotated in a helicoidal manner, through a small angle, about a radial axis perpendicular to the plane of the fibers. The layers are not necessarily discrete, although they are represented that way to facilitate the drawing. The lamellae become visible in sectioned chorion whenever the direction of the fibers is parallel to the plane of section; thus each lamella represents a 180° change in direction of the fibers (Diagram 1).

In electron micrographs of cecropia chorion approximately 15 fibers are visible in each lamella, implying that the angle through which the fiber direction changes from layer to layer is about 12°. As would be predicted from the geometry of the system, discontinuities or "fault lines" interrupt the lamellae, at irregular intervals.

Late in choriogenesis, the fibrous structure of the chorion lamellae is obscured by thickening and coalescing of adjacent fibers, and/or by impregnation with a matrix. This process first obscures the fibers nearest the oocyte, i.e., farthest from the secretory surface.

The lamellate chorion also contains air spaces. One type of space is the regular radial channel of each aeropyle. The future location of this channel is marked in early choriogenesis by the presence of a bundle of long villi, extending from the

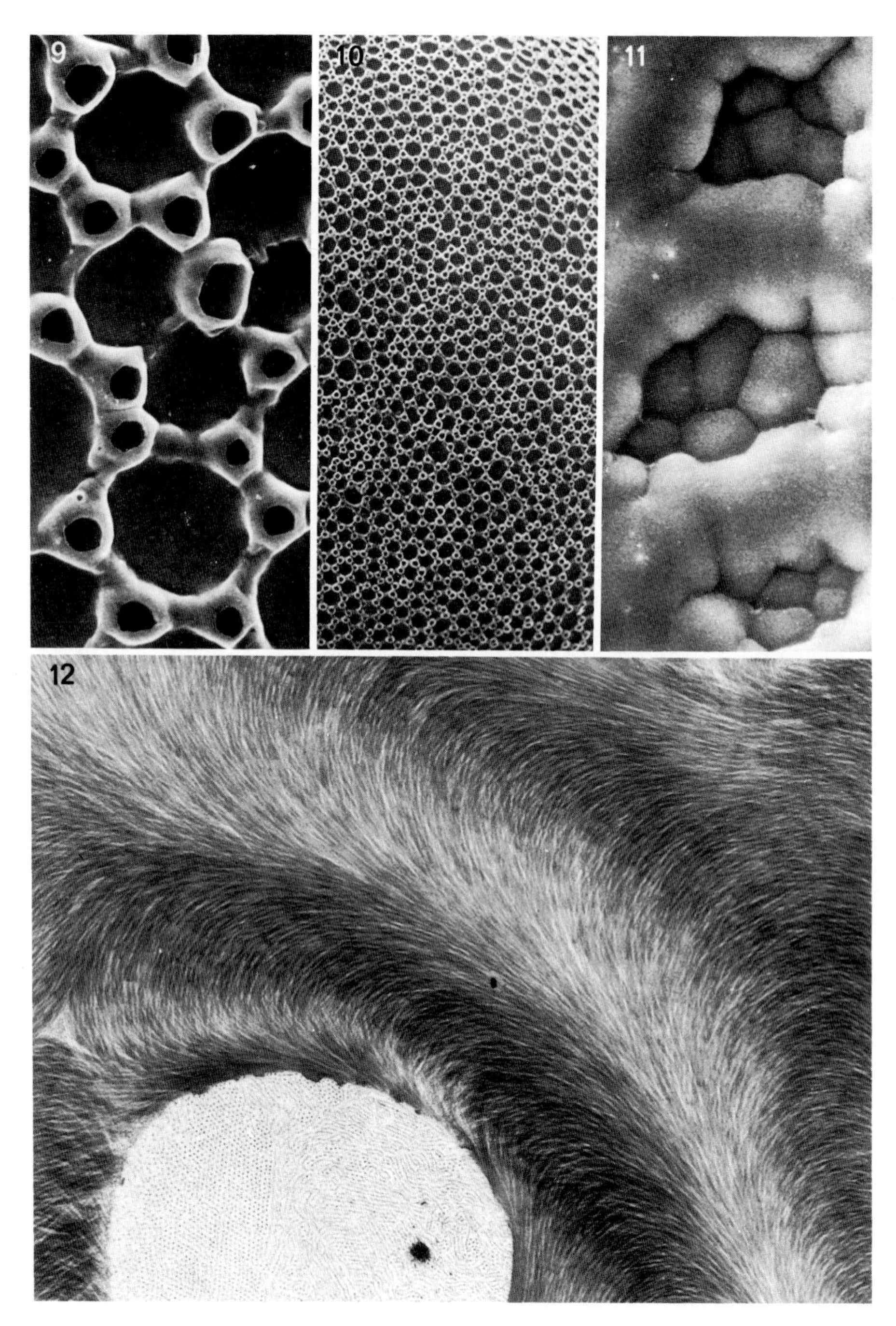

Fig. 9. Scanning electron micrograph of the mature chorion of pernyi. Except for the region around the micropyle the whole chorion is covered by tall aeropyles whose sides coalesce, often obscuring the polygonal follicle cell imprints ($\times 440$)

Fig. 10. Scanning electron micrograph of the pernyi chorion illustrating at low magnification a view of the flat side. The corresponding area of polyphemus chorion (Fig. 8) is devoid of tall aeropyles ($\times 44$)

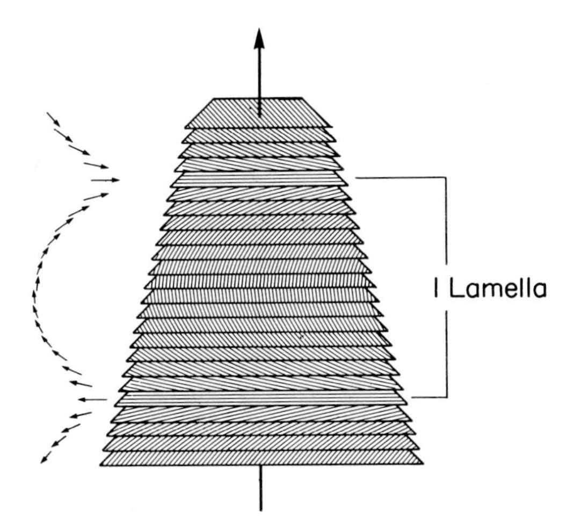

Diagram 1. Oblique section through stacked parallel layers of parallel fibers in Bouligand orientation. *Heavy arrow:* radial axis of follicle; *small arrows:* direction of fibers within each layer

follicle cell through the lamellate chorion, as far as the trabecular layer. These bundles occur at the junctions of three cells. Later in choriogenesis a cylindrical channel (1–2 µm in diameter) forms around the villi and is temporarily filled with loose flocculent material (Fig. 16). Toward the end of choriogenesis, localized secretion establishes the external rim of the aeropyle.

A second system of air spaces is an extensive network of irregular channels. Many small channels develop in the central third of the chorion, disrupting the lamellar organization into a distinct "holey layer". Larger channels (1–2 µm in diameter) radiate from this layer, some establishing communication with the space of the trabecular layer. The channels contain thick opaque threads (0.1 µm in diameter) and a nearly hexagonal array of fine filaments, 70 Å thick and spaced 200 Å apart (Fig. 12); both of these inclusions are oriented parallel to the axis of the channel. Smith et al. (1971) point out that this network of channels develops at some distance from the cytoplasm, possibly through a phase transition of previously secreted material. In this respect, as well as in the structure of the filaments, this system of spaces differs from the aeropyle channels, which are associated with villi. It is not known with certainty whether these two air systems of the lamellate chorion are confluent.

Fig. 11. Scanning electron micrograph of the chorion of *Bombyx mori* (strain GrB 703, heterozygote) showing the imprints of three follicle cells. Here the aeropyles are present only as small pores between the ridges and the knobs which fill the cell imprint (× 1540)

Fig. 12. Transmission electron micrograph of the lamellate chorion of cecropia (from Smith et al., 1971). The parabolic pattern formed by helicoidally oriented layers of parallel fibers is evident. A large air channel is seen, of the type which radiates from the central holey layer, containing a thick opaque thread (0.1 µm diameter) and a nearly hexagonal array of fine filaments; both of these inclusions run parallel with the long axis of the channel (× 29400)

D. Ultrastructure of the Bombyx Chorion

The chorion of Bombyx, while basically similar to that of cecropia, is thinner (20 to 25 µm), and contains three main morphologically distinct regions: the trabecular, inner lamellate and outer lamellate chorion (Fig. 13). The trabecular layer (1–2 µm thick) resembles that of cecropia, except that the trabeculae span a space between two fibrous spongy layers (Fig. 14) instead of terminating directly on the first layer of lamellate chorion.

The thick (15 to 20 µm) inner lamellate chorion contains 40–45 lamellae, made up of helicoidally oriented fibers as in cecropia. The first lamellae (toward the oocyte) are thinner. Extensive discontinuities of the lamellae occur, but the network of future respiratory channels is much less developed than in cecropia; the holey layer is restricted to about five lamellae in the inner quarter of the chorion (closest to the oocyte). The fibers which make up the lamellae progressively thicken and are eventually obscured, as in cecropia (Figs. 15–17). However, a distinctive feature of the Bombyx chorion is an extremely osmiophilic secondary matrix, which appears in mid-choriogenesis; it permeates the entire lamellate chorion beginning nearest the follicular cell (Fig. 18), and may even reach the cavities of the trabecular layer.

The outer lamellate layer of the Bombyx chorion also has no counterpart in cecropia. It is an electron-dense, extremely osmiophilic crust which stains as intensely as the secondary matrix (Figs. 13, 18). It is 3–4 µm thick, and contains 15–20 lamellae. While the osmiophilic lamellae are being laid down, their number varies in different parts of the chorion, indicating some regional differentiation. Fibers in Bouligand orientation are presumed to exist but cannot be seen because of the high density of the crust. This outer layer forms the knobs visible in scanning micrographs.

IV. Comparative Morphology of Insect Chorions

The morphology of insect chorions reflects diverse solutions to the common problem of providing strong and elastic mechanical protection while allowing respiration and penetration of sperm. The structure of the chorion differs, at least in detail, even among closely related species. Specialized life styles are accompanied by peculiarities in physiology and by distinctive morphology (e.g., see King et al., 1968).

A common form of mechanical support consists of fibers embedded in a matrix. As Slayter (1962) points out, two-phase materials which have fibers with high tensile strength in a weaker but more plastic matrix are stronger than either phase would be alone. The fiber phase gives strength, while the easily deformable matrix prevents the propagation of cracks from local imperfections by isolating the fibers from one another, and redistributing any stress evenly among the fibers. Two-phase systems of randomly oriented fibers in a matrix are found in grasshopper chorions (e.g., *Melanoplus differentialis*; Slifer and Sekhon, 1963), where the endochorion contains curved and twisted fibers, and in the eggshell of the dragonfly *Aeschna* (Beams and Kessel, 1969) where fibers are prominent in the exochorion.

A further elaboration of the two-phase system is parallel orientation of the fibers; in man-made composites, this may increase the impact resistance by a factor of two (Slayter, 1962). In Drosophila, the loose fibers of zone III (Quattropani and

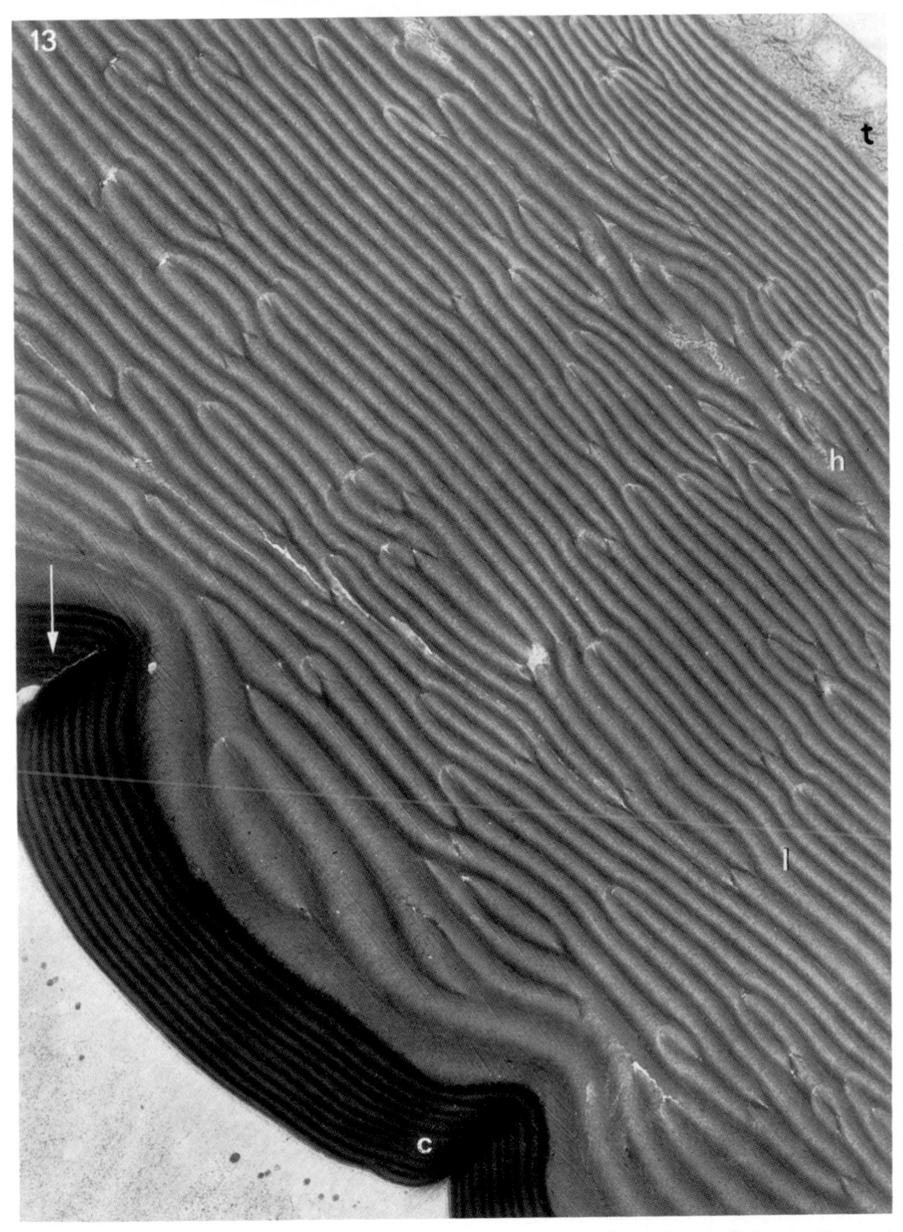

Fig. 13. Transmission electron micrograph of the chorion of Bombyx (wild type, Turtox). This chorion is from the 31st out of 43 postvitellogenic preovulatory follicles (position 0.720), and is almost fully mature. The trabecular layer (*t*), the inner lamellate layer (*l*), and much of the dense outer lamellate crust (*c*) have already been laid down. Lying closest to the oocyte, the 1–2 μm thick trabecular layer is a presumably respiratory structure formed by columns separating two spongy layers. The bulk of the chorion is in the 20 μm thick inner lamellate region, which shows many small scattered discontinuities as well as a holey region (*h*) in the inner quarter of the chorion. The outer lamellate layer forms the knobs visible in scanning electron micrographs and is occasionally traversed by an aeropyle *(arrow)*. Within each knob the lamellae have no discontinuities. They are uniform in thickness (~0.2 μm), comparable to the thinnest lamellae of the inner lamellate chorion. The apical cytoplasm of the follicular epithelial cell (*e*) shows osmiophilic secretory granules in the process of exocytosis (× 7560)

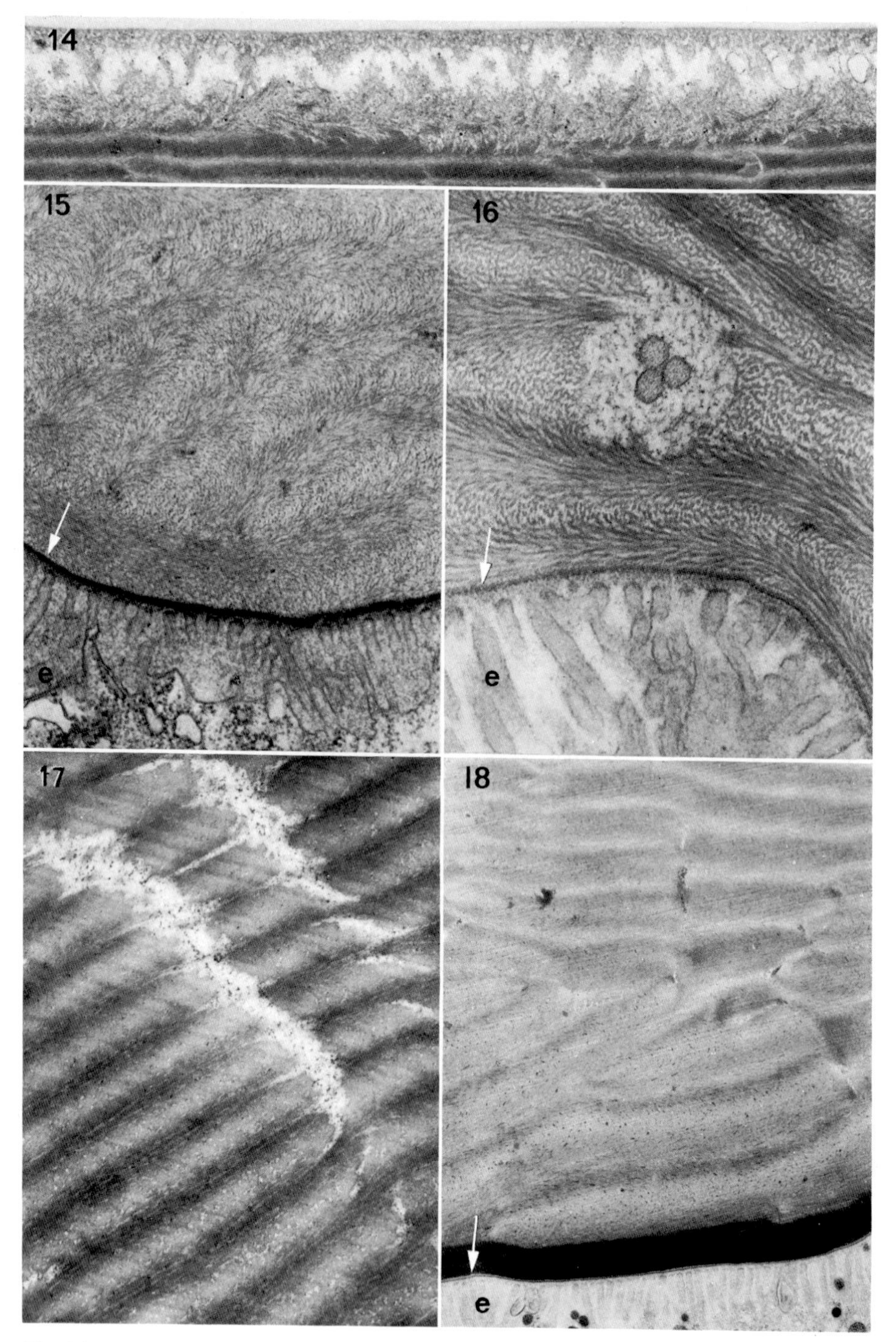

Fig. 14. Transmission electron micrograph of the chorion from the 25th out of 43 follicles (0.581) in Bombyx (wild type, Turtox). The 2 μm thick trabecular layer and two or three adjacent lamellae are shown (× 8820)

Anderson, 1969) are oriented in parallel (Margaritis, 1974). Helicoidal orientation of parallel fibers (see Sect. III.C. and Figs. 12, 15, 16) has been documented in Lepidoptera only, and may be of general occurrence in this order (Telfer and Smith, 1970; Furneaux and MacKay, 1972).

Chorions from a wide variety of insects, both aquatic and terrestrial, often contain crystalline proteins (Furneaux and MacKay, 1972). Typically, these proteins are found in contiguous small grains or crystallites, with a maximum size of 0.1 μm (comparable to the spacing of the microvilli of the secretory cells). The existence of grain boundaries implies that the crystals grow outward from crystal nuclei. Usually the crystals develop by secondary transformation of previously secreted material; in only one case (the cricket, *Acheta domesticus*) has addition of percursor material to growing crystallites been observed. The orientation of the crystallites is random in planes tangential to the chorion surface. The substructure of these crystals commonly shows a periodicity of approximately 50 Å; the subunits usually form layers parallel to the egg surface. In the grasshoppers, the crystalline proteins form long, irregular fibers with almost square cross-section (0.25 μm), rather than grains. The prevalence of the crystalline material and its location within the chorion are widely variable; no functional interpretation is as yet possible. Crystalline components have been observed in members of seven orders: Odonata, Orthoptera, Hemiptera, Neuroptera, Coleoptera, Hymenoptera and Diptera (Drosophila: L. Margaritis, personal communication).

Respiratory structures can be divided into three types with respect to their position in the chorion: an inner porous network next to the vitelline membrane (in the "endochorion"); central porous regions farther from the oocyte (in the "exochorion"); and surface structures (aeropyles and plastrons). The inner network often takes the form of a trabecular or colonnade layer (usually 1–2 μm thick or less); the spacing of the trabeculae may correspond to the periodicity of the microvilli in the secretory cell (Cummings, 1972). It is likely that this layer serves to

←

Figs. 15–18. Transmission electron micrographs showing a progression of stages in the deposition of the chorion in Bombyx (wild type, Turtox).

Fig. 15. shows a chorion from the 7th out of 34 follicles (0.206). In this early chorion the individual fibers can be seen clearly in helicoidal (Bouligand) orientation. At all stages the microvilli of the follicular epithelial cell (*e*) are apposed to the osmiophilic sieve layer (*arrow*) (× 22400)

Fig. 16. shows slightly older chorion from the 16th out of 43 follicles (0.372). At this point the chorion fibers have thickened but their orientation is still visible. Included in this field is an internal aeropyle channel containing three follicle cell microvilli surrounded by flocculent material (× 35000)

Fig. 17. shows chorion from the 25th out of 43 follicles (0.581). In this field, which includes future air spaces of the inner quarter, structure of the individual fibers has been obscured by coalescence and/or impregnation with a matrix (× 35000)

Fig. 18. shows the outer edge of the same chorion as the previous figure. Here, although the structure of individual fibers has already been obscured, their orientation can be inferred from distribution of the osmiophilic secondary matrix which has begun to permeate. At this stage, when osmiophilic outer lamellate layer is only 2–3 lamellae thick, the follicular cell is actively secreting osmiophilic components (× 10500)

equalize the oxygen concentration all around the oocyte, as well as for storage. A trabecular layer has been documented in Lepidoptera (Figs. 13, 14; Matsuzaki, 1968; Telfer and Smith, 1970; Cummings, 1972; Sakaguchi et al., 1973), in the Hemiptera (Hinton, 1969), in the Diptera (King and Koch, 1963; Quattropani and Anderson, 1969; Wigglesworth and Salpeter, 1962), and in the Hymenoptera (trabeculae approximately 20 μm high in *Bombus terrestris*; Hopkins and King, 1966).

It should be noted that, like the crystalline structures (Furneaux and MacKay, 1972), the central porous regions may develop by secondary reorganization of the secreted material (Smith et al., 1971), so that their absence cannot be established unless the structure of fully formed eggs is examined. Porous regions within the exochorion have been documented in the grasshoppers (Slifer and Sekhon, 1963) and in the silkmoths (Sect. III.). The surface structures are so diverse and their documentation so extensive that the reader is directed to the excellent reviews of Hinton (1969, 1970).

All insect eggshells have a micropyle, which allows the sperm to penetrate the oocyte. It is not clear to what extent the micropyle also functions in gas exchange (Beament, 1947). It is almost always distinct from the aeropyles, although in the moth *Cerura vinula* a common opening exists (Hinton, 1969).

V. Chemistry of the Silkmoth Chorion

The chorion is complex not only in terms of structure but also in terms of the multiplicity of its proteins. Moreover, its evolutionary diversity is evident biochemically as well as morphologically. Particularly exciting are the indications that many protein components in a single chorion may be closely related evolutionarily, and may serve partially overlapping functions. We consider the chorion system valuable not only for developmental analysis, but also for studying molecular evolution, and the relationship between structure and function in an interesting family of structural proteins.

As a point of departure, we will discuss the protein composition of the silkmoth chorion. We will then present the results of purification and characterization studies on individual proteins. Finally, in Section VI we will survey the information available on chorion proteins of various insect groups.

A. The Proteins of the Silkmoth Chorion

The silkmoth eggshell (families Saturniidae and Bombycidae) contains primarily protein (96% in Bombyx), and a very small amount of carbohydrate. In eight species examined, less than 1% of the shell consists of amino sugars (Kawasaki et al., 1971a, 1972; Smith et al., 1971). The bulk of the protein (85 to 99% in various species) is made soluble by treatment with denaturants plus reducing agents; the minor insoluble fraction contains essentially all the carbohydrate (Kawasaki et al., 1971a, 1972). The proteins solubilized by such treatment ("soluble proteins") are remarkable, both for their amino acid composition and for their small size (Kawasaki et al., 1971a, 1972; Paul et al., 1972a, b).

Table 1 includes the amino acid composition of the chorion in 10 Lepidoptera. Most notable is the high cys content (6.4 molar % in polyphemus, 12.0 in Bombyx), and the preponderance of the non-polar amino acids, pro, gly, ala, val, met, ile, leu, and phe (68.8 and 62.8 molar %, respectively), with gly and ala being particularly abundant (44.7 and 43 molar % for polyphemus and Bombyx, respectively). We may surmise that the relative insolubility of these proteins depends on hydrophobic interactions (as in gly+ala rich silk fibroins; Lucas et al., 1960) and on disulfide bonds (as in cys-rich vertebrate keratins; Fraser et al., 1972). This is confirmed by the requirement for both a strong denaturant (e.g., urea, guanidine hydrochloride, SDS) and a reducing agent (e.g., mercaptoethanol, DTT) for chorion dissolution (Kawasaki et al., 1971a; 1972; Smith et al., 1971; Paul et al., 1972b). Even 8 M guanidine hydrochloride fails to dissolve the chorion unless a reducing agent is added. However, a reducing agent only becomes necessary for dissolution late in chorion formation, suggesting that crosslinking of the sulfhydryl groups is a slow process (Sect. VII. C.). The composition of the minor residue which cannot be dissolved by known methods is given in Table 2. Apparently, it includes components substantially different from the soluble proteins. The latter are the only ones to be discussed hereafter.

When chorion proteins from polyphemus are electrophoretically separated according to molecular weight on SDS-polyacrylamide gels, a large number of constituent proteins become apparent. By mass, approximately 97% of the proteins have a molecular weight roughly between 7000 and 30000. For ease of reference the proteins have been assigned to four rather distinct molecular weight classes—A, B, C, and D (Fig. 19a), with respective average molecular weights of 9000, 13000, 18000 and 28000 (all ± 1000). A and B-type proteins by far predominate, accounting for 38 and 50% of the protein by mass, respectively. Class C is next in abundance, while D proteins are very minor constituents.

The protein classes differ from each other not only in molecular weight, but also in amino acid composition. Relative amino acid content can be evaluated conveniently by incubating chorionating follicles in culture with a mixture of two amino acids, one labeled with ^3H and the other with ^{14}C, and determining the $^3H/^{14}C$ ratio after fractionation of the proteins on SDS gels. The only assumption this technique makes is the reasonable one that all proteins are synthesized from the same intracellular amino acid pool. Figure 20 shows that A proteins are cysteine rich, relative to leucine. From similar experiments we know that relative to glycine the As are extremely methionine poor, the Bs are lysine poor, and the Cs are lysine rich. Direct chemical analysis of purified single proteins confirms these conclusions (Sect. V.B.). Although compositional heterogeneity exists within individual classes, the differences between classes are more pronounced.

Figure 19b shows polyphemus chorion proteins separated by isoelectric focusing in polyacrylamide gels. Once again the large number of proteins is apparent. The range of isoelectric points is approximately 4.0–7.5, with most of the proteins being clustered between 4.5 and 5.5 (data from focusing in liquid columns).

Maximum resolution of the proteins could be expected from a two-dimensional separation, using isoelectric focusing for the first dimension and an SDS slab gel for

Table 1. Compositions of insect eggshells (residues/100 residues recovered)[a]

	Order: Lepidoptera										Order: Diptera	Order: Coleoptera		Order: Orthoptera			Order: Odonata	
	Antheraea polyphemus[bh]	*Antheraea pernyi*[d]	*Antheraea mylitta*[d]	*Antheraea yamamai*[d]	*Philosamia ricini*[d]	*Philosamia pryeri*[d]	*Actias selene gnoma*[d]	*Bombyx mori*[d]	*Bombyx mandarina*[d]	*Malacosoma americanum*[bh]	*Drosophila melanogaster*[bk]	*Tenebrio molitor*[eh]	*Oryctes rhinoceros*[ij]	*Gryllus mitratus*[dg]	*Acheta domesticus*[c]	*Schistocerca gregaria*[i]	*Sympetrum infuscatum*[dg]	*Sympetrum frequens*[dg]
glu NH_2	0.0	0.0	0.0	0.0	0.1	0.0	0.0	0.0	0.0	N	0.6	0.0	N	0.6	N	N	0.3	0.0
gal NH_2	0.0	0.0	0.0	0.8	0.0	0.0	0.1	0.0	0.0	N	0.4	0.1	N	0.0	N	N	0.5	1.2
trp	1.0	0.8	1.2	0.7	0.7	0.9	1.1	1.4	1.3	0.7	1.0	0.1	N	2.5	N	N	0.9	0.7
lys	0.5	0.5	0.5	0.2	1.1	0.3	0.5	0.2	0.3	0.1	5.0	5.6	5.9	1.6	1.3	5.9	0.3	0.0
his	0.0	0.0	0.0	0.0	0.5	0.3	0.2	0.3	0.3	0.8	2.2	1.4	3.0	2.1	1.3	6.1	0.0	0.0
arg	2.3	2.4	2.0	2.2	2.4	2.5	1.6	2.6	2.7	1.8	3.8	4.1	3.0	3.3	2.4	5.1	0.7	0.8
cys	6.4	5.8	5.9	6.1	5.9	6.4	5.6	12.0	9.6	4.7	0.6	4.3	4.8	1.9	3.4	0.8	23.9	21.7
asx	3.7	3.6	3.5	3.6	5.5	5.4	4.3	4.3	4.4	5.7	5.9	9.5	9.3	7.0	7.4	9.0	3.1	2.6
thr	3.0	2.9	2.9	3.0	3.6	3.0	3.5	3.1	3.0	3.5	2.4	4.0	4.5	1.9	1.5	4.4	2.9	4.6
ser	3.7	3.7	3.6	5.1	4.5	3.7	3.6	3.5	3.3	4.9	8.4	8.0	8.8	29.3	34.7	5.2	7.1	5.4
glx	4.5	4.4	4.4	3.9	4.9	4.9	3.9	3.7	4.3	3.0	10.1	9.3	9.0	14.6	16.3	8.9	3.8	4.0
pro	4.4	3.8	3.9	3.8	4.2	3.8	3.8	4.0	3.8	4.2	10.9	9.9	7.1	4.9	2.2	7.2	7.9	7.8
gly	32.6	32.6	32.4	32.9	30.1	31.5	31.5	36.0	36.7	31.6	15.7	7.9	10.0	4.7	3.4	11.6	33.5	34.2
ala	12.1	12.7	12.9	12.4	13.5	12.7	12.6	7.0	7.1	12.1	14.7	6.8	6.8	4.3	2.6	7.3	3.4	4.2
val	6.5	6.6	6.3	6.4	6.2	5.6	5.6	5.9	5.5	6.9	6.1	7.3	7.8	10.9	14.7	7.2	1.3	1.0
met	0.4	0.3	0.5	0.5	0.5	0.4	0.3	0.1	0.1	0.1	0.5	1.1	0.1	0.2	0.1	0.3	0.5	0.3
ile	3.8	3.9	3.6	3.8	3.5	3.4	4.0	2.8	2.9	3.4	3.4	5.6	5.2	4.9	5.2	2.9	1.5	3.4
leu	7.6	7.2	6.9	6.8	7.1	6.9	8.2	5.1	5.8	7.6	4.9	8.9	7.1	3.7	3.3	9.7	4.7	4.5
tyr	6.4	7.4	7.0	6.7	6.4	6.5	7.1	6.3	7.1	6.8	3.8	3.2	4.7	1.4	0.9	5.6	3.8	3.8
phe	1.4	1.4	1.5	1.7	1.9	1.6	1.8	1.9	1.8	2.1	1.2	3.5	2.9	1.4	0.9	2.8	0.0	0.0
amide[e]	N	42.5	42.3	43.5	42.4	48.4	43.6	42.5	45.9	N	N	49.9	N	24.3	N	N	74.2	69.2
phosphate[f]	N	0.0	0.0	0.0	0.0	0.0	0.0	0.0	0.0	N	N	0.0	N	106.0	110.0	N	0.0	0.0

Table 2. Amino acid compositions of insoluble portion of silkmoth eggshells (residues/100 residues)

	Antheraea polyphemus[a]	*Antheraea pernyi*[b]	*Antheraea mylitta*[c]	*Antheraea yamamai*[b]	*Philosamia ricini*[b]	*Philosamia pryeri*[b]	*Actias selene gnoma*[b]	*Bombyx mori*[b]	*Bombyx mandarina*[c]
glu NH_2	N	1.3	0.7	0.7	0.6	0.2	0.4	0.5	0.3
gal NH_2	N	0.5	0.0	8.1	0.1	0.2	0.9	0.0	0.0
trp	N	1.5	2.0	0.9	1.3	1.3	0.9	0.7	1.0
lys	6.4	2.1	2.9	0.9	6.5	3.5	4.6	2.5	4.3
his	2.4	0.7	1.3	0.5	2.0	0.3	1.8	1.1	1.3
arg	3.8	3.0	2.9	2.9	3.4	0.8	3.1	2.5	2.9
cys	0.0	1.2	0.3	0.9	0.6	0.5	0.6	0.9	1.1
asx	10.7	7.3	9.4	5.1	10.2	4.0	9.8	9.5	9.5
thr	5.6	4.5	5.2	6.4	5.1	2.2	2.0	5.0	3.8
ser	7.6	7.8	9.1	18.3	7.2	4.2	10.4	6.6	7.1
glx	11.9	7.4	9.1	4.4	14.0	33.0	10.5	11.5	21.4
pro	2.5	5.0	5.5	3.2	5.3	2.7	5.6	10.1	8.2
gly	12.4	22.6	16.3	23.9	10.5	34.2	13.4	18.1	22.0
ala	10.1	9.9	10.1	9.0	7.6	4.7	7.1	8.3	4.7
val	8.5	6.3	6.2	4.6	5.5	2.2	4.6	5.9	3.3
met	1.2	1.3	1.3	0.9	2.0	0.3	2.8	1.3	0.4
ile	5.6	4.0	4.6	2.7	4.1	1.3	3.6	3.8	2.2
leu	8.6	7.4	7.2	4.2	7.0	2.3	5.8	7.0	3.3
tyr	0.0	4.1	3.6	2.2	3.9	1.3	3.3	2.2	2.2
phe	3.1	2.3	2.3	1.4	3.4	0.8	2.8	2.5	1.0
amide[a]	N	69.7	61.4	94.6	54.9	89.2	54.0	65.0	72.4

[a] Our data.
[b] Data of Kawasaki et al. (1971a).
[c] Data of Kawasaki (1972).
[d] Values are expressed as molar % of asx and glx.
N Not determined.

Footnote to table 1.

[a] Values are expressed in μmol/100 μmol recovered.
[b] Our data.
[c] Data of Furneaux (1970).
[d] Data of Kawasaki and collaborators (1971a, 1972, 1974).
[e] Values are expressed as molar per cent of asx and glx.
[f] Values are expressed as molar per cent of ser.
[g] Urea + dithiothreitol soluble proteins only.
[h] Guanidine-HCl + dithiothreitol soluble proteins only.
[i] Data of Furneaux and MacKay (1972).
[j] Includes only crystalline material of chorion.
[k] Zone II+III only.
N Not determined.

the second. We have performed SDS electrophoresis on individual fractions from gel or liquid column isoelectric focusing experiments (Sect. V.B. and X.B.). Moreover, we have used a two-dimensional gel directly (O'Farrell, 1975; Fig. 19c). The results suggest that more than 50 proteins exist in the silkmoth chorion.

It is important to know whether the striking heterogeneity of the chorion proteins is due to a multiplicity of genes, or to post-translational modifications. Such modifications are common in many systems. An extreme case are the small RNA viruses (e.g., polio virus and mengovirus; Summers and Maizel, 1968; Holland and Kiehn, 1968; Jacobson and Baltimore, 1968), all the proteins of which are apparently derived proteolytically from a single high molecular weight precursor. Less extensive proteolytic "processing" is exemplified by the activation of zymogens (Neurath, 1964), procollagen (Pontz et al., 1973) or proinsulin (Steiner and Oyer, 1967) or by the cleavage of N-terminal methionine (Waller, 1963; Wilson and Dintzis, 1970). Other post-translational modifications which would result in

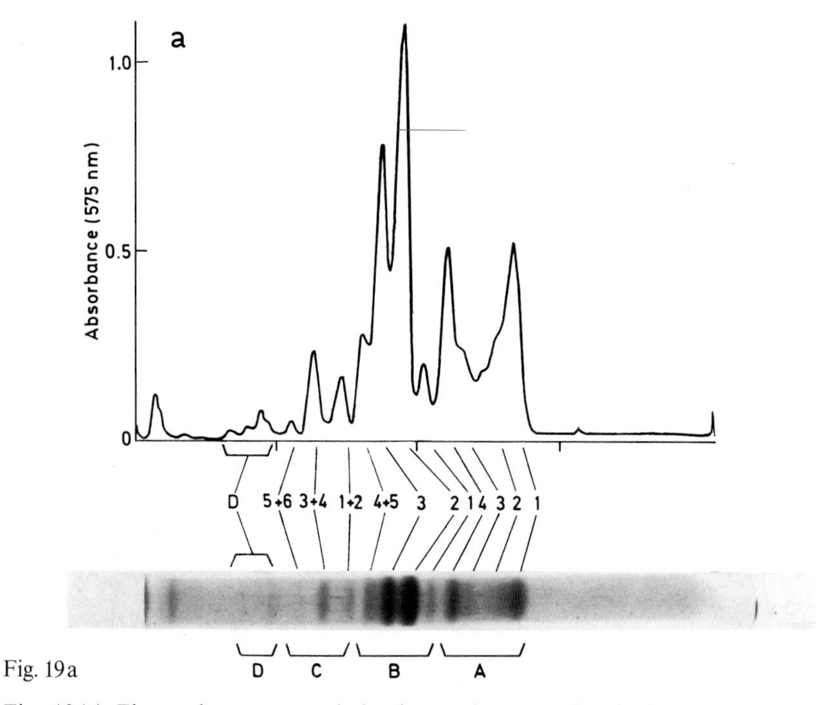

Fig. 19a

Fig. 19(a) Electropherogram and densitometric scan of polyphemus chorion proteins separated on as SDS-polyacrylamide gel (Paul et al., 1972a). Major groups of chorion proteins and their subclasses are labeled

(b) Isoelectric focusing pattern and densitometric scan of polyphemus chorion proteins separated on a slab polyacrylamide gel (ampholine range 4–6, acidic end on left)

(c) Two-dimensional fractionation of polyphemus chorion proteins by isoelectric focusing (first dimension) followed by electrophoresis on an SDS-polyacrylamide slab gel (second dimension). The polarity of IF is reversed relative to that in (b). The direction of SDS electrophoresis, from top to bottom, corresponds to the left-to-right direction in (a). The proteins were labeled in vitro with ^{14}C-iodoacetamide, and were detected by autoradiography; thus the cysteine-rich A proteins are overrepresented

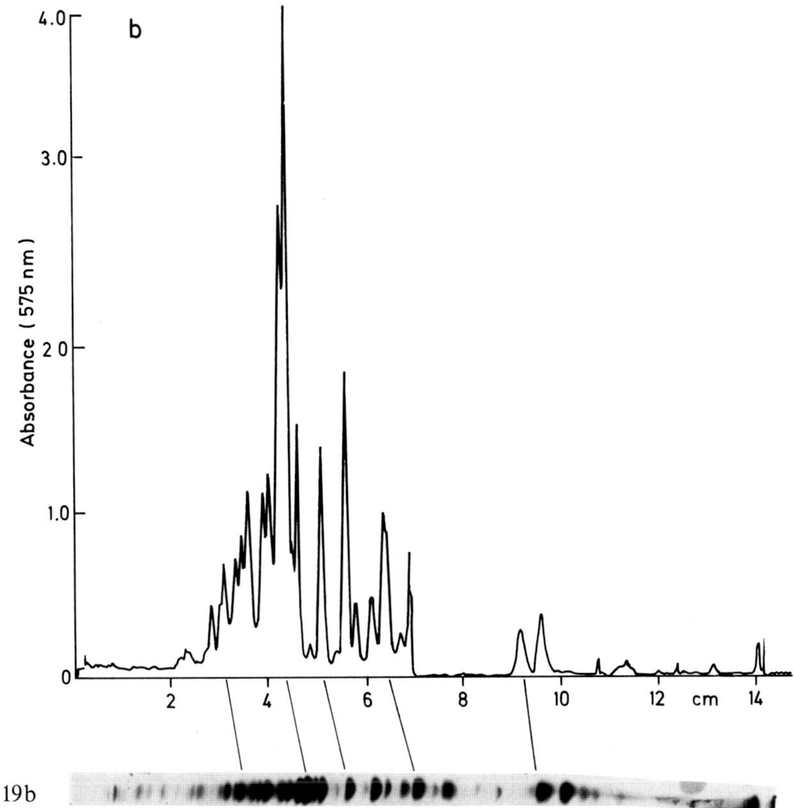

Fig. 19b

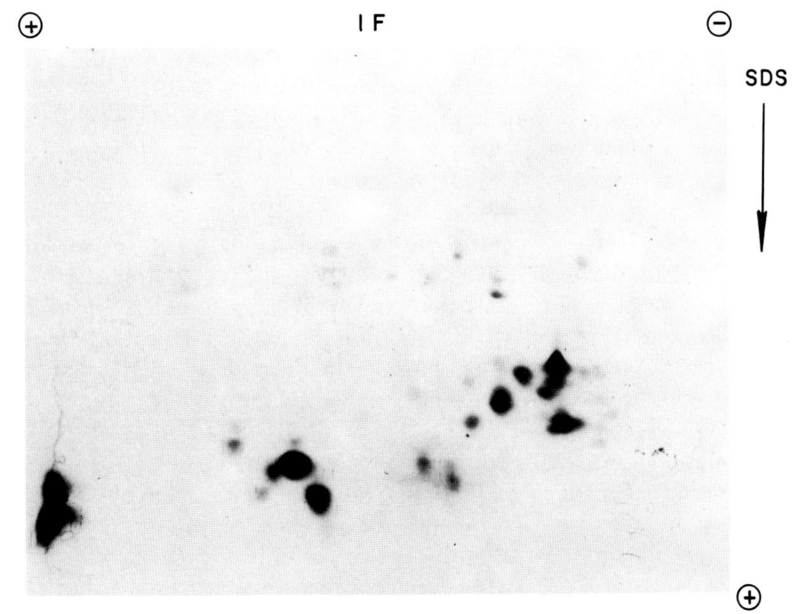

Fig. 19c

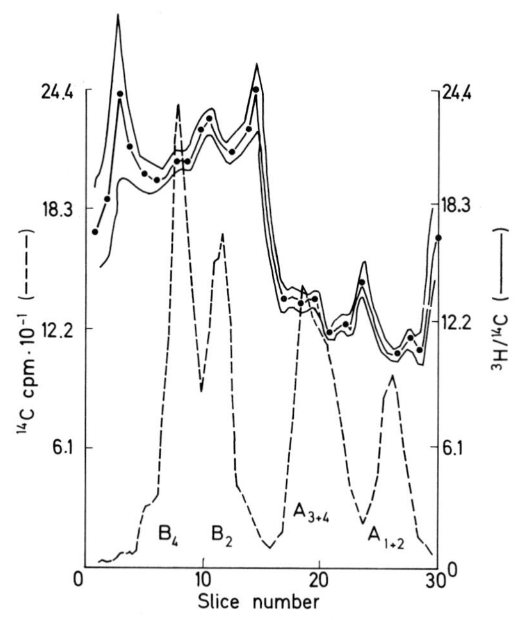

Fig. 20. Incorporation of cysteine and leucine in chorion proteins. A follicle 5 was labeled for 2 h in ^{3}H-leucine and ^{14}C-cysteine. After a 30-min chase in unlabeled medium to permit deposition of the labeled proteins, the chorion was separated from the epithelium, dissolved and electrophoresed on an SDS-polyacrylamide gel. The $^{3}H/^{14}C$ ratio (●——● shown here with the associated counting error) represents the relative content of leucine and cysteine in the respective protein subclasses. (See also Paul et al., 1972a)

molecular weight or charge changes are phosphorylation of serine (Furneaux, 1970; Kawasaki et al., 1971), methylation of lysine, arginine and histidine (Paik and Kim, 1971), acylation of N-terminal valine (O'Donnell et al., 1962; Narita, 1961), deamidation of glutamine and asparagine (Robinson et al., 1970), and glycosylation of asparagine, threonine and other residues (Gottschalk, 1972).

The evidence to date indicates that a substantial proportion of the heterogeneity of chorion proteins results from a multiplicity of structural genes. The distinct amino acid composition of individual proteins (Sect. V.B.), their characteristic kinetics of synthesis (Sects. VII.A., VIII, and XI.C.), and especially the homologous but distinct primary structure of the proteins partially sequenced to date (Sect. XIII), all indicate that the heterogeneity is due to a multiplicity of structural genes. It does not appear likely that chorion proteins are derived proteolytically from large precursors; the size distribution of nascent polypeptides (purified from peptidyl-tRNA) shows no evidence for precursors molecules larger than B proteins (Fig. 21a). Also, the polysomes of chorionating cells are inherently small (Paul et al., 1972a). SDS-electrophoretic comparison of "aged" and newly synthesized proteins in the same follicle was made by a double isotope method, using one isotope for pulse-chase and the other for secondary pulse labeling. The size distribution of proteins 26 min-old and 5 min-old on the average, was superimposable (Fig. 21b). Similar experiments failed to show size shifts between 45 min and 4 h from the beginning of synthesis (Paul et al., 1972a). However, processing of at least some chorion proteins

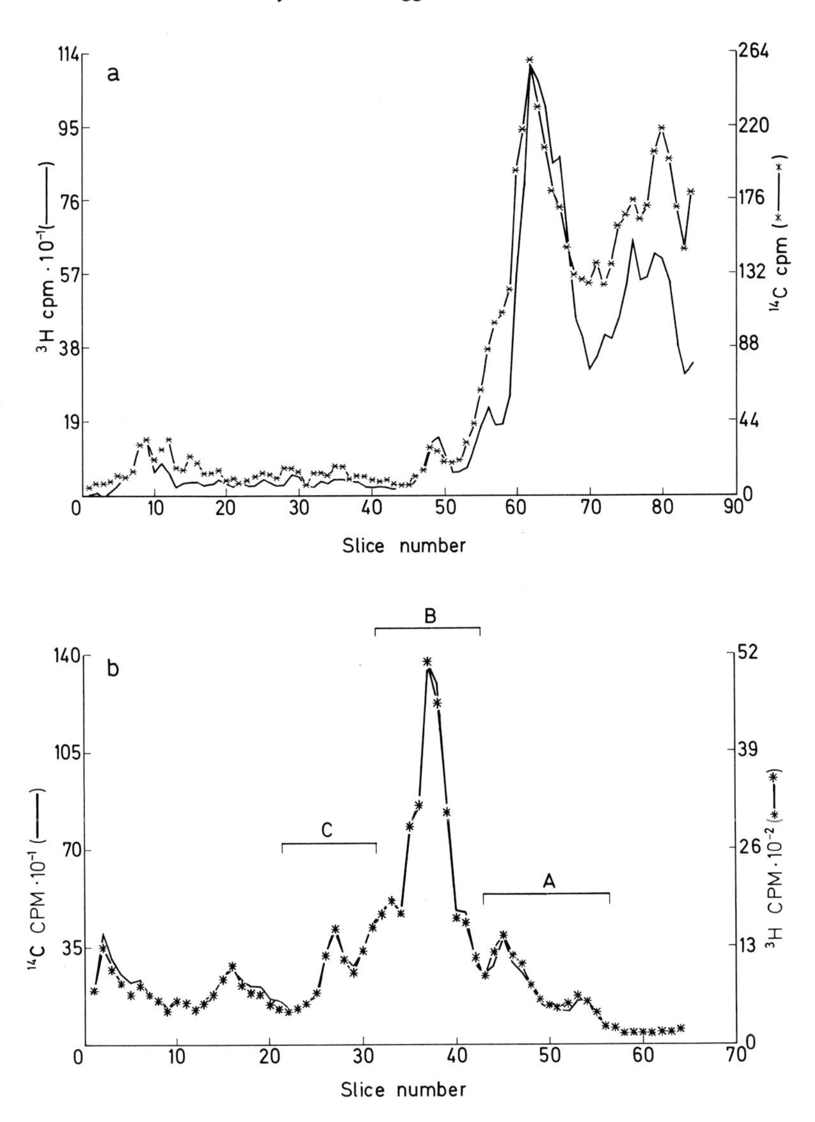

Fig. 21 (a) Comparison of the size distribution of nascent polypeptides and completed chains from polyphemus follicular cells (follicle positions 10–16). The cells were labeled in ^{3}H-glycine or ^{14}C-glycine in Grace's medium for 20 min. Synthesis was stopped by homogenization in 6 M urea, 1 M NaCl, 0.02 M sodium acetate (pH 4.2). ^{14}C-Peptidyl-tRNA was purified by binding to DEAE-cellulose at low ionic strength (0.1 M LiCl) and eluting at high ionic strength (0.4 M LiCl; Slabaugh and Morris, 1970). ^{14}C-peptidyl-tRNA (deacylated; ⋆——⋆) and ^{3}H-completed chain follicular proteins (——) were mixed and electrophoresed on an SDS-polyacrylamide gel, which was then sliced and counted

(b) Comparison of the size distribution of "aged" (——) and newly synthesized (⋆——⋆) follicular proteins. A follicle 4 was incubated in ^{14}C-leucine in Grace's medium for 12 min, chased in unlabeled medium for 10 min and incubated in medium containing ^{3}H-leucine for 10 min. Sample was dissolved, electrophoresed on an SDS-polyacrylamide gel, sliced and counted. (From Paul et al., 1972a)

appears to take place (Sect. XIII). Although the SDS pattern remains unchanged during chase incubations, the IF pattern does not. At least the major B proteins become slightly more acidic (by 0.2 pH units) following their synthesis, and this correlates with their acquisition of a blocked N-terminus: A proteins are not similarly modified. Moreover, experiments with cell-free protein synthesis suggest that the proteins translated from chorion mRNAs are of slightly higher molecular weight than the proteins observed in vivo; this is consistent with current ideas concerning the precursors of secretory proteins (Blobel and Dobberstein, 1975). In summary, all the evidence indicates that the chorion proteins are encoded in many structural genes, although at least some of these proteins are modified post-translationally.

B. Purification and Characterization of Individual Chorion Proteins from Silkmoths

In conjunction with our developmental, evolutionary and structure-function studies, we have begun to isolate from polyphemus, pernyi and Bombyx individual chorion proteins; we plan to characterize them in terms of amino acid composition and primary sequence.

The protein classes can be distinguished by the solubility of their derivatives, as well as by molecular weight (Fig. 19a) and relative amino acid content (Fig. 20). The A proteins from polyphemus can be purified quantitatively by a simple fractionation based on differential solubility of the carboxamidomethyl derivatives prepared by reaction of the reduced proteins with iodoacetamide (Fig. 22). The same fractionation scheme, applied to the aminoethyl derivatives prepared by reaction with ethyleneimine, purifies quantitatively the Cs from the more soluble As and Bs.

Further fractionation of the A proteins takes advantage of their relatively widely distributed isoelectric points. Isoelectric focusing of the A proteins from polyphemus in a sucrose gradient results in the fractionation shown in Figure 23. Individual peaks from this fractionation can be collected and analyzed on SDS gels. We have detected at least 12 distinct A proteins by this approach. Judging by the minor contaminants in the A fraction (Fig. 22b), as well as by the two-dimensional gel (Fig. 19c), Bs and Cs tend to cluster over a narrower, more acidic pH range than the As. When selected fractions from the first isoelectric focusing separation (Fig. 23a) are pooled and focused on an even narrower pH gradient (Fig. 23b), two A proteins appear to be purified to homogeneity. A similar approach has yielded homogeneous preparations of one A and one B protein from pernyi.

We have also made progress toward purifying polyphemus B proteins. If proteins dissolved in guanidine hydrochloride are diluted with water, a precipitate forms which is enriched for B_{3+4} (Fig. 24a). Depending on the amount of water added, the supernatant can be extensively enriched for B_2 relative to B_3 (Fig. 24b). When the As are removed from this supernatant by the procedure already described (Fig. 22), a fraction remains which is highly enriched for B_2 (Fig. 24c) and which, upon isoelectric focusing, yields a high purity preparation of B_2.

The amino acid content of these purified proteins from polyphemus and pernyi is shown in Table 3. Most notable is that all the purified As are enriched (50–84%)

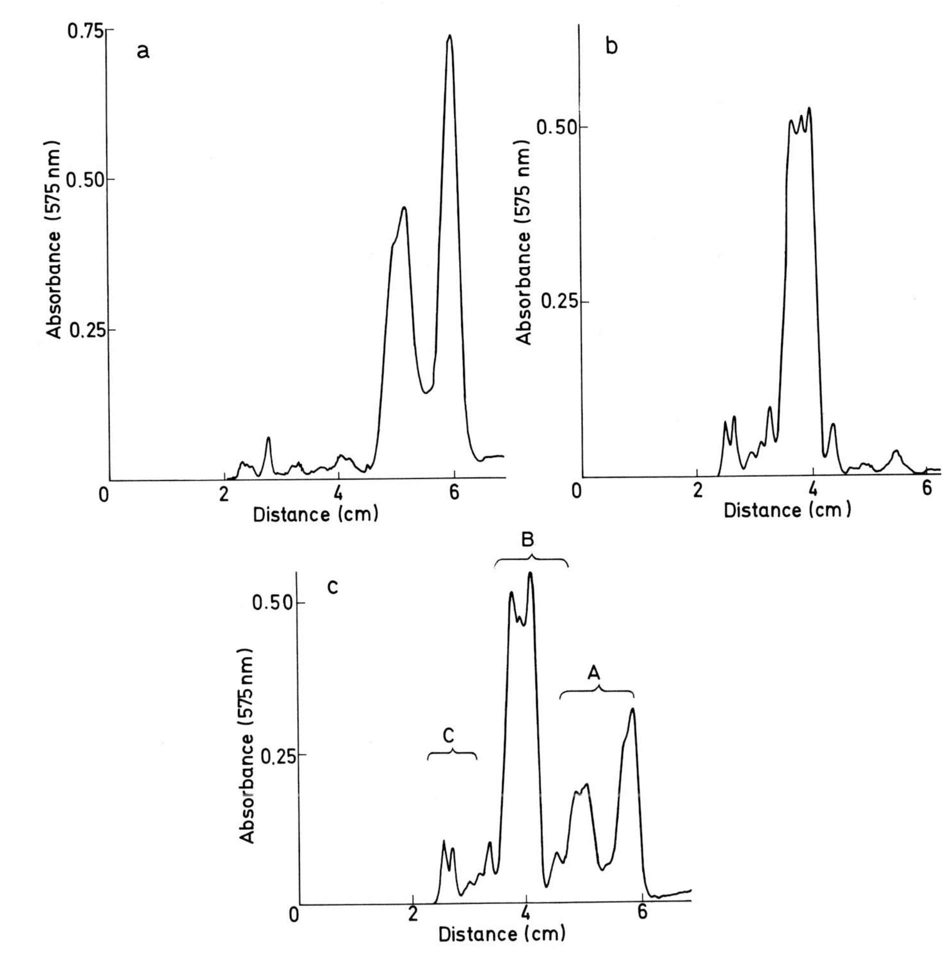

Fig. 22a–c. Fractionation of solubilized, carboxamidomethylated polyphemus chorion proteins by dialysis into 0.3 M guanidine-HCl, 0.1% mercaptoethanol (brought to pH 3.5 with acetic acid). The fractions were analyzed on SDS polyacrylamide gels. (a) Soluble fraction (predominantly As). (b) Insoluble fraction (Bs and Cs). (c) Unfractionated proteins

for cysteine relative to the pernyi B protein (cf. Fig. 20) and that the three polyphemus As completely lack methionine (cf. Sect. V.A.). An important conclusion from Table 3 is that all three polyphemus A proteins isolated (d_1, f, i) appear to be products of distinct genes. This follows from differences, beyond the margin of error, in the content of trp, arg, cys, asx, thr, pro, gly, ala, leu, and tyr. At the same time, the striking similarities among these proteins lead one to suspect that they are evolutionarily related. These proteins also show considerable similarity to the pernyi A and even B proteins.

Sequence analysis of the purified proteins is now beginning. The polyphemus A protein, fraction f (Fig. 23a and Table 3) has been sequenced from the amino terminus, for approximately half the total length (Table 4). Even by itself, this

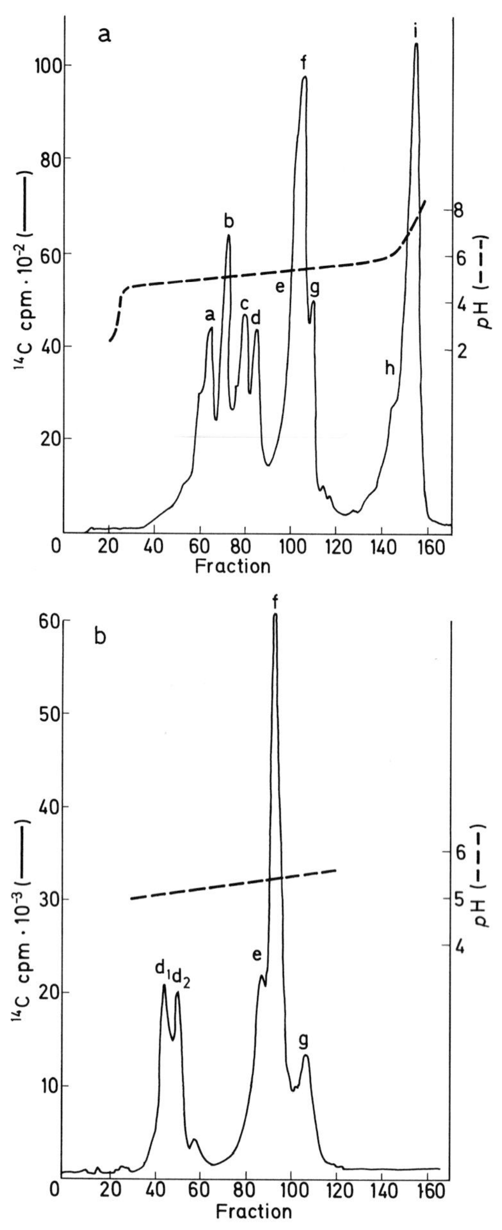

Fig. 23a and b. Isoelectric focusing of polyphemus A proteins in a sucrose gradient. (a) Total A proteins (see Fig. 22a). (b) Rerunning of fractions d and f from Figure 23a

sequence is of some interest. First, it contains several internal repeats. The tetrapeptide gly-leu-gly-tyr is repeated three times (residues 20 to 23, 25 to 28, and 35 to 38), and the related tetrapeptide ala-leu-gly-tyr occurs once (residues 30 to 33). The tetrapeptide pro-ala-cys-gly (residues 14 to 17) is repeated near the middle of the molecule (residues 45 to 48). Repeating sequences are common in structural

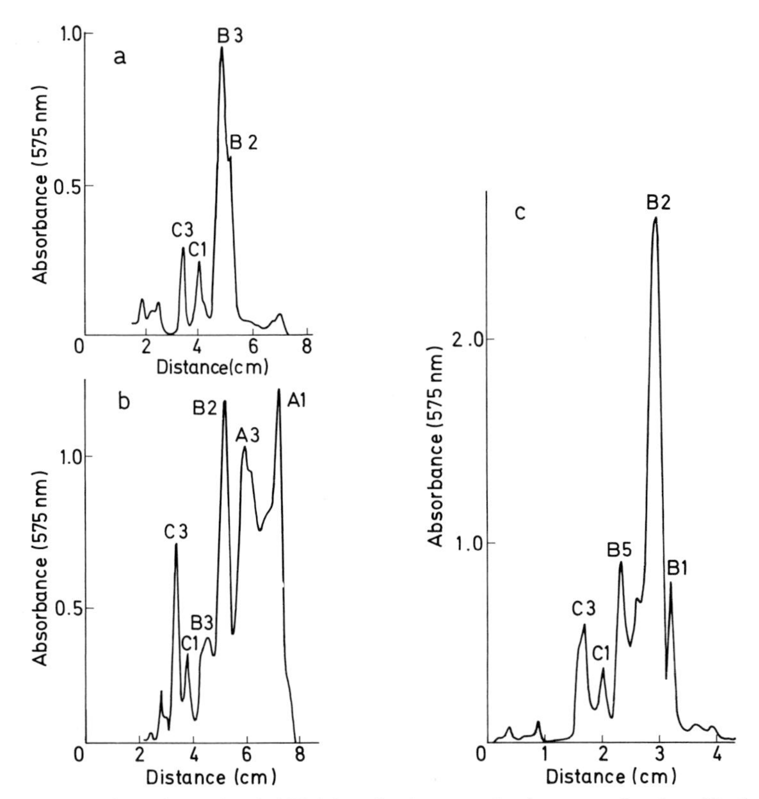

Fig. 24a–c. Fractionation of solubilized polyphemus chorion proteins by dilution of a solution of the proteins (in 3.6 M guanidine-HCl, 0.018 M dithiothreitol, 0.064 M Tris-HCl, pH 8.5). (a) The precipitate after dilution with 2 parts H_2O. (b) The supernatant after addition of 14 parts H_2O. (c) The precipitate formed when the supernatant of part b was dialyzed into 0.3 M guanidine-HCl, 0.1% mercaptoethanol, pH 3.5, adjusted with acetic acid (cf. Fig. 22b)

proteins. For example, fibroin from *Bombyx mori* contains extensive repetitions of the dipeptide gly-ala, and the tetrapeptide gly-ala-gly-tyr (somewhat reminiscent of one of the chorion repeats) is also repeated. Collagen has a repeat of gly-pro-hydroxypro. Freezing-point-depressing glycoproteins from certain fish repeat the sequence ala-ala-thr-O-galactosyl-N-acetyl-galactosamine. The high sulphur fraction of the α-keratins of sheep's wool contains a repeat of cys-cys-glu-pro-val (Ycas, 1972).

Sequence repetitions may indicate evolutionary derivation by partial gene duplication. It will be interesting to determine whether the repeating sequences also occur in other polyphemus A proteins, in the Bs and Cs from polyphemus, and in the chorion proteins from other species as well. The evolution of chorion proteins is of substantial interest since their characterization to date suggests that they may be encoded in "informational multigene families" (Hood et al., 1975). Such a family is defined as a closely linked series of genes, which share nucleotide sequence homology and have identical or overlapping phenotypic functions. Multigene systems can be expected to permit the evolution of genes in clusters (coincidental

Table 3. Amino acid compositions of purified polyphemus and pernyi chorion proteins

	polyphemus A proteins fraction i, Fig. 23	polyphemus A protein fraction f, Fig. 23	polyphemus A protein fraction d_1, Fig. 23	polyphemus proteins, mostly A, fraction c, Fig. 23	polyphemus proteins, mostly A, fraction b, Fig. 23	purified pernyi A protein	purified pernyi B protein
trp	0.0	0.8	0.8	0.8	0.8	0.0	0.0
lys	0.9	0.9	1.0	0.7	0.9	1.1	0.4
his	0.0	0.0	0.0	0.1	0.0	0.0	0.0
arg	2.1	2.5	1.7	2.0	1.0	1.6	1.9
cys	9.8	9.4	10.0	8.4	10.3	9.5	5.6
asx	1.3	3.6	2.1	3.3	2.9	3.0	2.9
thr	3.4	2.7	3.1	3.0	3.8	2.8	1.5
ser	1.2	3.6	3.5	4.7	3.4	3.6	4.6
glx	4.0	3.6	4.5	4.3	3.1	3.8	4.4
pro	1.9	3.6	4.0	4.4	4.8	3.3	3.0
gly	39.4	31.5	30.9	28.6	29.7	34.6	39.2
ala	10.8	12.9	14.1	14.7	15.3	14.0	10.4
val	7.3	7.1	7.3	7.3	6.9	6.3	6.6
met	0.0	0.0	0.0	0.2	0.0	0.3	0.5
ile	4.4	3.5	4.0	4.4	3.8	3.0	3.1
leu	5.6	7.4	6.8	6.3	6.1	4.8	7.4
tyr	6.9	6.1	5.4	5.5	6.2	7.3	7.0
phe	1.0	0.9	0.9	1.2	1.0	1.1	1.6

Values are expressed in μmol/100 μmol with trp, thr and ser corrected for incomplete recovery

Table 4. Partial amino acid sequence of purified polyphemus chorion protein f, Fig. 23

1 5 10 15
H_2N-val-cys-arg-gly-gly-leu-gly-leu-lys-gly-leu-ala-ala-pro-ala-cys-gly-cys-gly-

20 25 30 35
gly-leu-gly-tyr-glu-gly-leu-gly-tyr-gly-ala-leu-gly-tyr-asp-gly-leu-gly-tyr-gly-

40 45 50
ala-gly-trp-ala-gly-pro-ala-cys-gly-?-tyr-gly-gly-glu

evolution), and the relatively rapid creation, expansion and contraction of gene families. It has been postulated that multigene systems play an important role in eukaryotic development and evolution (Hood et al., 1974).

A second feature of interest is that the sequenced residues appear to fall into functional domains. Much of the interior of the sequence consists of hydrophobic residues, among which are interspersed the few acidic residues present in this molecule. Two of the three basic residues are found near the N-terminus (arg3 and lys9), and another arginine is found in the C-terminal third of the polypeptide. The cysteine residues are also asymmetrically distributed: three of the eight are found near the N-terminus (cysteines 2, 16, and 18). This cluster of cysteine residues is then followed by a 28 residue internal stretch (residues 19 to 46) lacking cysteine. Domains are interesting, not only from a structure-function viewpoint, but also in terms of the possible occurrence of constant and variable (or hypervariable) regions in the chorion proteins, analogous to those seen in immunoglobin chains.

VI. Comparative Chemical Studies on the Insect Chorion

Chemical, as well as morphological, studies show that the insect chorion has evolved markedly, probably more rapidly than would be expected from common rates of protein divergence (Dickerson, 1971) and the estimated phylogenetic distances. While the eggshells are grossly different in different orders, substantial differences also exist at the family level. In the best-studied group, the silkmoths, macromolecular differences are obvious between species within a genus and even between strains of the same species.

A. Solubility

The solubility of the insect chorion varies widely, depending on the type of bonding which must be disrupted for dissolution. As already discussed, the silkmoth chorion is almost fully dissolved by a combination of denaturants and reducing agents (Kawasaki et al., 1971a, 1972). In similar solvents, we have been able to dissolve a substantial portion of the chorion of two hemipterans (the bugs *Oncopeltus fasciatus* and *Pyrrhocoris apterus*) and one dermapteran (the earwig *Forficula auricularia*). By contrast, the shell of a mature dipteran egg (*Drosophila melanogaster*) is insoluble, although it can be dissolved at earlier stages (Sect. XII.E.). Among the Orthoptera, the cricket, *Gryllus mitratus*, has a partially soluble chorion, the insoluble portion containing dityrosine and trityrosine crosslinks (Kawasaki et al., 1971b); the cockroach chorion decreases in solubility as tanning increases (highly soluble in untanned *Leucophaea maderaea*, partially soluble in *Periplaneta americana*, and almost completely insoluble in fully tanned *Blatella germanica*). Most of the chorion from two Odonata (dragonflies) is soluble, although the much larger vitelline membrane is tanned and thus insoluble (Kawasaki et al., 1974). The eggshell of the mealworm beetle *Tenebrio molitor* (Kawasaki et al., 1974) is soluble in guanidine-HCl and DTT, but insoluble in urea and DTT, suggesting substantially more hydrogen bonding in this eggshell. As in the cricket eggshell, the insoluble portion of the mealworm eggshell also contains dityrosine and trityrosine crosslinks.

B. Composition

The overall macromolecular composition also varies among different groups. A high protein content appears to be general for insect eggshells (e.g., estimated as

90% in a grasshopper and a beetle; Furneaux and MacKay, 1972). Carbohydrates and lipids are also present, and may be quantitatively important in some cases (e.g., *Rhodnius*, Beament, 1946; Drosophila, Sect. XII.C.).

The amino acid compositions (Table 1) manifest drastic differences between eggshells from different orders. Important differences are even apparent in interfamily comparisons.

Relative to the Lepidopteran chorion, the eggshells of Drosophila and the mealworm, *Tenebrio molitor*, are considerably richer in the basic amino acids and asx, ser and pro, but much poorer in gly, ala and cys. The cricket eggshell is grossly deficient in gly and ala, relatively low in cys and tyr, rich in asx, glx, and val, and strikingly enriched in ser; much of the latter is phosphorylated, providing the possibility of Ca^{2+} crosslinks (Kawasaki et al., 1971b). The composition in another orthopteran, the grasshopper *Schistocerca*, is very different from that of the crickets, and relatively similar to the composition in a scarab beetle, *Oryctes*; both *Schistocerca* and *Oryctes* eggshells are rich in crystalline material. The dragonfly chorion is unusual in being extremely cysteine-rich and quite poor in arg, ala, and val (Kawasaki et al., 1974).

Among the moths, all members of the family Saturniidae so far examined have chorions of similar amino acid composition. Bombyx (family Bombycidae) is enriched in cys and shows less ala but more gly, as compared to the Saturniids.

C. Size

SDS-electrophoretic profiles provide an indication of the evolutionary plasticity of the chorion (Table 5 and Fig. 25). The Lepidoptera and Odonata have almost exclusively small proteins (93–99% by mass below 30000 daltons for Lepidoptera; average molecular weight of 13500 for Odonata); other insects have mostly larger proteins (only 5–50% below 30000). Large differences in the molecular weight distribution are also observed between families (cf. the abundance of C-type proteins in the moth family Sphingidae, and the differences among cockroaches and bugs).

Within the moth superfamily, Bombycoidea, the SDS profiles show certain constant features: predominance of B-type proteins, abundant As, minor levels of Cs, and traces of Ds. Bombyx is unusual in having greater amounts of protein of the D size range; these are Bombyx-specific proteins quite unlike the Ds from polyphemus and they are called Hc (high cysteine, since they are extraordinarily rich in cys; Sections III.A. and VII.D.). Despite the overall similarity, detailed differences are apparent even between species of the same genus; the Cs appear to be the most variable class. Many of the species differences appear to be qualitative (cf. the As in polyphemus and pernyi), especially when one takes into account additional properties of the proteins. For example, of the two major B bands the one that appears to be more soluble and to be synthesized earlier in development has the lower molecular weight in polyphemus, but the higher molecular weight in pernyi (Paul et al., 1972b).

The most surprising differences are those observed among inbred strains of *Bombyx mori* (Fig. 26; see also Sect. XI.B.).

Table 5. Percentage of soluble insect eggshell proteins in different molecular weight classes

Class: Insecta		<30000		
Order: Lepidoptera	>30000 (%)	Cs and Ds (%)	Bs (%)	As (%)
Superfamily: Bombycoidea				
Family: Saturniidae				
Antheraea polyphemus	3	9	50	38
Antheraea pernyi	1	8	51	40
Antheraea mylitta	1	11	50	38
Hyalophora cecropia	1	10	50	39
Hyalophora rubra	1	13	50	36
Family: Bombycidae				
Bombyx mori (strain 703, wildtype)	5	17	55	23
Family: Lasiocampidae				
Malacosoma americanum	2	7	60	31
Superfamily: Sphingoidea				
Family: Sphingidae				
Manduca sexta	4	54	8	34
Order: Orthoptera	>30000 (%)		<30000 (%)	
Suborder: Ensifera				
Family: Gryllidae				
Acheta domesticus	92		8	
Suborder: Blattaria				
Family: Blaberidae				
Leucophaea maderae	50		50	
Family: Blattidae				
Periplaneta americana	92		8	
Order: Dermaptera				
Family: Forficulidae				
Forficula auricularia	95		5	
Order: Hemiptera				
Family: Lygaeidae				
Oncopeltus fasciatus	60		40	
Family: Pyrrhocoridae				
Pyrrhocoris apterus	82		18	
Order: Diptera				
Family: Drosophilidae				
Drosophila melanogaster	77		23	

All data except those for *Drosophila melanogaster* are taken from the densitometric scans of SDS-polyacrylamide gels shown in Figure 25. Data for *Drosophila melanogaster* is taken from an SDS gel similar to that shown in Figure 52.

D. Charge

When soluble chorion proteins from different insects are separated by isoelectric focusing in polyacrylamide gels, considerable differences become apparent (Fig. 27).

In Lepidoptera the majority of the proteins fall within the narrow limits of pH 4.5 to 5.5; nevertheless, no two species have qualitatively identical profiles in this region. In addition, large differences are apparent in a normally minor class of proteins (pI > 7). These qualitative and quantitative differences among silkworm

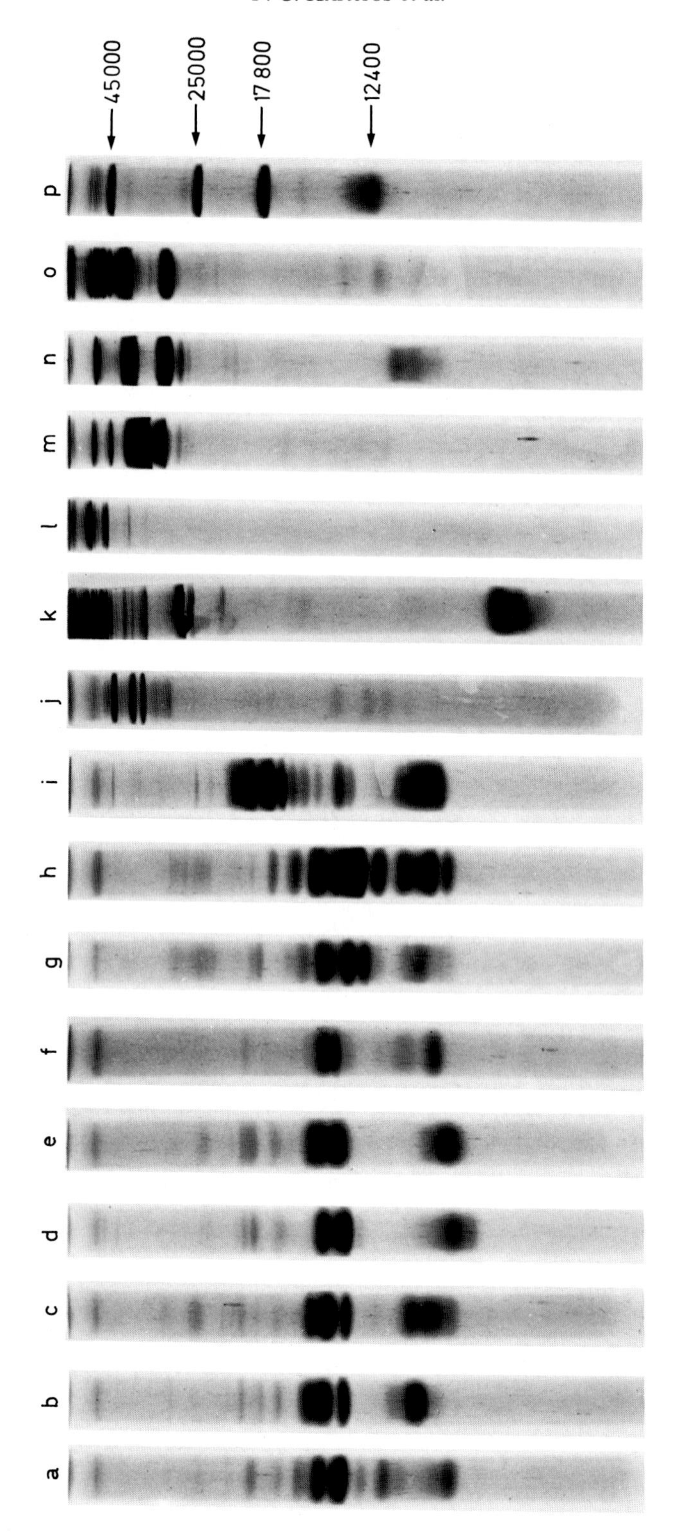

Fig. 25

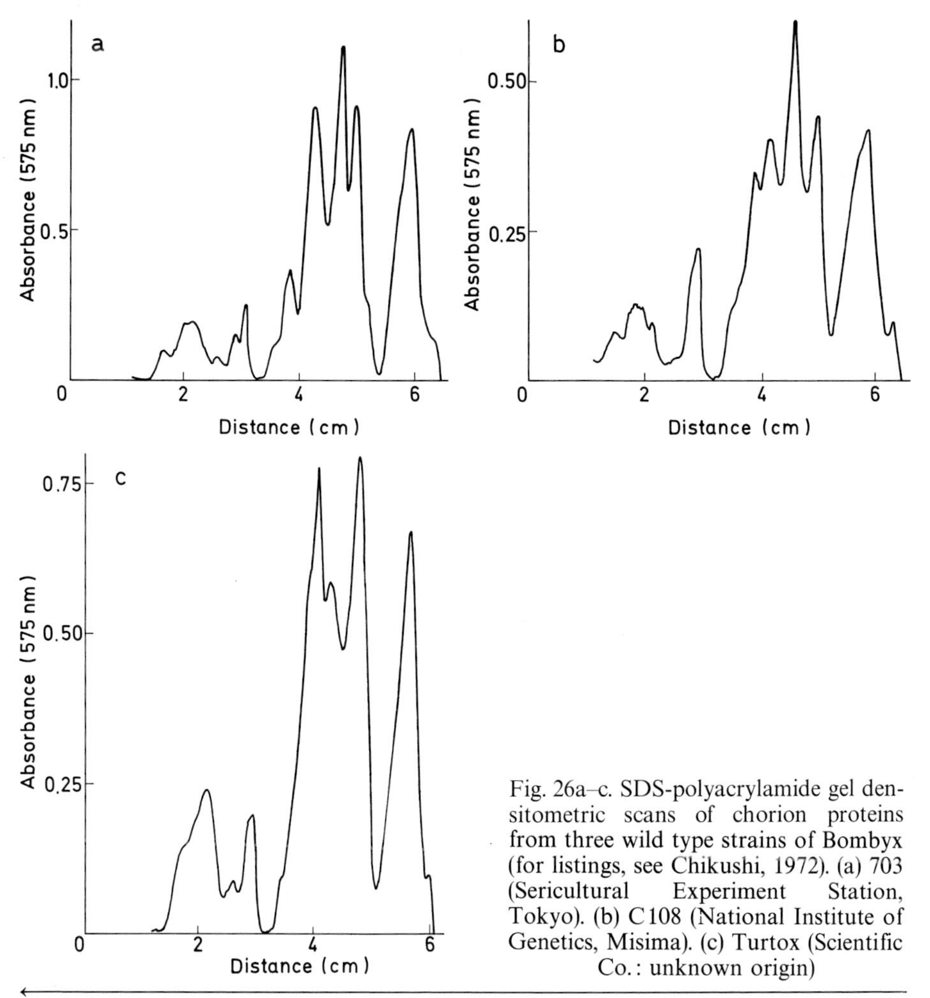

Fig. 26a–c. SDS-polyacrylamide gel densitometric scans of chorion proteins from three wild type strains of Bombyx (for listings, see Chikushi, 1972). (a) 703 (Sericultural Experiment Station, Tokyo). (b) C108 (National Institute of Genetics, Misima). (c) Turtox (Scientific Co.: unknown origin)

←

Fig. 25a–p. Electropherograms of insect eggshell proteins. Eggshells were dissolved (some partially, some fully) in 7.2 M urea, 0.36 M Tris-HCl (pH 8.5) and 30 mM dithiothreitol. The proteins were carboxamidomethylated, SDS was added and the samples were electrophoresed in parallel on SDS polyacrylamide gels

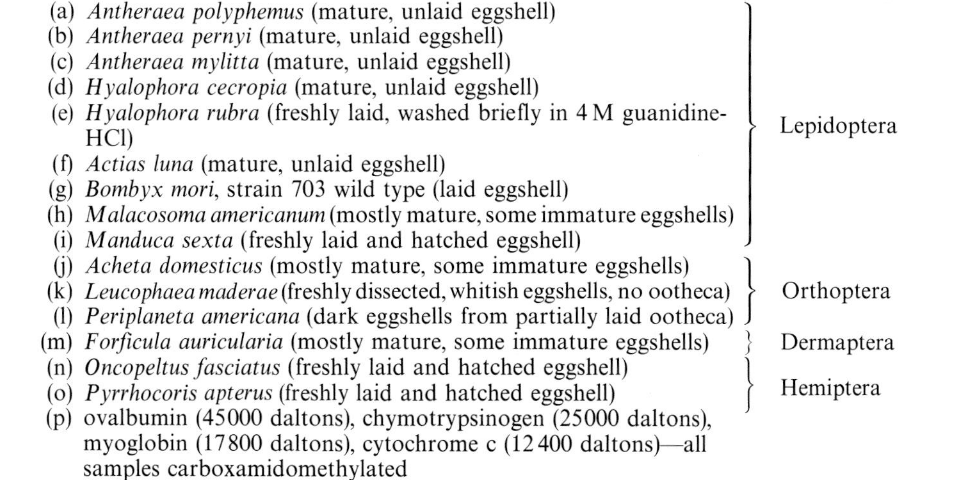

(a) *Antheraea polyphemus* (mature, unlaid eggshell)
(b) *Antheraea pernyi* (mature, unlaid eggshell)
(c) *Antheraea mylitta* (mature, unlaid eggshell)
(d) *Hyalophora cecropia* (mature, unlaid eggshell)
(e) *Hyalophora rubra* (freshly laid, washed briefly in 4 M guanidine-HCl)
(f) *Actias luna* (mature, unlaid eggshell)
(g) *Bombyx mori*, strain 703 wild type (laid eggshell)
(h) *Malacosoma americanum* (mostly mature, some immature eggshells)
(i) *Manduca sexta* (freshly laid and hatched eggshell)
} Lepidoptera

(j) *Acheta domesticus* (mostly mature, some immature eggshells)
(k) *Leucophaea maderae* (freshly dissected, whitish eggshells, no ootheca)
(l) *Periplaneta americana* (dark eggshells from partially laid ootheca)
} Orthoptera

(m) *Forficula auricularia* (mostly mature, some immature eggshells) } Dermaptera

(n) *Oncopeltus fasciatus* (freshly laid and hatched eggshell)
(o) *Pyrrhocoris apterus* (freshly laid and hatched eggshell)
} Hemiptera

(p) ovalbumin (45000 daltons), chymotrypsinogen (25000 daltons), myoglobin (17800 daltons), cytochrome c (12400 daltons)—all samples carboxamidomethylated

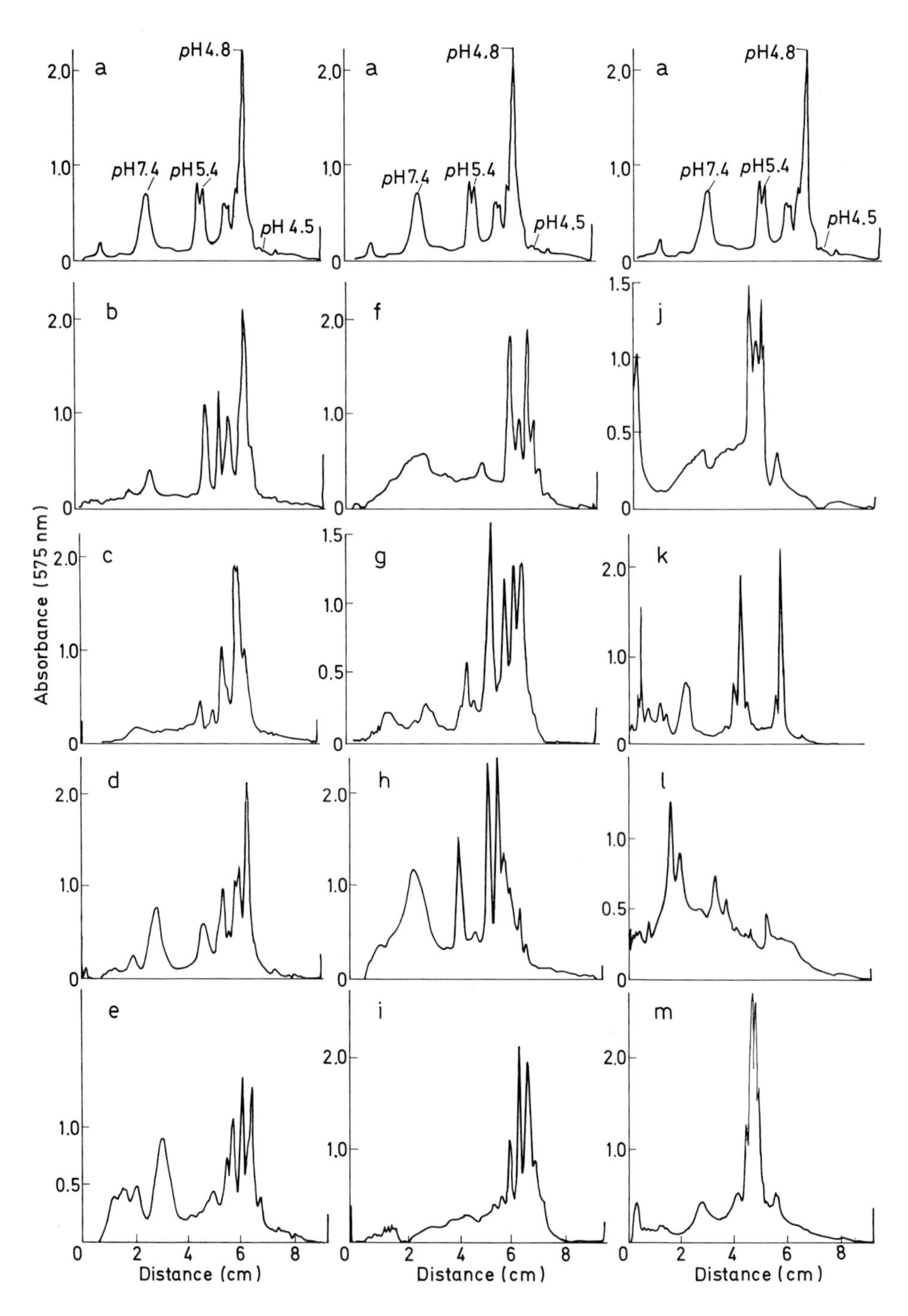
a
pH4.8
pH7.4
pH5.4
pH 4.5
b
c
d
e
f
g
h
i
j
k
l
m
Absorbance (575 nm)
Distance (cm)

Fig. 27

species are suggestive of rapid chorion evolution. Similarly, inbred races of Bombyx, with normal chorion appearance, can be distinguished both qualitatively and quantitatively by the isoelectric focusing profiles of their chorion proteins (see Sect. XI. B.; Fig. 49).

The sphingid moth, *Manduca sexta*, which is characterized by the prevalence of C rather than B proteins in SDS gels, is less distinctive in terms of the isoelectric focusing profile; this is explainable by the similarly acidic nature of Bs and Cs (see Sect. V. B.). The profiles of three nonLepidopterous insects, by contrast, show proteins of widely differing pI. In *Periplaneta*, the basic (and high molecular weight; Fig. 25) proteins are most abundant—or at least they are preferentially soluble. Some of the pI values may be affected by post-translational modifications of the proteins (e.g., phosphorylation of *Gryllus* proteins; Kawasaki et al., 1971b).

VII. Morphogenesis of the Silkmoth Chorion and Functional Analysis of Its Proteins

Beyond protein synthesis, gene expression entails functional interactions of proteins within supramolecular structures. We are interested in exploring not only the evolution and developmental regulation of the chorion proteins, but also their significance in the extracellular morphogenesis of a most elaborate protein assembly.

A. Chorion Proteins are Produced Asynchronously

The profile of chorion proteins synthesized and secreted by the follicular epithelium changes as a function of time (Paul et al., 1972a; cf. Sect. VIII). Figure 28 shows SDS electrophoretic profiles of accumulated proteins from progressively more mature polyphemus eggshells. In each case, the percentage of the chorion mass already deposited is indicated as a staging parameter. Early chorion is seen to

Fig. 27a–m. Isoelectric focusing-polyacrylamide gel densitometric scans of insect eggshell proteins. Eggshells were dissolved without SDS as described in Figure 25. The ampholyte range was pH 2.5—10

(a) *Antheraea polyphemus*	Lepidoptera
(b) *Antheraea pernyi*	
(c) *Antheraea mylitta*	
(d) *Hyalophora cecropia*	
(e) *Hyalophora rubra*	
(f) *Actias luna*	
(g) *Bombyx mori* (strain 703, wild type)	
(h) *Malacosoma americanum*	
(i) *Manduca sexta*	
(j) *Acheta domesticus*	Orthoptera
(k) *Leucophaea maderae*	
(l) *Periplaneta americana*	
(m) *Forficula auricularia*	Dermaptera

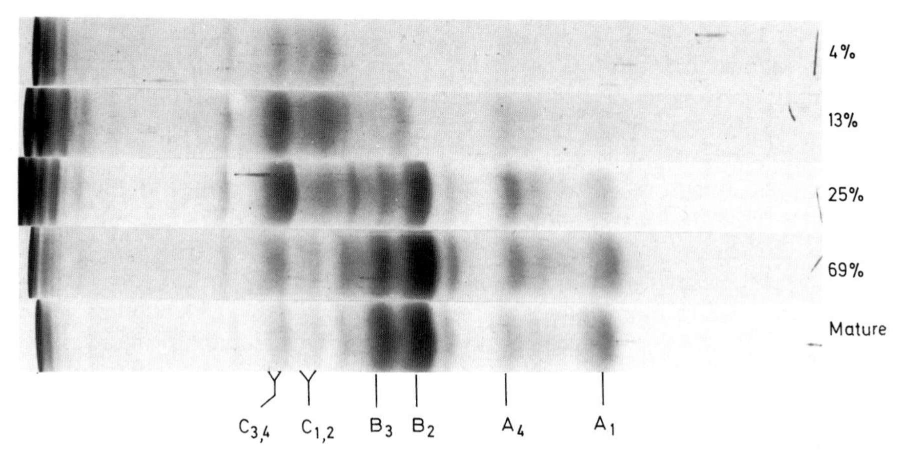

Fig. 28. The protein composition of progressively older eggshells in polyphemus. Chorions from a single ovariole were dissected rapidly in 7% n-propanol. The dry weight of each chorion was determined and expressed (numbers on the *right*) as a percentage of the mature dry weight (472 μg in this animal). Shown are electropherograms of the accumulated proteins, analyzed on SDS-polyacrylamide gels (aliquots of 1, $^1/_2$, $^1/_3$, $^1/_8$, and $^1/_{12}$ of a chorion for positions 2, 3, 4, 7, and 11, from *top* to *bottom*, respectively)

consist almost exclusively of C proteins. Substantial amounts of A and B proteins are present by the time 25% of the chorion has been deposited, and these two classes predominate thereafter. In further detail, and correcting for the fraction of chorion analyzed in each gel, we may note considerable asynchrony within each class. Thus, $C_{3,4}$ accumulates later than $C_{1,2}$. A_4 and B_2 are relatively early proteins, whereas A_1 and B_3 accumulate late in choriogenesis. Pulse labeling experiments reveal that this asynchrony in accumulation is the result of differences in the proteins being synthesized at various stages (Paul et al., 1972a; and Sect. VIII).

Figure 29 shows the total proteins synthesized by progressively older follicles in Bombyx. In their broad outline, these "synthetic profiles" are typical of all silkmoth follicles. As choriogenesis starts, the synthesis of high molecular weight "cellular" proteins is suppressed and chorion C proteins begin to be produced, followed by a changing population of A and B proteins. At the peak of choriogenesis, synthesis of proteins other than chorion is negligible; chorion production in Bombyx accounts for more than 90% of total leucine incorporation, >93% of cysteine incorporation. A peculiarity of Bombyx is the late production of a class of cysteine-rich proteins (see Sect. VIII. D.). At the end of choriogenesis, synthesis of nonchorion proteins predominates again, but now proceeds at a very low absolute rate.

B. Chorion Morphogenesis: Apposition vs. Intercalation

Given the sequential synthesis of chorion proteins, it might seem obvious to identify the proteins synthesized at each developmental stage with the chorion structures closest to the secretory cells at that time. That would indeed be the case if the chorion grew only by apposition, layer by layer. However, we now know that this is not the case. Proteins which make up a substantial proportion of the mass of the chorion reach their final destination by permeating through the previously laid down layers. Thus, much of the morphogenesis of the chorion takes place at some distance from the secretory cells, and involves the interaction of previously laid down material with freshly secreted components. These interactions include substantial reorganization of "old" chorion layers. We shall call this morphogenetic mechanism intercalation.

Intercalation also occurs in plant cell walls, where the cellulose component is deposited by apposition while the hemicelluloses permeate the preexisting wall (Ray, 1967). Similarly, chondrocytes secrete a protein, thought to be tropocollagen, which diffuses through the cartilage matrix to polymerize into striated collagen fibrils at some distance from the cell (Revel and Hay, 1963).

The occurrence of intercalation can be inferred from the fact that the chorion attains its definitive thickness at a time when less than half of its protein has been deposited. We can define follicular stages by reference to the amount of chorion already secreted: stages I, II, III ... for chorion which is 0–10%, 10–20%, 20–30% ... of the final weight, respectively (Sect. VIII.A.). Figure 30 shows that the chorion has nearly reached the final thickness, 32 μm in this case, by the end of stage IV. Since the surface area of the chorion does not change beyond that time, it would seem that the majority of the chorion protein is deposited by intercalation in a previously laid down framework. This conclusion agrees with the ultrastructural observations that the total number of lamellae is laid down quite early and that the fibers thicken and are progressively obscured by a matrix in both cecropia and Bombyx (Sects. III.C. and III.D.). Similarly, it agrees with the observation of a permeating osmiophilic matrix in Bombyx (Sect. III.D.), and with the secondary "phase transition" which forms the network of channels in the cecropia chorion (Sect. III.C.).

Autoradiographic studies of chorion secretion in polyphemus provide direct evidence for intercalation. At certain developmental stages, newly secreted protein quickly permeates throughout the older layers. At stage III, permeation is already evident at 10 min following an amino acid pulse, when secretion is just beginning; the autoradiographic grains are essentially uniform throughout the chorion by 30 min, when secretion is nearly complete (Fig. 31). At somewhat later developmental stages (e.g., stage VII), permeating proteins are distributed within the chorion in a stable gradient, most dense nearest the follicular cells. Finally, at late developmental stages (e.g., stage X_c; Fig. 31), permeation is negligible and the newly synthesized proteins are deposited in a thin layer, nearest the cells.

C. Intracellular Transport, Secretion, and Insolubilization of Chorion Proteins

The follicular cells are unusually favorable material for the study of secretory kinetics since the secreted products represent nearly all the protein synthesized and, following secretion, are immobilized in a compact extracellular layer. The kinetics with which the newly synthesized proteins are transported through the cytoplasm, secreted at the apical surface, and deposited at their final positions are being investigated by a combination of autoradiographic (Figs. 31 and 32) and biochemical procedures.

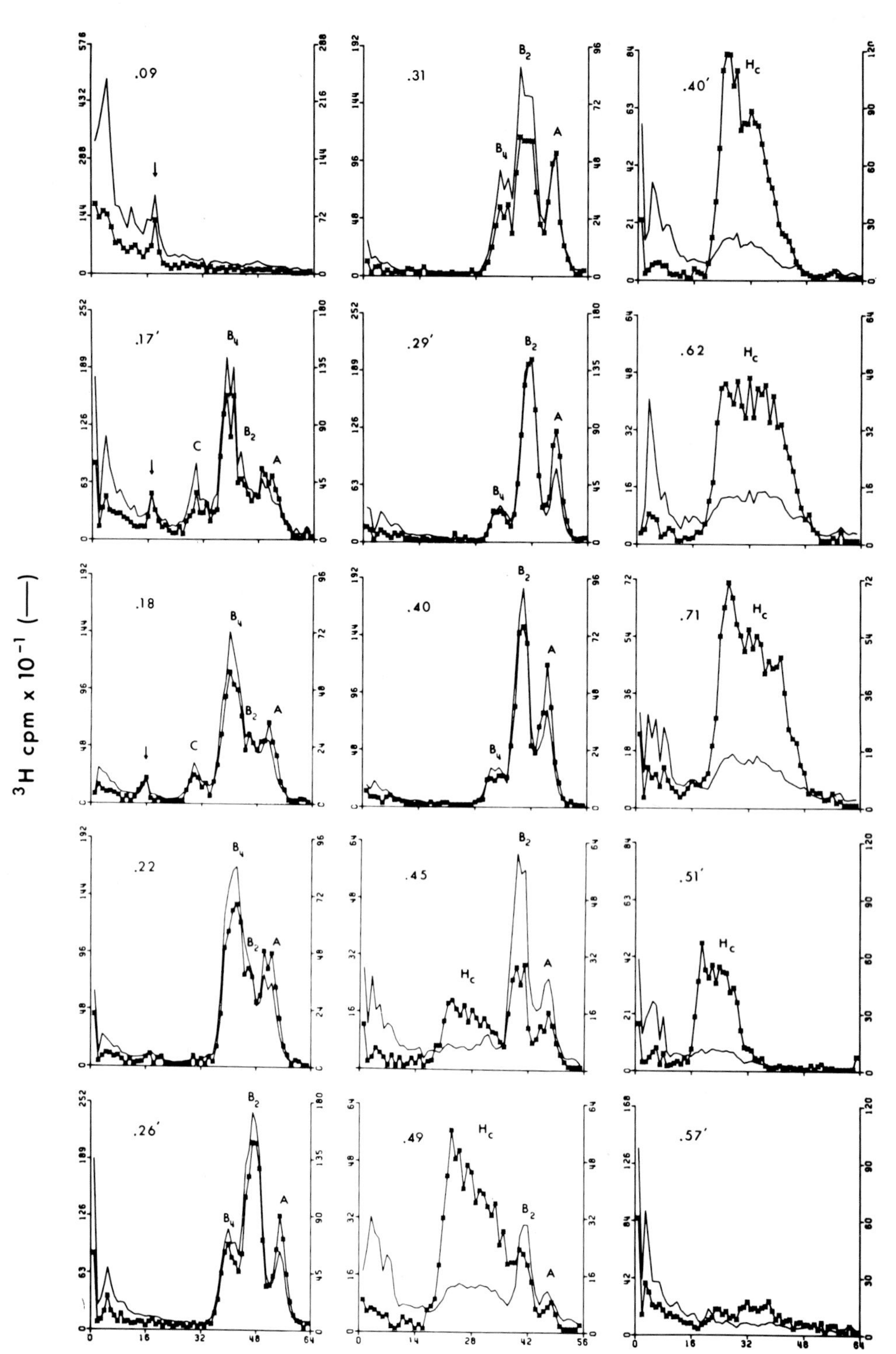

.09
.31
B2
B4
A
.40′
Hc
.17′
B4
C
B2
A
.29′
B2
B4
A
.62
Hc
.18
B4
C
B2
A
.40
B2
B4
A
.71
Hc
.22
B4
B2
A
.45
B2
Hc
A
.51′
Hc
.26′
B2
B4
A
.49
Hc
B2
A
.57′
3H cpm x 10-1 (—)
35S cpm
SLICE NUMBER

Fig. 29

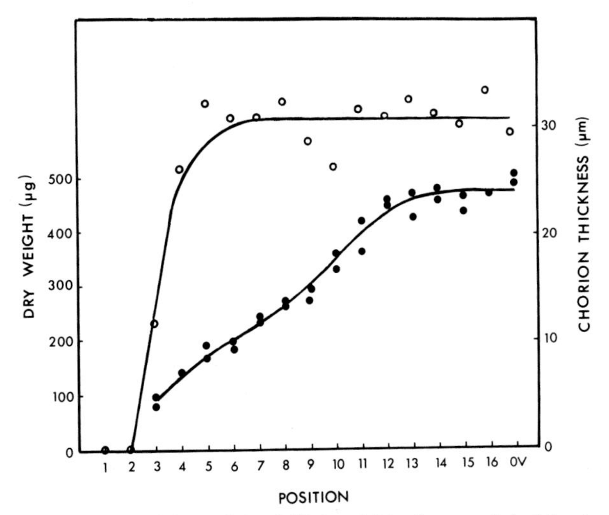

Fig. 30. Correlation of the position of the follicle within the ovariole (abscissa) with chorion thickness and the amount of protein accumulated (dry weight). Three parallel ovarioles were dissected from a day 15 polyphemus and the follicles were cut and washed free of yolk. The chorions were dissected from two of the ovarioles (in 1% H_2O_2, pH 5) for determination of dry weight. Follicles of the third ovariole were fixed, embedded in Araldite and sectioned, after careful orientation, for determination of chorion thickness. The chorion reaches its final thickness long before it acquires its final dry weight, as might be predicted by the occurrence of intercalation

Using very brief pulses of ^{3}H-glycine it is possible to follow the movement of the newly synthesized polyphemus chorion proteins. Immediately after a one minute pulse the proteins are evenly distributed throughout the cytoplasm, but by the end of a two to four-minute chase the proteins are organized into numerous aggregates. These correspond to Golgi zones which are present throughout the cytoplasm, indicating that the chorion proteins are secreted via the well documented pathway, ER→Golgi→secretory vessicle (the involvement of Golgi is morphologically

←

Fig. 29. The changing protein synthetic profiles of progressively older follicles in Bombyx (*top* to *bottom*, *left* to *right*). Results from two animals are combined (wild type, Turtox). The fractional position of the follicle within the ovariole, between the end of vitellogenesis and ovulation, is indicated for each panel (total positions 45 and 35 for animals I and II, plain and prime numbers, respectively). The follicles were labeled in a medium containing a mixture of ^{3}H-leucine and ^{35}S-cysteine (for 75 and 140 min, animals I and II, respectively), and were then cut in half to wash the follicular epithelial cells free of oocyte material. The epithelia were dissolved in electrophoresis sample buffer (Paul et al., 1972a), carboxamidomethylated, and aliquots were analyzed on SDS-polyacrylamide gels ($^1/_2$ and $^1/_3$ of a follicle, animals I and II, respectively). The major components of newly synthesized proteins are identified: A, B_{2+3} (abbreviated B_2), B_{4+5} (abbreviated B_4), C, high cysteine proteins (Hc) and an early high molecular weight component (*arrow*). Comparison of the two animals shows that fractional position only approximately measures developmental maturation. Note that both radioactivity scales are reduced 3× for follicle 0.09, and that for follicles older than 0.40 the ^{3}H-scale is expanded 3× (1.5× for 0.57′) and the ^{35}S scale is reduced 1.5×. When allowance is made for the different scales for older follicles, the high cysteine proteins are seen to have a cys/leu ratio substantially higher than that of even the A proteins (see also Sect. VII.D.)

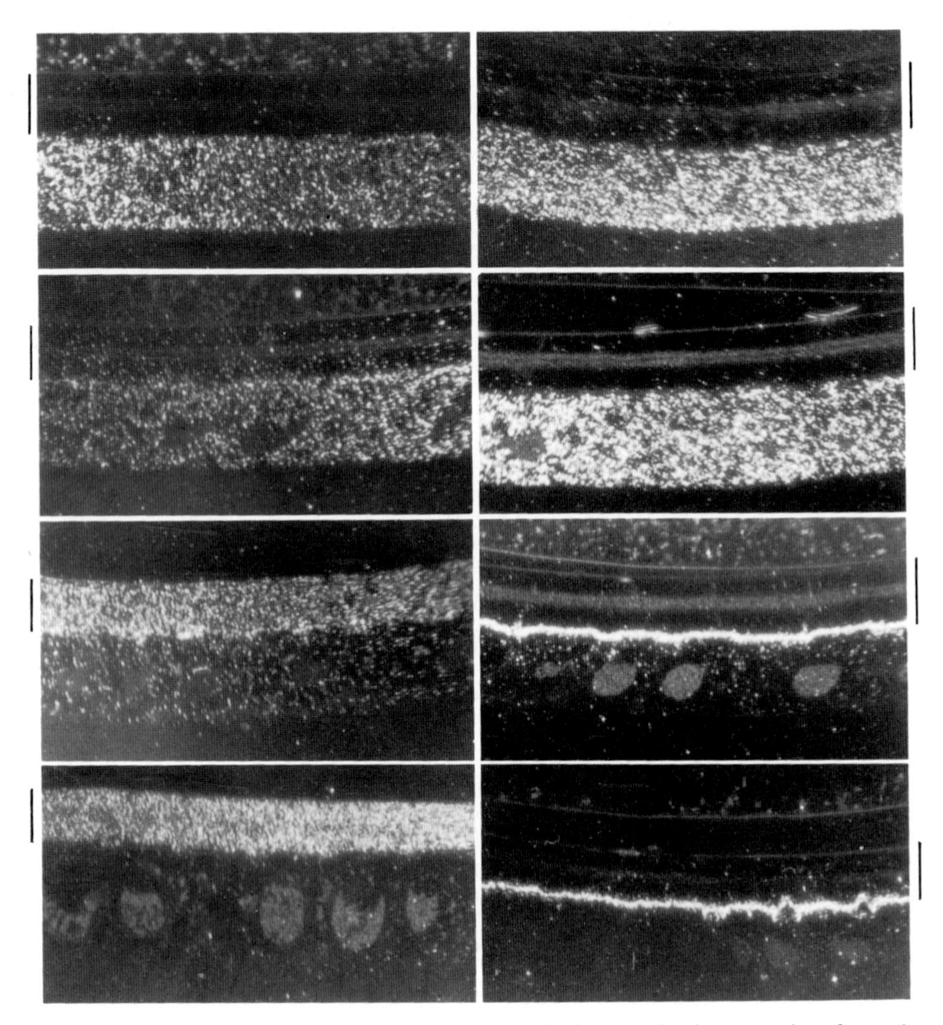

Fig. 31. Autoradiograms showing secretion of polyphemus chorion proteins, from the epithelial cells to the overlying extracellular eggshell (indicated by *vertical bar*). Two different developmental stages (see Fig. 37) are shown, stages III (*left*) and Xc (*right*); they illustrate intercalation and apposition of newly synthesized proteins, respectively. Follicles were labeled with ^{3}H-glycine for 6 min and chased in nonradioactive medium prior to fixation, for progressively longer times (0, 10, 30, 200 min for *left*, *top* to *bottom*; 0, 10, 60, 400 min for *right*, *top* to *bottom*). Secretion is essentially complete by 30 to 60 min, and the pattern of protein distribution (intercalation or apposition) remains stable thereafter. *Double bright line* visible in the middle of the chorion is the holey layer. Autoradiograms exposed for four weeks, photographed by darkfield in the Leitz MPV-I microscope

obvious in the case of the putative high cysteine components of Bombyx, which can be traced easily because of their osmiophilia).

After a 10–15 min chase, label begins to accumulate in the apical border of the cells. By 30 min, secretion into the chorion is essentially complete. The lag time (ca. 10 min) between aggregation and movement to the apical border presumably

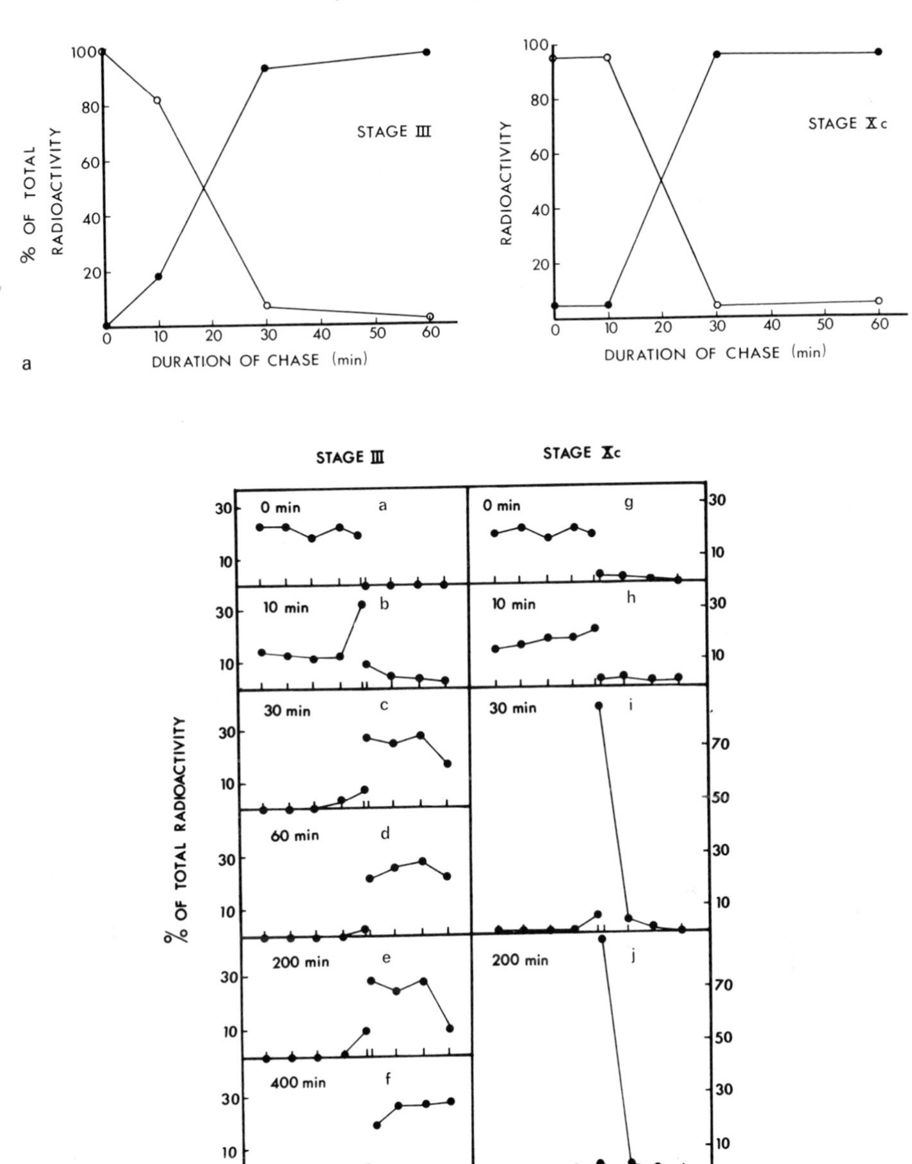

Fig. 32a and b. Quantitative analysis of autoradiographic grain distribution, from the same experiment as Figure 31. Autoradiograms were illuminated by epi-illumination and the light reflected by the grains was measured using the Leitz MPV-I microscope. Light intensity was shown to be proportional to grain density. Measurements in the cytoplasm and the chorion were taken, beginning at the border between the two and progressing in 10 μm intervals. (a) Sum of measurements in cytoplasm (○) and chorion (●), relative to total, shown as a function of chase duration. (b) Measurements in individual regions of cytoplasm (from basal to apical, regions E_1 to E_4 and epithelial border, B_E) and chorion (chorion border, B_C and regions C_1 to C_3); expressed relative to total of all measurements

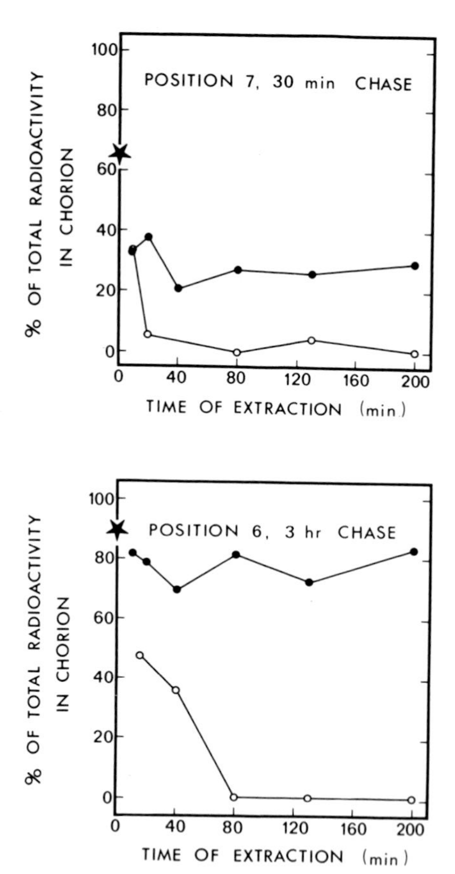

Fig. 33. Extraction of newly synthesized, secreted proteins from the chorion. Midchoriogenic follicles were pulse labeled with ^{3}H-leucine for 4 min and chased in nonradioactive medium for 30 min (*top*) or 3 h (*bottom*). Follicles were then cut, washed free of yolk, frozen, thawed and maintained in distilled water (●——●) or 1% SDS (○——○) at 24° for indicated times of extraction. Chorion was then dissected, and cell remnants plus soluble proteins were combined. TCA-precipitable radioactivity was determined for each of these two fractions; radioactivity remaining in the insoluble chorion is shown as percent of sum of both. Asterisk: % of total labeled protein present in chorion before extraction, as estimated by quantitative autoradiography on follicles from another ovariole of the same material

reflects the time needed for "packaging" the proteins within the Golgi system (Jamieson and Palade, 1967a, b)[3]. By contrast, both the intracellular and the extracellular movement of the proteins appear to be rapid, since transient gradients in distribution are not prominent either in the subapical cytoplasm or in the chorion (Fig. 32b). The time needed for secretion of half the labeled protein is 15 to 30 min (Fig. 32a). Eventually, more than 95% of the incorporated glycine appears in the chorion.

[3] It should be noted that the bulk of the silkmoth chorion is probably not glycoprotein (Table 1). In one experiment, we found that glucosamine was efficiently incorporated in the polyphemus chorion, but was not associated with any of the soluble proteins.

Table 6. Resistance of chorion to extraction by SDS in the absence of a reducing agent

Follicle stage	% resistance[a]
mid-choriogenic follicle (position 6)	10%
late-choriogenic follicle (position 10)	43%
last pre-ovulatory follicle (position 13)	63%
first post-ovulatory follicle (position 14)	80%
fourth post-ovulatory follicle (position 17)	97%

[a] Chorions were extracted with 1% SDS, 0.01 M Tris pH 8.4, under nitrogen, first for 14 days at room temperature with shaking and then for one hour at 100° C. Resistance was estimated as the % of the chorion protein which was insoluble under these conditions but was dissolved by subsequent treatment with electrophoresis sample buffer (Paul et al., 1972a), which includes 1% mercaptoethanol.

The autoradiographic kinetics were supplemented with the results of biochemical experiments. The cells and chorion can be separated by dissection, and each part can then be assayed for total TCA precipitable radioactivity and for the profile of labeled proteins it contains. In this method, only that radioactivity which remains associated with the dissectable chorion is scored as extracellular. As a result, the half-secretion time estimated in this way is slightly higher than by the autoradiographic method. After 10 to 50 min of chase, the intracellular and extracellular labeled chorion proteins have similar profiles, suggesting that there are no gross differences in the kinetics of secretion of particular proteins.

Although the proteins reach their extracellular site within the chorion rapidly, they become cemented only gradually. Thus, newly secreted proteins are easily lost during dissection (Fig. 33), but can be fixed in place if the chorion is treated with 95% ethanol or 1% H_2O_2(pH 5). In midchoriogenesis, distilled water can extract nearly 60% of the proteins secreted during a chase of 30 min, but only 12% of those that have been secreted by 3 h. The secreted proteins remain soluble in 1% SDS (without reducing agent) for a long time. After 30 min of chase, extraction is complete with 20 min of SDS treatment at room temperature, but after 180 min of chase, complete extraction requires more than an hour of SDS treatment. When a series of staged chorions is extracted with 1% SDS (Table 6), the results indicate that the chorion proteins become refractory to vigorous extraction under nonreducing conditions primarily toward the end of choriogenesis.

The extracellular morphogenesis of the polyphemus chorion involves construction of an initial fabric and its elaboration by a combination of intercalation and apposition. The localization of the proteins is attained by (1) rapid movement of the secreted proteins to their respective sites; (2) a relatively slow initial immobilization and (3) a final cementing of the chorion structure with disulfide bonds, much of which is accomplished at and after ovulation.

D. The Distribution of Particular Proteins Within the Chorion

Despite the complications introduced by the interesting process of intercalation, we have begun to identify the spatial distribution of specific proteins within the chorion. The clearest case is the distribution of the high cysteine proteins in Bombyx.

Kawasaki et al. (1971a, 1972) purified a high cysteine fraction (30 molar %) from Bombyx chorion by extracting with a low concentration of thiol reagents. The high cysteine proteins are obtained as an insoluble residue, which can then be solubilized at a higher thiol concentration. They are reported to have an average molecular weight higher than that of the lower cysteine fraction (18800 vs. 12300). Our developmental studies show that the high cysteine proteins are synthesized toward the end of choriogenesis (Fig. 29; Sect. VII.A.). At first their synthesis coincides with some A and B synthesis, but soon they become virtually the only chorion proteins synthesized. They are strikingly evident in radioactivity profiles as a heterogeneous group of relatively high molecular weight proteins, intensely labeled with cysteine but poorly so with leucine (cysteine to leucine ratio 5.4 × higher than for As, which in turn have a ratio 1.6 × higher than for Bs and 2.0 × higher than for Cs; Fig. 29).

At the time when high cysteine proteins begin to be synthesized, the intensely osmiophilic outer lamellate layer of the Bombyx chorion begins to be deposited (Fig. 18). We ascribe the strong osmiophilia of this layer to the high sulfur content (Rogers, 1959). A mature Bombyx chorion treated gently with a modification of the Kawasaki et al. (1971a) low thiol procedure yields a thin insoluble shell which can be identified under the electron microscope with the outer lamellate chorion, by virtue of its thickness, number of lamellae and staining intensity. This shell is considerably enriched in high cysteine proteins (Fig. 34b).

At the time when the outer layer is being deposited, material of similar staining properties, the secondary Bombyx matrix (Sect. III.D.), penetrates the previously laid down chorion. We believe that this material is also high cysteine protein, based on (1) its staining properties (2) its appearance at a time when autoradiography reveals permeation of cysteine rich material into the chorion and (3) the presence of substantial amounts of high cysteine protein in the low thiol-extractable fraction (Fig. 34a), which presumably corresponds to the inner lamellate and trabecular layers. It would be interesting to know whether the same proteins are found in both the outer shell and dense matrix, or whether only a subset of the high cysteine proteins (perhaps together with special A and B proteins) are used in the outer crust.

By mass alone, as well as by consideration of the developmental synthetic profiles, we may safely identify the bulk of the fibers and the occluding "ordinary" matrix of the lamellate chorion in all silkworms as B and A proteins. Since the cysteine content of the As (ca. 10%) is somewhat greater than that of the Bs (ca. 5%), it may be that the fibers are mostly Bs and the matrix mostly As, by analogy with mammalian keratins (Fraser et al., 1972), where cysteine poor fibers (ca. 7% in sheep's wool) and a cysteine rich matrix (ca. 18% in wool) have been identified. The relative insolubility of most Bs and the solubility of As argue in the same direction. On the other hand, feather keratin proteins appear to be rather uniform in cysteine content (7.3–8.2% in partially fractionated fowl feather rachis proteins) despite their having fibers embedded in a matrix. In this case it is thought that there are cysteine-rich portions of molecules which form matrix, and cysteine poor portions of the same molecules which form fibers. The partial sequence of the polyphemus chorion protein shown in Table 4 demonstrates a nonuniform cysteine distribution consistent with this latter model. In any case, autoradiography of mid-choriogenic

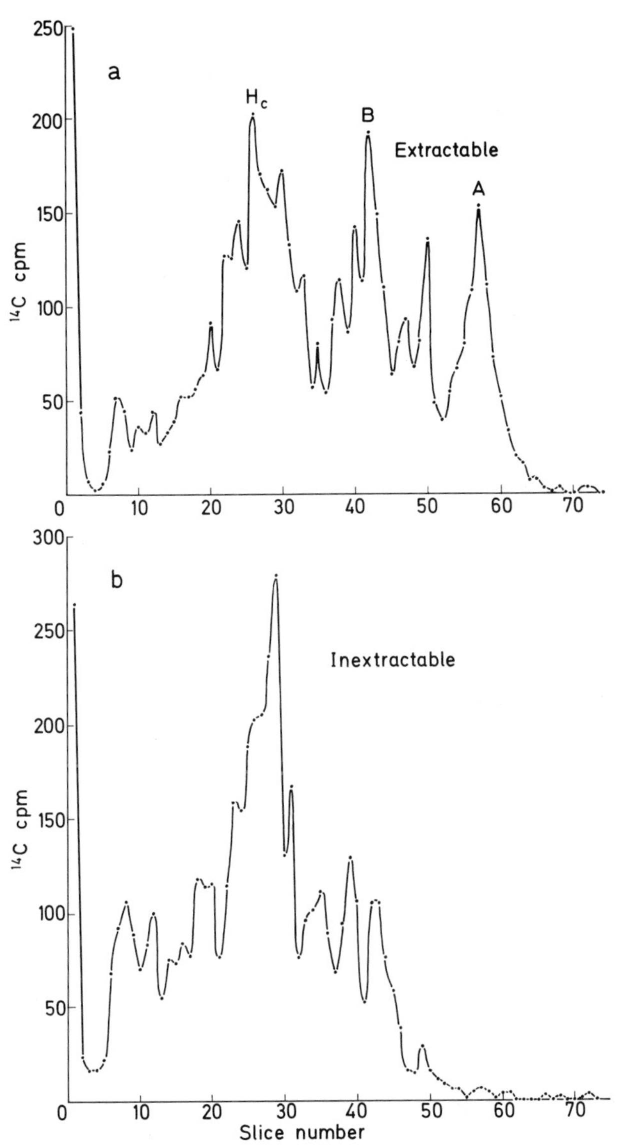

Fig. 34a and b. Differential solubilization of *Bombyx mori* eggshells (strain obtained from Turtox). Eggshells were shaken in 6 M guanidine-HCl, 0.18 M Tris-HCl (pH 8.5), 0.75 mM Na_2 EDTA and 10 mM dithiothreitol for 108 min at room temperature (method modified from Kawasaki et al., 1971a, 1972). The solubilized fraction (a, Extractable) was carboxamidomethylated and electrophoresed on an SDS-polyacrylamide gel. The residue was dissolved by increasing the dithiothreitol concentration to 50 mM (b, Inextractable), and then processed as for (a)

polyphemus follicles, separately pulse chased with ^{3}H cysteine and ^{3}H methionine, reveals a similar permeation of the two labels. This indicates that both the Bs and the methionine free As produced at that time intercalate into the previously laid down chorion.

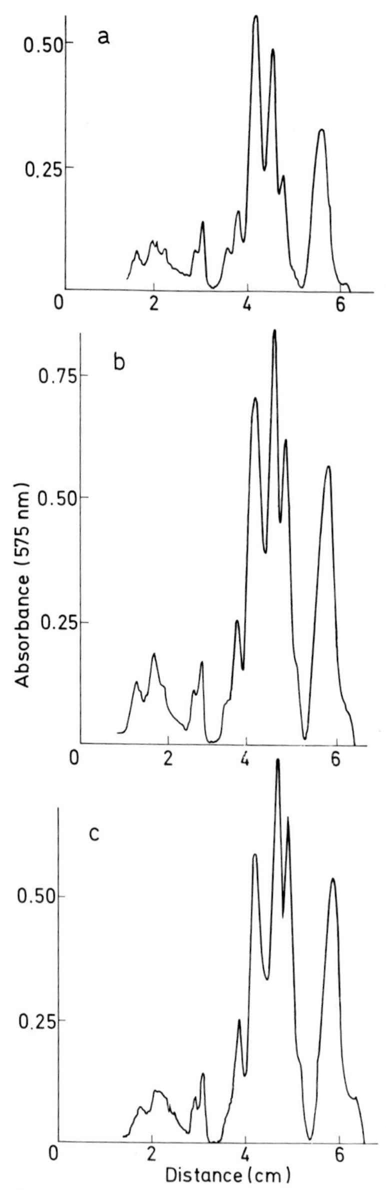

Fig. 35a–c. Regional differences in the protein composition of the Bombyx chorion. Densitometric scans of SDS-polyacrylamide gels are shown for three regions (strain 703 wild type, Sericultural Experiment Station, Tokyo). (a) Micropyle. (b) Edge (equivalent to stripe and tall aeropyle region in polyphemus). (c) Flat sides

The Cs are the earliest chorion proteins to be produced (Figs. 28 and 29) and the trabecular chorion is the first layer to be formed. The conclusion that the trabecular chorion is enriched in Cs was verified by direct electrophoretic analysis of the innermost layer, scraped with a razor blade from polyphemus chorion. The scrapings consisted of 26% Cs by weight. It is not known whether the non-C material of the scrapings is a contaminant or exists within the actual trabecullar layer as a result of intercalation. Additional scraping experiments clearly indicated that small amounts of C proteins are also present within the A and B-rich lamellate chorion. It is noteworthy that C proteins again become prominent in the synthetic profiles of late polyphemus follicles which are elaborating the external cups and expanded cavities of the aeropyles. The external part of these aeropyles appears to be made of lamellate helicoidally oriented fibers identical to the rest of the lamellate chorion. The cups are filled with flocculent material similar in appearance to that which fills the internal aeropyle channels in polyphemus, cecropia and Bombyx (Fig. 16). It is possible that Cs are generally associated with structures within which air spaces develop secondarily (trabecular layer, channel system in lamelate chorion, aeropyles).

In addition to having distinct layers, the chorion contains morphologically recognizable surface regions—for example, in polyphemus, the micropyle, the stripe, the tall aeropyle zone, and the flat sides. It is not surprising that minor compositional differences can be detected between such regions both in polyphemus and in Bombyx (Fig. 35). All the protein subclasses recognizable on SDS gels are present in each region; the differences may be mostly quantitative. It would be desirable to have isoelectric focusing analyses of accumulated and newly synthesized proteins from the various regions at the various developmental stages, in order to appreciate the full extent of cellular heterogeneity in the follicular epithelium.

We can safely exclude the possibility that each follicular cell produces a single protein, and that the developmental synthetic program of the epithelium as a whole results from temporally coordinated periods of activity of the individual cells. We have already noted that some proteins are deposited by apposition rather than intercalation; this implies that at least part of the complexity of the chorion associated with each cell is the result of secretory activity of the cell itself. Restricted circumferential mobility of the proteins is also indicated by the morphology of the chorion mutant, "Bird's eye," in Bombyx (Sect. XI.A.). This mutant has opaque chorion, with the exception of a transparent region in the middle of the flat side. Careful examination of the border between opaque and transparent regions reveals a cellular mosaic: individual polygons of the chorion are all-or-none in opacity.

Follicular cell differentiation has a spatial as well as a temporal dimension. However, at the present level of analysis, the temporal component is much more striking, and the epithelium of each follicle can be used as a unit for the study of sequential protein synthesis.

VIII. Programmed Changes in the Synthesis of Chorion Proteins During Differentiation

We have already indicated that chorion proteins are produced asynchronously. Indeed, the developmental interest of this system is based on the fact that the follicular cells synthesize a limited number of specific proteins according to a strict developmental timetable. In this section we will document in some detail the synthetic program in polyphemus, and will demonstrate that it can proceed autonomously (Paul and Kafatos, 1975).

A. Synthetic Stages in Polyphemus

In order to define the stages, ovarioles from a large number of animals were briefly labeled with radioactive leucine, either in situ (by injection of the isotope into the animal) or in vitro (by addition of the isotope to Grace's tissue culture medium, into which the ovarioles were transferred shortly after dissection). Individual labeled follicles were collected, and record was kept of their numerical position in the ovariole relative to the last vitellogenic follicle (position 0; see Sect. II.C). The follicles were cut in half with iridectomy scissors and the yolky oocyte was gently washed away. The remaining follicular cells and chorion (if any) were dissolved with an SDS-mercaptoethanol buffer in toto, and aliquots were subjected to electrophoresis in SDS polyacrylamide gels. In some cases, follicles from parallel ovarioles of the same animal were dissected into the cellular layer and the extracellular chorion; the dry weight of the chorion was then determined and expressed as a function of position within the ovariole.

The number of follicles between the end of vitellogenesis and ovulation is somewhat variable, so that the follicles cannot be staged with great accuracy by absolute numerical position alone. Fractional numerical position is a more accurate developmental parameter. The fraction of chorion mass already deposited gives additional staging information. The most direct staging criterion is the synthetic activity of the cells, i.e., the electrophoretic profile of newly synthesized proteins.

We adopted the convention of expressing absolute and fractional numerical positions with arabic numerals. Stages determined by dry weight or by synthetic profile were expressed with roman numerals. The dry weight of the accumulated extracellular chorion proved to increase almost linearly with numerical position, between approximately 10% and 90% of the final weight (Fig. 36). Therefore, we defined synthetic stages II to IX merely by correlating synthetic profiles with chorion dry weights. Accumulation of early and late chorion proceeded considerably slower relative to numerical position (and, therefore, relative to time). Accordingly, we subdivided stages I and X as follows. All available synthetic profiles prior to stage II, or after stage IX, were ordered in a progression. Taking into account relative frequencies, ease of recognition, and the results of timing experiments (Sect. VIII.B.), five stages were defined for the initial and four for the final period. Stage 0 was defined as the period before any synthesis of C, B, or A proteins is detectable. Thus, 18 synthetic stages were defined in all (Fig. 37a). The 17 chorionating stages are represented by an average of 12 follicles per ovariole, and thus not all stages are found in a single animal (Fig. 37b).

The most striking changes in protein synthesis take place at the beginning and end of choriogenesis. Synthesis of C proteins begins, shifts quantitatively among several subclasses, and then subsides to a very low level (stages Ia to III); near the end of choriogenesis, synthesis of Cs again becomes apparent and ultimately predominates (stages Xa to Xd). From the rates of isotope incorporation into total

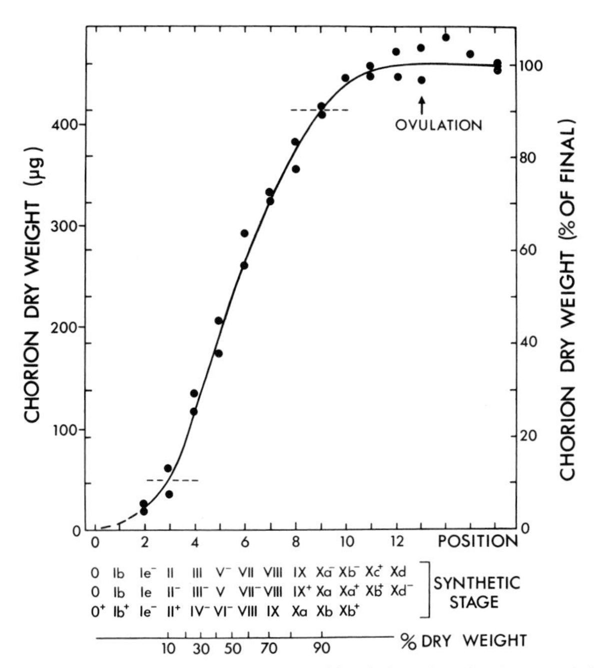

Fig. 36. Correlation of protein synthetic stage of individual polyphemus follicles with their position in the ovariole (abscissa) and the amount of chorion protein already secreted (ordinate). Three ovarioles from a day 16 female were labeled with ^{3}H-leucine for 1 h upon dissection, and labeled proteins of each of the follicles were subsequently analyzed on an SDS-polyacrylamide gel to determine the synthetic stage of the follicle (see Fig. 37). Chorion was dissected (rapidly, in water) from individual follicles in two parallel ovarioles from the same animal, and chorion dry weight was determined. (From Paul and Kafatos, 1975)

chorion and individual peaks, as well as from the accumulation of chorion mass as a function of stage, we estimate that the absolute rate of C synthesis may increase only slightly in late choriogenesis; C synthesis becomes prominent largely because A and B synthesis is turned off rather abruptly. Definite conclusions must await careful determinations of absolute rates of synthesis, taking into account the specific activity of the intracellular precursor pool (Regier and Kafatos, 1971). Only after absolute kinetics of synthesis are known for individual proteins will it be possible to evaluate whether the changes in chorion protein synthesis are gradual enough to be driven by contemporaneous transcription, without gene amplification or redundancy (Sect. I.C.), and without accumulation of temporarily inactive mRNA.

As expected from their final abundance, synthesis of A and B proteins is dominant for most of choriogenesis (Id to Xb), including the periods of most active chorion deposition (II to IX). Synthesis of nonchorion proteins is minimal throughout choriogenesis. Each of the 18 stages can be recognized by a unique protein synthetic profile. Although some of the synthetic changes are subtler than others, choriogenesis entails a continuously changing pattern of specific protein synthesis. With SDS fractionation, identical patterns are obtained with pulses varying from 2 min to several hours, indicating that the changes do not reflect posttranslational modifications (see also Fig. 21).

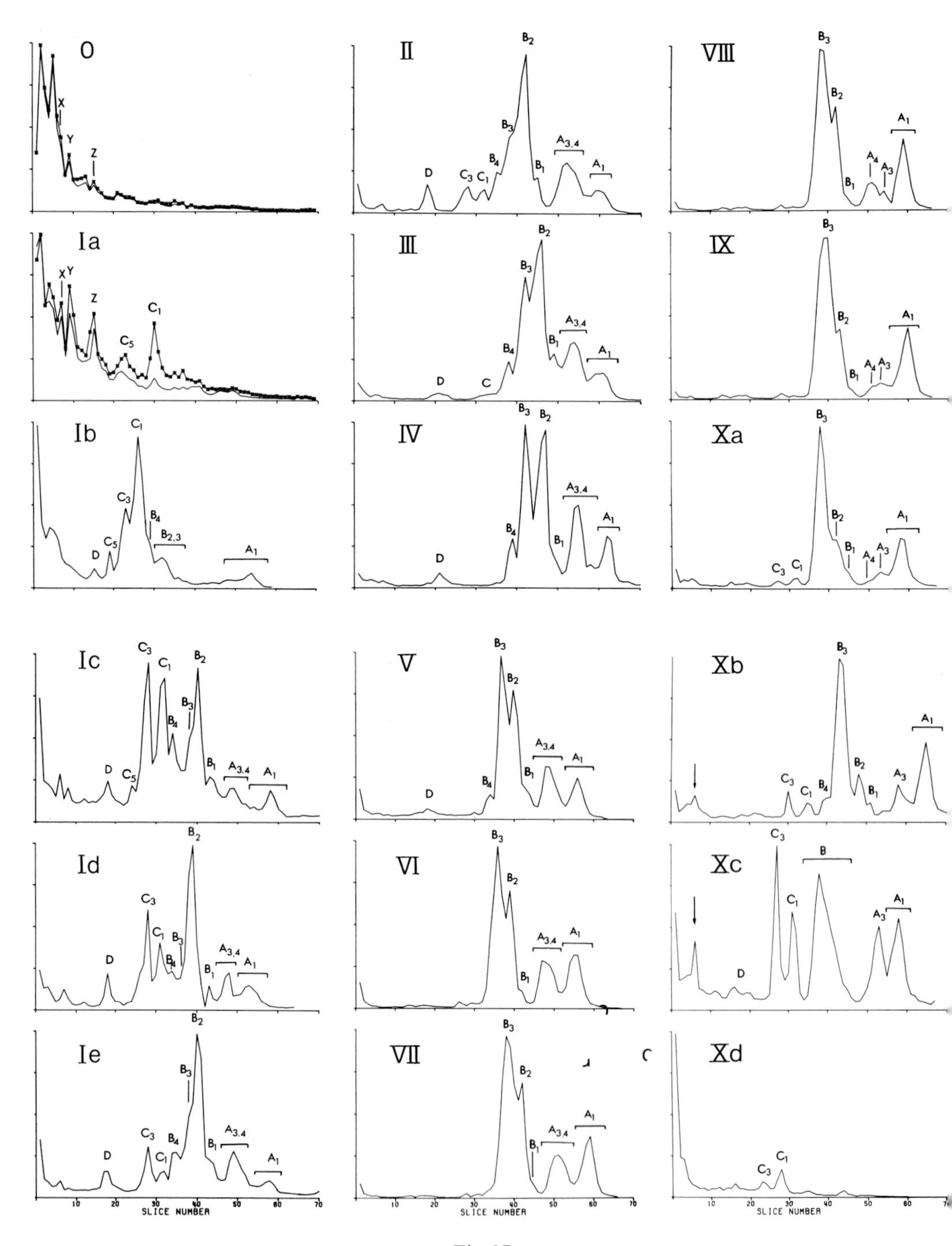

Fig. 37a

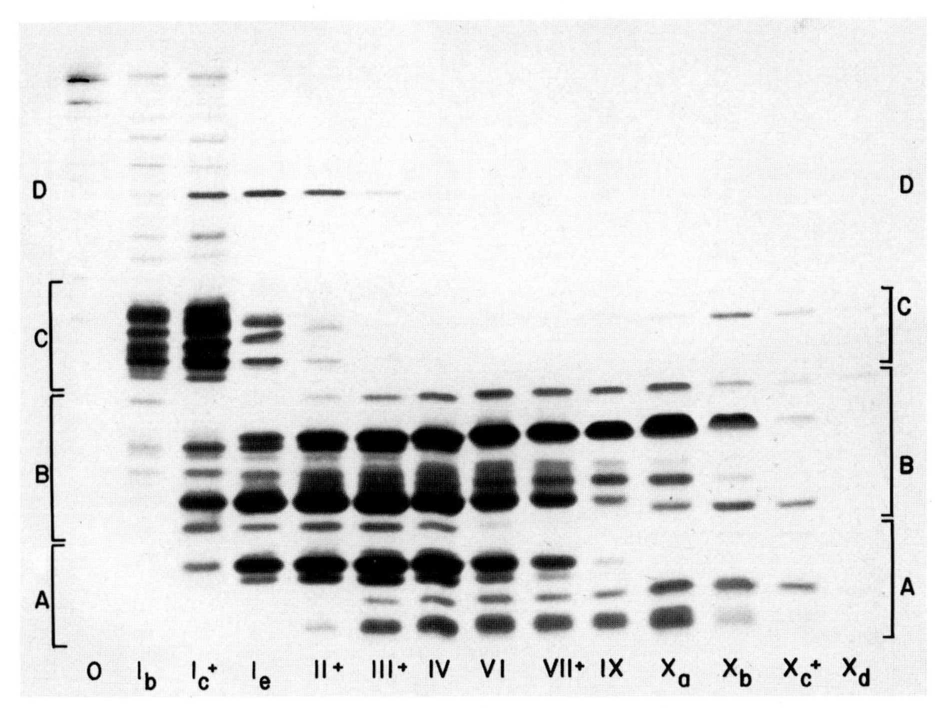

Fig. 37b

Fig. 37. (a) Protein synthetic stages in polyphemus. Follicles were labeled with ^{3}H leucine briefly, soon after dissection, the oocyte was washed away, and the follicular epithelial cells plus chorion were dissolved in SDS-urea-mercaptoethanol and analyzed by electrophoresis on cylindrical SDS-polyacrylamide gels. Mobility from *left* to *right*. Radioactivity profiles shown are characteristic of the indicated synthetic stages (0 to X_d). Ordinate scale was 2 to 32X 10^3 ^{3}H cpm per division, depending on the conditions of the experiment; we estimate that the absolute rate of chorion synthesis is nearly constant between stages II and IX (from the nearly linear increase in chorion dry weight, Fig. 36), and somewhat lower on either side of this period. Chorion subclasses are identified as in Paul et al. (1972a); for convenience, $A_{1,2}$, $B_{4,5}$, $C_{1,2}$, $C_{3,4}$, and $C_{5,6}$ are abbreviated as A_1, B_4, C_1, C_3, and C_5, respectively. Characteristic high molecular weight components of unknown function are also indicated (*X, Y, Z, arrow*). Follicles 0 and I_a were labeled with a mixture of ^{3}H-leucine (———) and ^{14}C-glycine (×——×) to detect early synthesis of the glycine-rich C proteins. Each stage lasts approximately 3 h (see Tables 7, 8) (From Paul and Kafatos, 1975). (b) Protein synthetic stages in polyphemus, analyzed by electrophoresis on an SDS-polyacrylamide slab gel, followed by autoradiography. Note the increased resolution, relative to (a). All samples were from a single animal. Stages determined by comparison to (a)

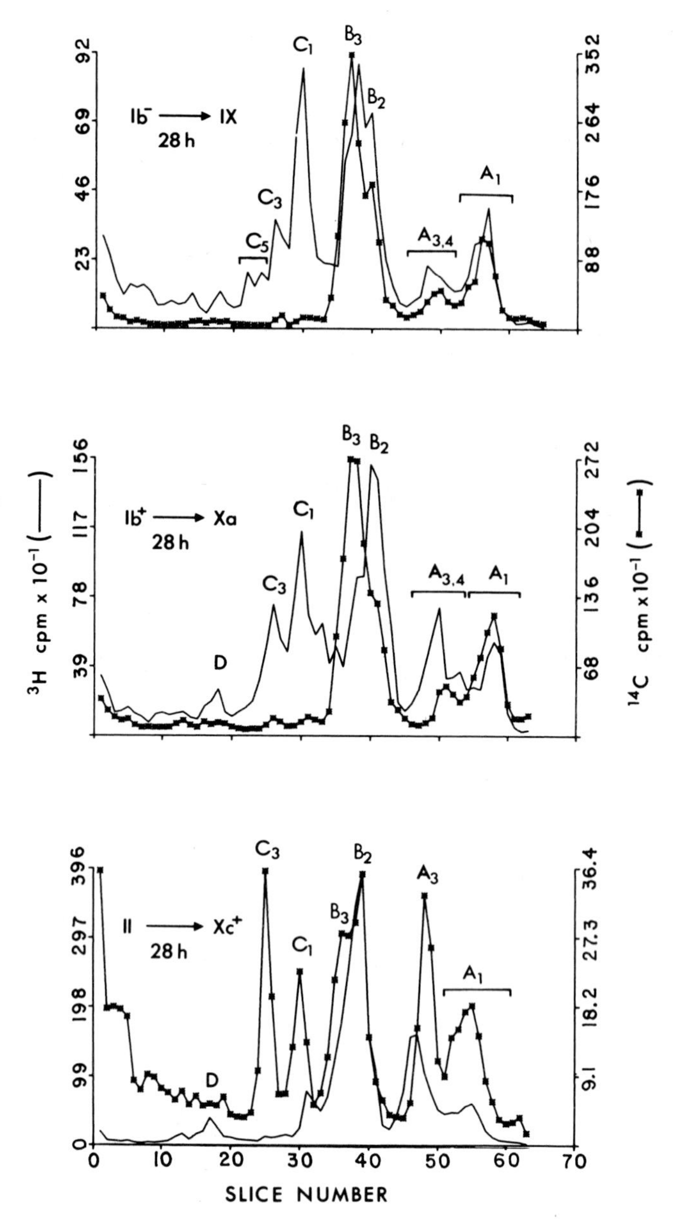

Fig. 38. Maturation of polyphemus follicles in situ (*left*) and in vitro (*right*). Each panel represents synthetic profiles of a single follicle, labeled at the beginning of the experiment with ^{3}H-leucine and at the end with ^{14}C-leucine. Notations at the *upper left* indicate the initial synthetic profile (^{3}H), the final synthetic profile (^{14}C), and the time interval between the midpoints of the two isotope pulses (h). For in situ experiments, ^{3}H pulse duration was

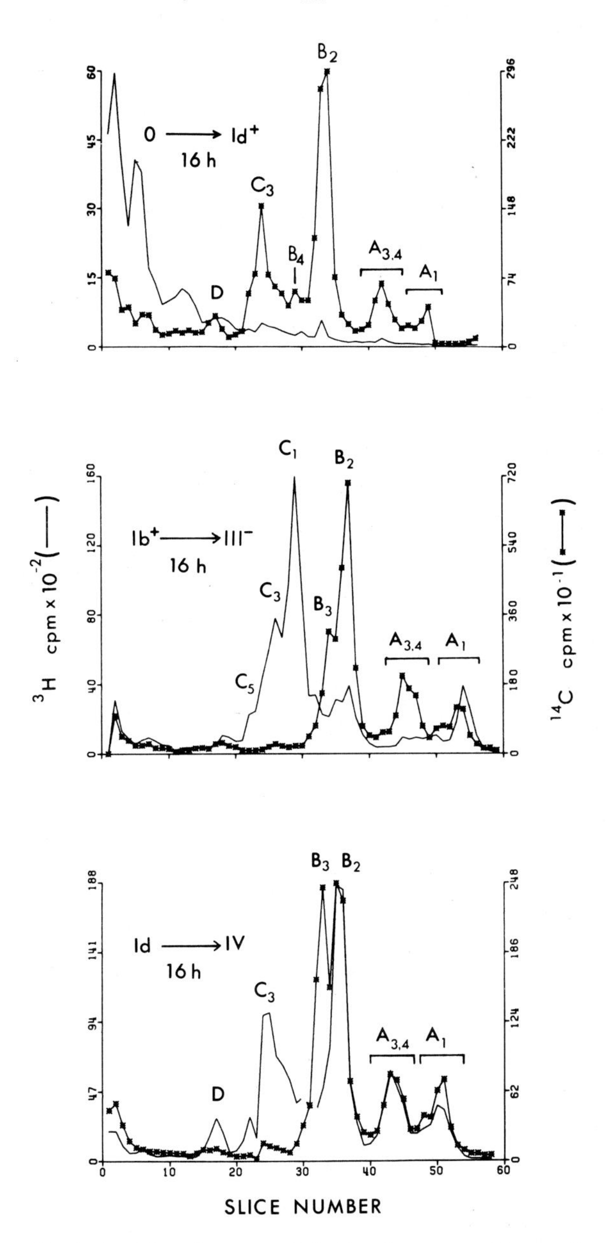

approximated as 2 h, the half-life of amino acid in blood; to correct for continued low-level ^{3}H incorporation during "chase," ^{3}H profile was interpreted with special attention to early features (e.g., the profiles of Cs in the Ib^{-} and Ib^{+} follicles). In in vitro experiments (*right*) chase was better defined and synthetic profiles easier to interpret. See also Tables 7, 8. *Left* and *top right panels* from Paul and Kafatos (1975)

B. The Time Axis of the Developmental Program

Having defined the stages, we sought to determine their normal duration. The timing experiments were simple in concept: follicles were labeled with a pulse of ^{3}H-leucine at the beginning of the experiment, maintained in the absence of isotope for a known period of time, and then labeled with a pulse of ^{14}C-leucine. Upon electrophoresis of the total proteins on an SDS gel, the ^{3}H profile revealed the synthetic stage at the beginning of the experiment, and the ^{14}C profile the synthetic stage of the same follicle at the end of the experiment. Knowing the total duration, we could calculate the average duration of the stages traversed by the follicle between the two isotope pulses.

In order to minimize any possible interference with the normal process of choriogenesis, a series of timing experiments were performed in situ: the first isotope was injected into the developing moth, and the animal was left unperturbed until the end of the experiment, when the ovarioles were dissected and labeled with the second isotope in vitro (Fig. 38, left). Timing information could be obtained from this experiment, even though the first isotope was not delivered in a strict pulse: the concentration of label in the blood was monitored during the experiment and proved to decline fairly rapidly to approximately 10% of the initial level, with an initial half-life of 2 h. Thus, the ^{3}H profile largely reflected the proteins synthesized at the time of the injection. Using follicles labeled by the first isotope during early stages (which are usually unmistakeable), and a long experimental period, timing could be evaluated with some confidence. An average stage duration of 3.0 h was estimated from 11 analyses in two animals (range 2.4 to 4.0 h; Table 7).

Similar results were obtained in vivo. In this type of experiment, portions of ovarioles were dissected, labeled briefly in vitro with ^{3}H-leucine, and then transplanted into the haemocoel of a host. After a period of several hours, the transplant was recovered and labeled in vitro with ^{14}C-leucine. With this design, there was no ambiguity regarding the initial synthetic stage of the follicle, and the duration of the experiment could be shorter, permitting determination of the duration of a small number of stages, rather than of an overall average. Again, the duration seemed reasonably uniform for different stages and equal to that determined by in situ experiments (9 analyses in 3 animals; average duration 3.1 h, range 2.3 to 4.0 h; Table 7).

C. Choriogenesis Can Begin and Proceed in vitro According to the Normal Developmental Program

In vitro timing experiments (Table 8) clearly revealed that the synthetic program for chorion proteins is autonomous. Follicles kept in isolation in a defined, protein free tissue culture medium (Grace, 1962) can initiate chorion production. Moreover, they can proceed through the biochemically defined synthetic stages at the same rate as they do within the animal. Even the stripe, which is detectable as an opaque line in late polyphemus eggshells, appears after in vitro incubation of young follicles.

Table 7. Duration of stages in individual follicles, in situe and cultured in vivo

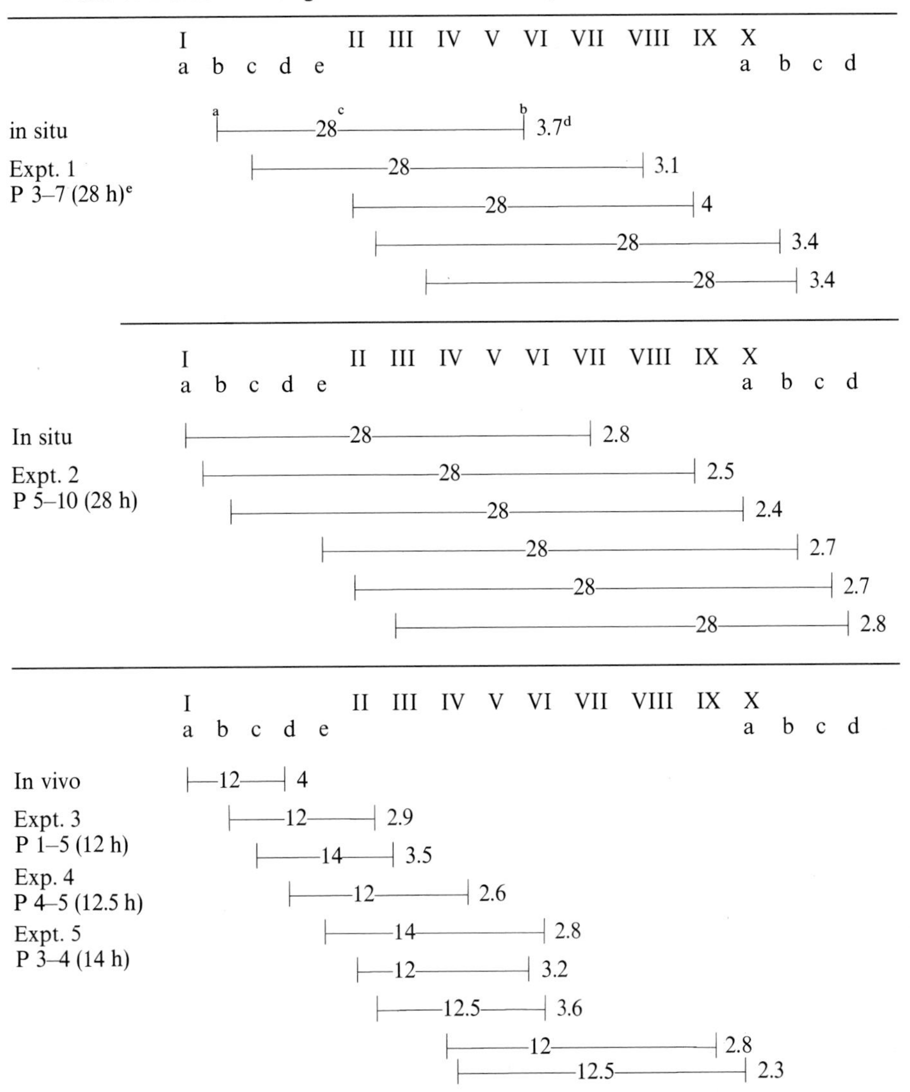

[a] Synthetic profile at beginning of experiment (^{3}H).
[b] Synthetic profile at end of experiment (^{14}C).
[c] Duration of experiment, h.
[d] Average duration of the stages traversed, h.
[e] Position of the follicles at dissection, and duration of the experiment.
For further details, see the text; from Paul and Kafatos (1975).

In these experiments labeling with the two isotopes and the intervening "chase" were all performed in Grace's tissue culture medium without hemolymph. Essentially normal stage duration was observed for all chase periods (8 to 24 h), regardless of the initial stage of the follicle (Fig. 38, right). Moreover, *initiation* of choriogenesis was clearly detected using stage 0 follicles (Table 8). In all in vitro experiments (36 analyses, using follicles from 7 animals), the average stage duration was 3.0 h (range 2.1 to 4.0 h).

A significant observation was that the follicles do not synthesize normal amounts of protein in vitro, although they change their relative rates of protein synthesis with the same kinetics as they do in vivo (Table 9). For example, a follicle

Table 8. Duration of stages in individual follicles cultured in vitro[a]

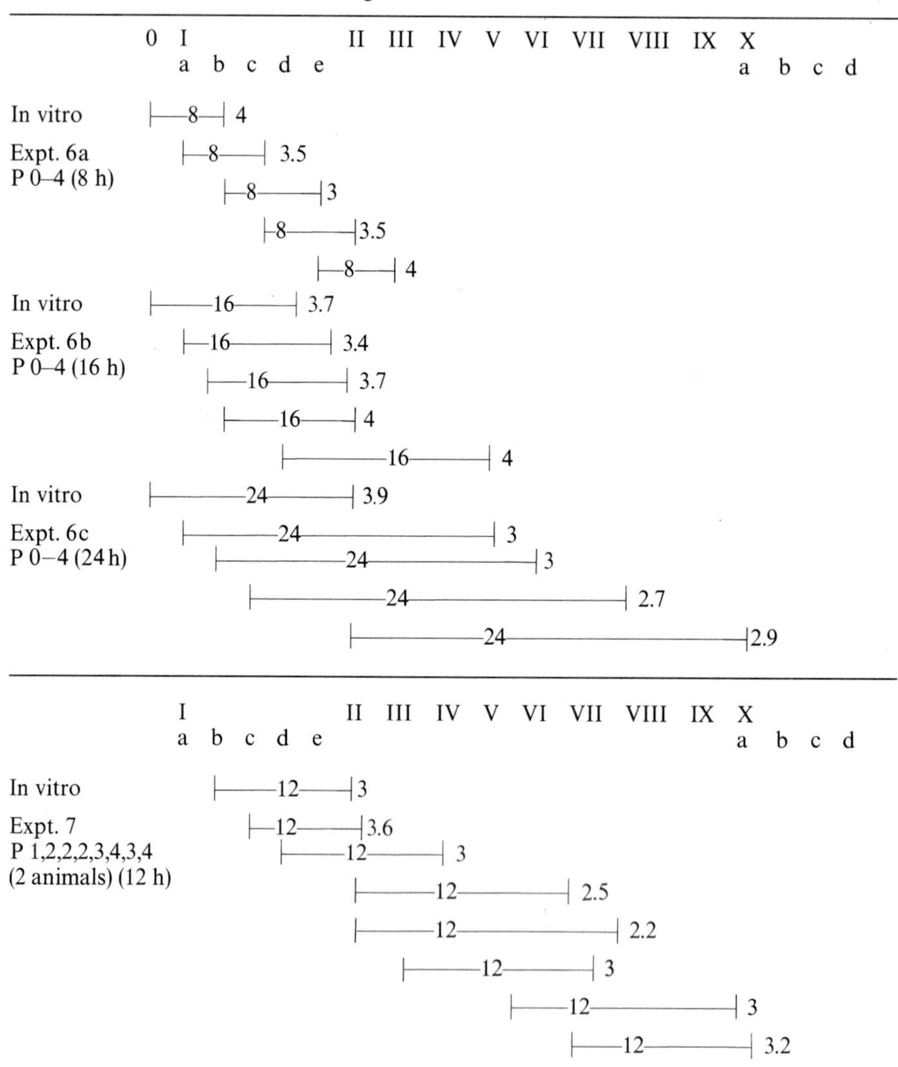

Tab. 8 (cont.)

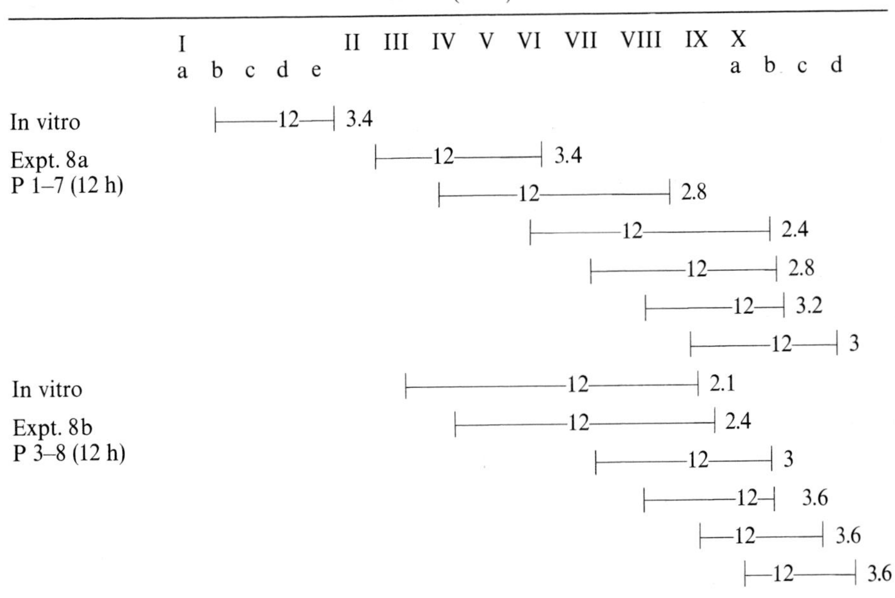

[a] For explanation, see Table 6 and the text; from Paul and Kafatos (1975).

Table 9. Chorion production and developmental maturation of cultured follicles

Position at	Chorion weight, % of 460 μg (initial stage→final stage)			
t=0 h	t=0 h	t=8 h	t=16 h	t=24 h
P_1	— (I_a)	– ($I_a \rightarrow I_c^+$)	— ($I_a \rightarrow II^-$)	10% ($I_a \rightarrow V$)
P_2	1% (I_b)	— ($I_b^+ \rightarrow I_e$)	9% ($I_b^- \rightarrow II$)	22% ($I_b \rightarrow VI$)
P_3	2% (I_c)	4% ($I_d^- \rightarrow II$)	21% ($I_b^+ \rightarrow III^-$)	24% ($I_c \rightarrow VIII$)
P_4	7% (I_e^-)	15% ($I_e \rightarrow III$)	25% ($I_d \rightarrow IV$)	37% ($II^- \rightarrow X_a^+$)

Six ovarioles from a single animal were put in culture. Three were pulsed for 30 min with ^{3}H-leucine to permit determination of the synthetic stage at time zero, and were then cultured in nonradioactive Grace's medium. At 8, 16, and 24 h one ^{3}H-labeled string was labeled for 30 min with ^{14}C-leucine and was then frozen together with one of the three unlabeled strings. Initial and final synthetic stages were determined from the ^{3}H and ^{14}C profiles, respectively, and the chorion dry weights from the extracellular chorions dissected from the unlabeled string frozen at the corresponding time. Chorion weights are expressed as percent of the normal final dry weight, 460 μg. The weights at t=0 h are estimates summarizing our knowledge of the normal relationship between chorion weight and synthetic stage, and the stages indicated are the average of the three determinations (^{3}H). (From Paul and Kafatos, 1975).

required 24 h to go from stage II to stage Xa in vitro, as expected, but in so doing deposited only approximately 120 μg of chorion (expected: 390 μg). These results indicate that the program of chorion protein synthesis in polyphemus is insensitive not only to cues from outside the follicle (e.g., hormones; Sect. IX.A.) but also to feedback from the absolute amount of chorion produced.

IX. Regulation of Choriogenesis

In the preceding sections, we discussed the structure and function of the chorion, the mechanism of its morphogenesis, and the changes in specific protein synthesis which make that morphogenesis possible. We showed that the protein synthetic changes characteristic of choriogenesis are programmed, and can be implemented autonomously by the follicles. We now turn to a consideration of the mechanisms regulating this program of cell-specific protein synthesis. In this section, we will discuss hormones and other molecules which may play a role as intercellular or intracellular signals. In the next section, we will discuss various aspects of macromolecular metabolism, from translation to DNA accumulation, with a view toward determining biochemically at what levels of intracellular information flow choriogenesis is regulated.

A. Hormonal Control

While it is not known whether hormones have any direct role in choriogenesis, hormones have been shown to participate during earlier stages of oogenesis (Engelmann, 1970). In many insects juvenile hormone is necessary for the production of yolk proteins by the fat body and/or their uptake by the follicle. In some cases neurosecretory hormones of the brain may also play a role, for example, by affecting the basal metabolic stage of the animal or by regulating the secretion of juvenile hormone by the corpora allata.

However, environmental and hormonal control of oogenesis has been nearly abolished in Bombyx and the saturniid moths, which are nonfeeding and short-lived. Only the onset of maturation is affected by the endocrine state; although yolk protein is already present in the blood at the beginning of the pupal-adult transformation, vitellogenesis begins only if ecdysone acts in the *absence* of juvenile hormone. Once the uptake of yolk has begun, the subsequent maturation of follicles is carried out autonomously, according to an endogenous timetable, in which the presence or absence of juvenile hormone appears to be without consequence (Williams, 1959).

While spermiogenesis in the silkmoths depends on a macromolecular factor (MF) which is made available to the spermatocytes through the action of ecdysone (Kambysellis and Williams, 1971), it is not known whether MF plays any role in oogenesis. Nor is it known what mechanism ensures the sequential entry into the vitellarium of individual follicles, many of which were formed long before, in late larval life. It does seem probable that in the saturniids, once a follicle has entered the vitellarium, it matures autonomously. Even so, it is not unlikely that intercellular communication *within* the follicle ensures its orderly and timely maturation.

In the sphingid moth, *Manduca sexta* (Nijhout and Riddiford, 1974) juvenile hormone is required for some late aspects of follicle development. In normal development, the corpora allata begin to release juvenile hormone about the time the adult female emerges; thus follicular development occurs up to this time in the absence of juvenile hormone. Examination of ovarioles at this time reveals that in the most advanced follicles vitellogenesis has proceeded to the stage just prior to

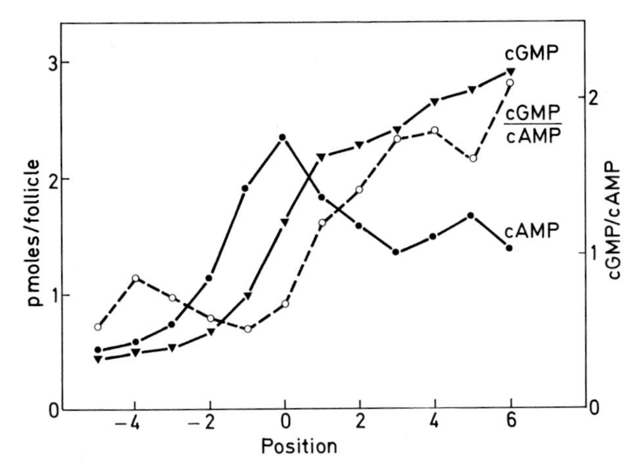

Fig. 39. Cyclic nucleotide content of polyphemus follicles, as a function of position within the ovariole. Measurements performed on the complete follicle (including the oocyte). Four to eight follicles of identical position were pooled from a single animal and immediately homogenized at 4° C in 6% TCA. After a brief, low speed centrifugation the supernatant was recovered and extracted several times with ethyl ether. Traces of ether were evaporated under a stream of N_2. Aliquots of the extract were made 0.05 M in Na acetate buffer, pH 6.2 and were used directly for radioimmunoassays of cAMP and cGMP. Both assays were specific, with a sensitivity of 0.25 pmol. Results are averages of duplicates or triplicates, except for positions −5 to 0

degeneration of the nurse cells. Vitellogenesis proceeds normally up to this stage even in allatectomized animals; however, once this stage is reached the follicles continue to develop only if exogenous juvenile hormone is administered, and degenerate if it is withheld. It should be noted that in normal moths, the subsequent maturation of follicles in emerged females (containing juvenile hormone) occurs sequentially rather than in groups, suggesting that the hormone acts only at a particular phase of an endogenous developmental program. The sensitive period seems to begin slightly more than a day before ovulation, since one egg per ovariole is ovulated by 27 h post emergence. Finally, the follicles become insensitive to juvenile hormone withdrawal, approximately 12 h before ovulation. The exact relationship between the end of the hormone sensitive period and the onset of choriogenesis remains to be determined.

B. Cyclic Nucleotides

Preliminary measurements of cyclic-AMP and cyclic-GMP levels in the maturing silkmoth follicle revealed repeatable changes correlated with its progression from vitellogenesis to terminal growth and choriogenesis. Analyses have been performed only on the follicle as a whole, and so it is not known whether the cyclic nucleotide changes are correlated with follicular cell maturation, oocyte maturation, or intercellular communication.

Positionally staged follicles from a single polyphemus were homogenized in TCA. The homogenate was extracted with ethyl ether to remove the TCA, and aliquots were used directly for determination of cAMP and of cGMP by separate radioimmunoassays (Steiner et al., 1972). Figure 39 shows results of one such experiment. The cAMP level in the follicle rises steeply in late vitellogenesis, reaches a peak during terminal growth and then declines to an intermediate level. Six repeats showed that the cAMP peak usually occurs at position +1; a second very minor peak is observed reproducibly between positions +4 and +6. By contrast, the cGMP level rises later (by one follicle position; 2 repeats), reaches a high level at the beginning of choriogenesis, and continues to increase slowly thereafter. As a result, the cGMP/cAMP ratio nearly triples between the end of vitellogenesis and follicle +3 or +4.

It has been reported (Kuo et al., 1971) that in the majority of tissues of Lepidoptera (cecropia, polyphemus and *Manduca sexta*), cGMP-dependent kinases predominate, and that the relative abundance of the cAMP and cGMP kinases varies between tissues. The tissue differences suggest that the variation of the cGMP/cAMP ratio in the follicle, as a function of developmental stage, may be related to a specific follicular function, rather than to general metabolism.

X. Chorion mRNAs and Their Metabolism

Our quantitative understanding of choriogenesis is still too incomplete to permit strong inferences about the levels at which it is regulated. For example, we do not yet know the absolute kinetics of synthesis for individual proteins, and hence cannot evaluate whether these kinetics are compatible with contemporaneous transcription on a single copy of each gene per genome (Sect. I.). Nevertheless, we have chosen initially to investigate mRNA synthesis and accumulation as two aspects of information flow which are likely to be important in regulating choriogenesis. In additional studies, we have measured the translational productivity during development, and have begun to study the DNA of the follicular cells.

A. Translational Productivity During Follicular Maturation

After terminal growth, there is a massive upsurge of chorion protein synthesis. One may ask whether this is accompanied by an increase in the productivity[4] of translation, i.e., the number of polypeptides produced per minute per molecule of polysome-bound mRNA (Paul et al., 1972a). In general, the productivity of a template-copying process is controlled by the rates of initiation and chain elongation. It can be measured as the rate of chain elongation (codons/min) divided by the average spacing of the ribosomes on the mRNA (codons/ribosome)—irrespective of the size of the message (Kafatos, 1972). At 25° C, the time required for translation of an average chorion protein was measured as 2.0 min by an adaptation of the technique of Fan and Penman (1970); since the average size of the nascent

[4] The term efficiency is often used by other workers for what we call productivity.

chorion proteins was estimated as 150 amino acids, the calculated rate of chain elongation is 75 codons/min. The average size of the polysomes is four to five, indicating an average spacing of 30–37 codons/ribosome on the monocistronic mRNAs. Thus, the translational productivity for chorion is 2.1 to 2.6 polypeptide chains per minute per active mRNA (Paul et al., 1972a)—a not unreasonable value for eukaryotic systems at 25° (Kafatos, 1972a; Kafatos and Gelinas, 1974).

In the same experiments, the rate of chain elongation in late vitellogenic follicles was estimated as 110 codons per minute—close to the value for chorionating follicles. The polysomes of prechorionating follicles (which are synthesizing relatively large proteins, 58000 daltons average) are considerably larger than those of chorionating follicles—suggesting that the spacing of the ribosomes is in the same range. Thus, it appears that the translational productivity does not change drastically with the onset of choriogenesis. Control through changes in translational productivity may not be generally significant in differentiation (Kafatos, 1972a; Kafatos and Gelinas, 1974).

This conclusion, however, has no bearing on other types of translational control. For example, measurements of translational productivity would not reveal a population of preformed but inactive messengers. For a more meaningful inquiry into the molecular control of choriogenesis, it is necessary to shift our attention from the protein to the nucleic acid level.

B. Isolation and Characterization of Chorion mRNA

The high degree of specialization of the follicular cells, the unusually small size of the chorion proteins, and the low levels of endogenous RNase make isolation of chorion mRNA relatively simple (Gelinas and Kafatos, 1973). When RNA is extracted from either whole cytoplasm or polysomes, and is then fractionated on two successive sucrose gradients, a 7-14S RNA "cut" sedimenting between tRNA and 18S rRNA can be recovered and shown to be highly enriched in chorion message. When this RNA is prepared from pooled chorionating follicles labeled with ^{32}P-phosphate in organ culture or in vivo (3–48 h), and is further fractionated by electrophoresis on 6% polyacrylamide gels, autoradiography reveals two highly labeled broad zones, at 8 and 9S (Fig. 40). The lower molecular weight zone 1 is less intensely labeled than the higher molecular weight zone 2; in favorable preparations, zone 3, a high molecular weight shoulder of zone 2 can also be distinguished. Our working assumption is that zones 1, 2, and 3 are messages for chorion protein classes A, B, and C, respectively. In recent work, purified chorion mRNA has been prepared by the following steps: Mg^{2+} precipitation of ribonucleoprotein complexes (mostly polysomes; Palmiter, 1974) from a crude follicular homogenate, extraction of the RNA, binding to oligo-dT cellulose, and sucrose gradient fractionation of the bound fraction.

Identification of the zones as chorion mRNA is supported by the following lines of evidence:

1. They are the predominant labeled components in RNA preparations which can be translated in cell-free systems to yield polypeptides precipitable by chorion-specific antibody. Zones 1 and 2 can be excised individually from the polyacryl-

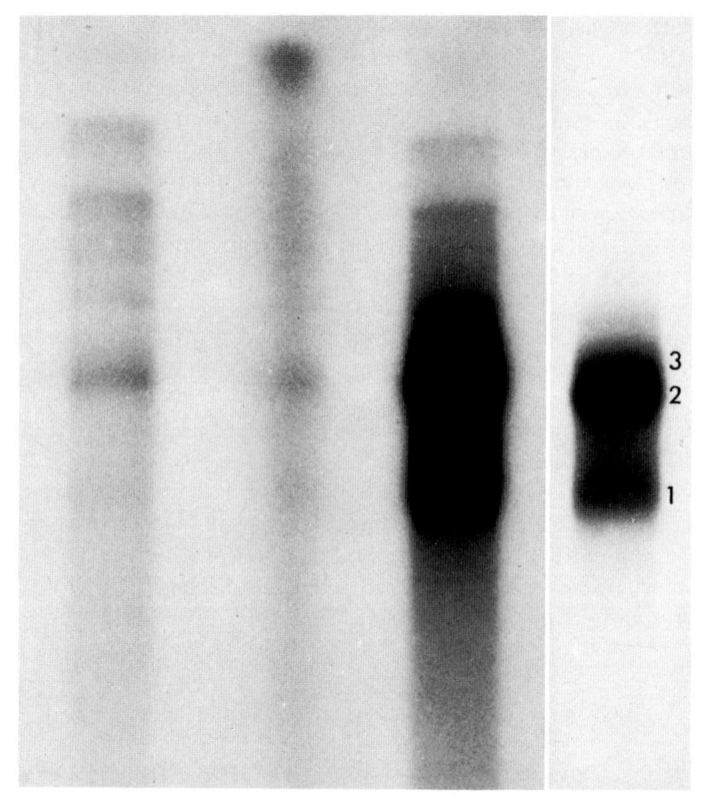

Fig. 40. Purification and electrophoretic fractionation of polyphemus chorion mRNA. One hundred chorionating follicles were pooled from animals labeled in vivo with ^{32}P-phosphate. Polysomes were collected in a sucrose gradient and disrupted in a buffer containing 0.5% SDS and 1 mM EDTA. RNA was centrifuged in two consecutive sucrose gradients; 7–14S fractions were collected and passed through a nitrocellulose filter at high ionic strength. Filter was washed at high ionic strength, and bound RNA was finally eluted with 0.5% SDS at low ionic strength. Direct effluent, high salt wash and low salt eluate were collected and analyzed by electrophoresis on a 6% slab polyacrylamide gel (migration from top to bottom). RNAs were detected by autoradiography. *Left* to *right*: effluent, wash, eluate, eluate photographed with a shorter exposure to permit resolution of heavily labeled bands. *Zones 1*, *2*, and *3* are the putative chorion mRNAs (from Gelinas and Kafatos, 1973; see also Vournakis et al., 1974)

amide gel and translated into products in the size range of A and B chorion proteins, respectively (Sect. X. C.).

2. Their size correlates in detail with the size of the chorion proteins. In polyphemus, their size is that expected of monocistronic mRNAs coding for chorion proteins (Gelinas and Kafatos, 1973): rabbit globin message (approximately 650 nucleotides; Gaskill and Kabat, 1971) comigrates with zone 2, the presumed mRNAs for B proteins (average size approximately 150 amino acids). In *M. sexta*, a sphingid whose chorion proteins are more than 5000 daltons larger, on the average, than those of polyphemus, the size of the putative chorion mRNAs is correspondingly larger (by 50000–100000 daltons; Gelinas and Kafatos, 1973).

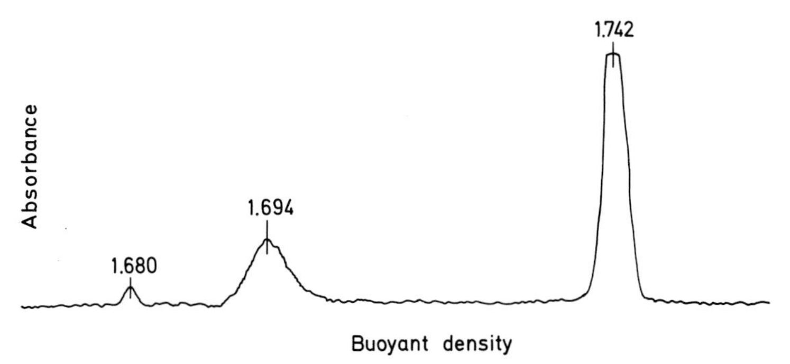

Fig. 41. Calculation of base composition of pernyi DNA, based on buoyant density of follicular DNA, centrifuged to equilibrium in neutral CsCl in the Spinco Model E ultracentrifuge. DNA was purified from whole nonchorionating follicles (from germarium to the stage of nurse cell degeneration) by a modified Marmur procedure. The G+C content of follicular DNA (35%) was calculated (Mandel et al., 1968) from the buoyant density ($\varrho = 1.694$) relative to phage SPO1 DNA ($\varrho = 1.742$). Species which bands at $\varrho = 1.680$ is presumably mitochondrial DNA, with the same density to that of Drosophila (Peacock et al., 1974). The calculated main band base composition is in agreement with estimates obtained from DNA melts

3. These RNAs are almost exclusively associated with the translational machinery of chorionating follicles (97% in polysomes and ribosomes combined; Gelinas, 1974; Gelinas and Kafatos, 1977). They can be released from polysomes by EDTA or puromycin treatments, in the form of RNP particles. They are absent from the polysomes of prechorionating follicles (Gelinas and Kafatos, 1973).

4. The putative mRNAs contain poly (A), and can be purified by binding to oligo d(T) cellulose or to nitrocellulose filters (Fig. 40; Sect. X. D.).

5. They are moderately rich in G+C, as expected from the amino acid composition of the chorion proteins (Sect. V. A.). ^{32}P-labeled pernyi mRNA, purified by oligo (dT) cellulose binding and sucrose gradient centrifugation, showed the following isotopic composition, after alkaline hydrolysis, paper electrophoresis, and correction for the poly (A) content (12.5% of As): 26.7% A, 27.6% U, 25.8% G, 19.9% C, i.e. 45.7% G+C. Similar results were obtained with total polyphemus chorion mRNA (Vournakis et al., 1974), and with polyphemus bands 1 and 2 recovered from a slab gel. By composition, the main band DNA in saturniids is approximately 35% G+C (in cecropia 35.2%, Wyatt and Linzen, 1965; in pernyi 35%, Fig. 41).

C. Translation of Chorion mRNA in Cell-Free Systems

Chorion mRNAs have been translated successfully in two cell-free systems, from wheat germ and from Krebs II ascites cells. The product has been identified as chorion through its precipitability by chorion-specific antibody, as well as through its characteristic size and isoelectric point distribution. Identification of individual chorion proteins among the cell-free products is difficult, because of the multiplicity of the components and the limited faithfulness of the cell-free systems, and the probable occurrence of protein precursors (Sect. XIII).

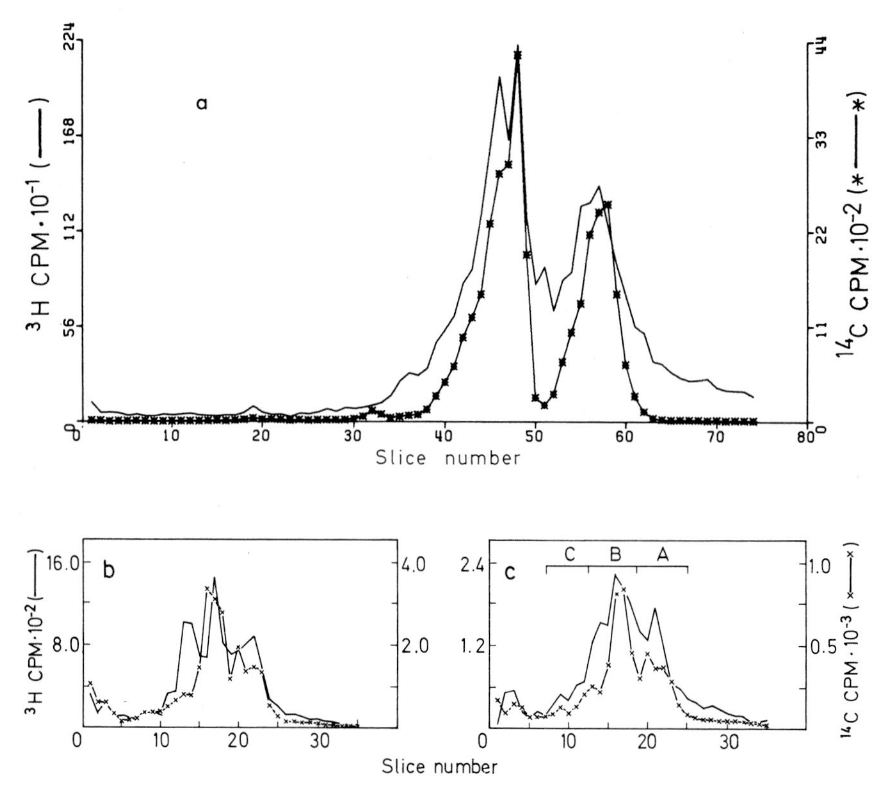

Fig. 42a—c. Cell-free translation of partially purified chorion mRNAs from pernyi (a) and polyphemus (b, c). Aliquots of ^{3}H-leucine labeled wheat germ translation products were dissolved and analyzed on SDS-polyacrylamide gels (Paul et al., 1972a), either directly (a) or after immunoprecipitation with chorion-specific rabbit antibody (b, c). Authentic ^{14}C-labeled chorion was added as an internal standard, prior to electrophoresis. In experiment shown in (a), 0.5 µg Mg^{2+}-precipitated (Palmiter, 1974) oligo dT-cellulose bound RNA from chorionating follicles was translated in presence of high K^+, 120 mM (which, while suboptimal in terms of total incorporation, improves fidelity of translation by decreasing premature chain termination; Schmeckpeper et al., 1974). In experiments shown in (b) and (c), RNA extracted from polysomes and monosomes of chorionating follicles, respectively, was used. The translation products were identified as chorion by their size distribution and their recognition by chorion antibody. (a): courtesy N. Rosenthal

Figure 42 shows the ^{3}H-labeled cell-free products from wheat germ extracts stimulated by the addition of partially purified chorion mRNA. In the experiments shown in Figure 42a, the fidelity of cell-free translation was improved by the presence of high K^+. Figures 42b and 42c show translation products after immunoprecipitation with chorion-specific antibody. In all experiments, the products were mixed with ^{14}C-authentic chorion prior to electrophoresis on an SDS gel designed to fractionate small chorion-sized proteins. The presence of chorion polypeptides in the cell-free products is indicated by their immunoprecipitability and chorion-specific size distribution.

Gel isoelectric focusing was used to evaluate the isoelectric point distribution of the cell-free product of an ascites extract which was stimulated by the addition of

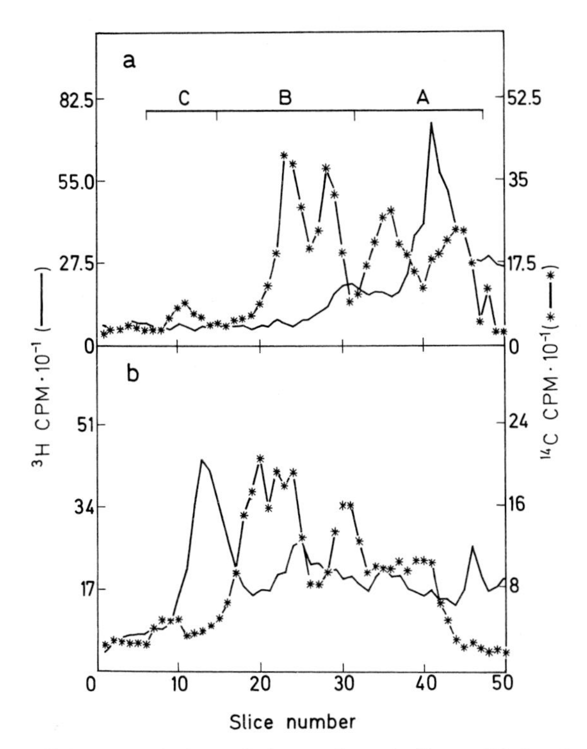

Fig. 43a and b. Cell-free translation of electrophoretically purified polyphemus chorion mRNA. 50 µg of 7–14S polysomal RNA from chorionating follicles (see Fig. 40), including a small amount of ^{32}P-RNA, was fractionated on a 4 to 8% polyacrylamide slab gel. Zones 1 and 2 were located by autoradiography, 1 mm gel slices were excised, and the RNA was extracted from the gel (Gelinas, 1974). The extracted RNA was ethanol precipitated in the presence of yeast tRNA as carrier, and was used in the wheat germ cell-free translation assay. The cell-free product was displayed by electrophoresis in SDS-polyacrylamide gels in the presence of authentic ^{14}C-chorion standard. RNA from the leading part of zone 1 (a) yielded products in the size range of A proteins, and RNA from the high molecular weight side of zone 2 (b) yielded products in the size range of large Bs or small Cs (see also Sect. XIII)

chorion mRNA, extracted from polysomes and purified by two sequential sucrose gradient centrifugations. The ^{3}H-labeled product contained acidic material, extensively overlapping the distribution of an authentic ^{14}C-chorion marker.

Figure 43 shows the cell-free products from wheat germ extracts, stimulated by RNA from the leading part of zone 1 and from the high molecular weight half of zone 2. Zone 1 stimulates the synthesis of A-sized polypeptides, whereas the sample of zone 2 used stimulates the synthesis of polypeptides at just above the B-protein size range.

In most assays, especially with the wheat-germ system, the in vitro product fails to parallel exactly the internal standard (e.g., Fig. 42). Most of the discrepancy is probably due to post-translational processing which fails to occur in the cell-free systems, although a second source of the discrepancy may be unequal translation of different messengers by the cell-free system.

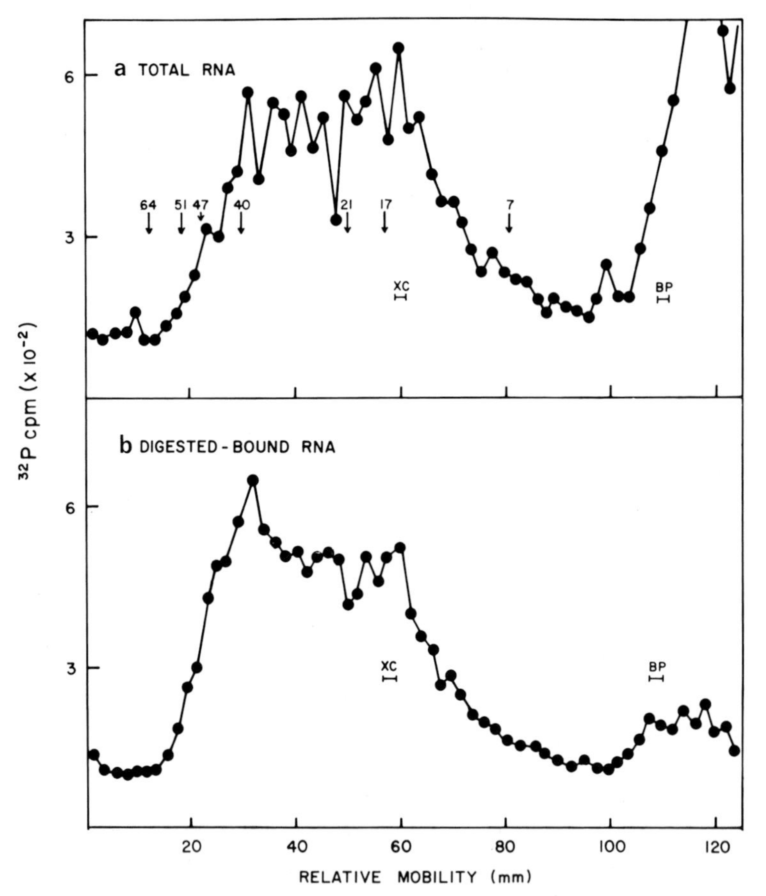

Fig. 44a and b. Size distribution of poly (A) fragments from polyphemus chorion mRNA. 7 to 14S polysomal RNA from chorionating follicles, labeled in vivo with ^{32}P-phosphate, was purified by sucrose gradient centrifugation and digested with a mixture of RNase $A+T_1$. The RNase resistant fragments were ethanol precipitated, either directly (a) or after binding to oligo (dT)-cellulose (*b*). The poly (A) size was estimated by electrophoresis (migration from *left* to *right*) in 15% polyacrylamide – 7 M urea gels standardized with polyribonucleotides of known polymer size *(arrows)* and the dyes, xylene cyanol (*XC*) and bromophenol blue (*BP*). The mono- and oligo-nucleotide front migrated with *BP*. In this experiment, the weight average size of the poly (A) was estimated as 28 to 29 nucleotides (a and b, respectively). (From Vournakis et al., 1974)

D. The Poly (A) of Chorion mRNA

The chorion messages contain poly (A) of a size sufficient to permit their binding to oligo d(T) cellulose and to nitrocellulose filters (Vournakis et al., 1974). In experiments using sucrose gradient purified polysomal ^{32}P-mRNA, the bound labeled material consists largely of zones 1, 2, and 3 (Fig. 40), while the nonbound fraction contains much of the contaminants and very little material of the same size as zones 1 and 2.

The size of the poly (A) was estimated by a combination of careful nuclease digestion and electrophoretic mobility experiments. The poly (A) proved to be unusually small in mRNA molecules labeled with a 12–36 h in vivo exposure to ^{32}P-phosphate: 11 to 14% of the adenines and 3.5 to 5.2% of the phosphates remained acid precipitable after treatment with RNase A and T_1, which should solubilize all but poly (A) sequences. From the size of chorion proteins, and the percent nuclease resistance of the adenines (together with the base composition of the RNA), it can be calculated that the weight average of the poly (A) length per molecule should be a minimum of 20 nucleotides [assuming no untranslated regions in the mRNA, other than poly (A)]. Electrophoresis indicated that the nuclease resistant poly (A) fragments (Fig. 44) have an average size of approximately 30 nucleotides (28 to 39 weight average in all experiments; 19 to 34 number average). The poly (A) nature of the nuclease resistant fragments was confirmed directly by base composition analysis (approximately 99% adenine). The percent nuclease resistance was the same for total sucrose gradient- plus nitrocellulose-purified polysomal message, and for electrophoretically purified zone 2 material, eluted from the acrylamide gel. Chorion messages are synthesized by chorionating follicles in culture. The poly (A) size is slightly larger (weight average 47 nucleotides) after a short in vitro labeling period (3 h).

E. Developmental Studies on Chorion mRNA Synthesis and Accumulation

Once chorion mRNAs became assayable by electrophoresis and by cell-free translation, it was possible to inquire when they are produced and accumulated in polyphemus, relative to the period of chorion protein synthesis (Gelinas, 1974; Gelinas and Kafatos, 1977). The messages cannot yet be assayed with a resolution comparable to that of the proteins. However, at least in its gross features it appears that choriogenesis may be "message driven". Cell-free chorion translation activity becomes detectable in parallel with the in vivo initiation of chorion protein synthesis. Newly synthesized, electrophoretically recognizable chorion mRNA appears in the cytoplasm throughout choriogenesis, and never before. Synthesis of the high molecular weight message class, zone 3, is only detectable in early choriogenesis, when synthesis of the high molecular weight C proteins is predominant.

Figure 45 shows the cell-free products of wheat germ extracts stimulated with total RNA from follicles 0 and +3. The total product from follicle +3 RNA contains material of the size expected for all three chorion protein classes, A, B, and C; no such material is detectable above background in the product from follicle 0 RNA (Fig. 45a). Immunoprecipitation confirms that the follicle +3 product is chorion polypeptides, and fails to demonstrate any chorion components in the follicle 0 product, even though the background is drastically reduced (Fig. 45b). We may conclude that no translatable chorion mRNA is detectable at position 0, i.e., in the last prechorionating follicle.

In evaluating this experiment, it is important to bear in mind some qualifications. Cell-free translation may not detect some possible message precursors, i.e., any molecules which require "capping" or other covalent modification before they

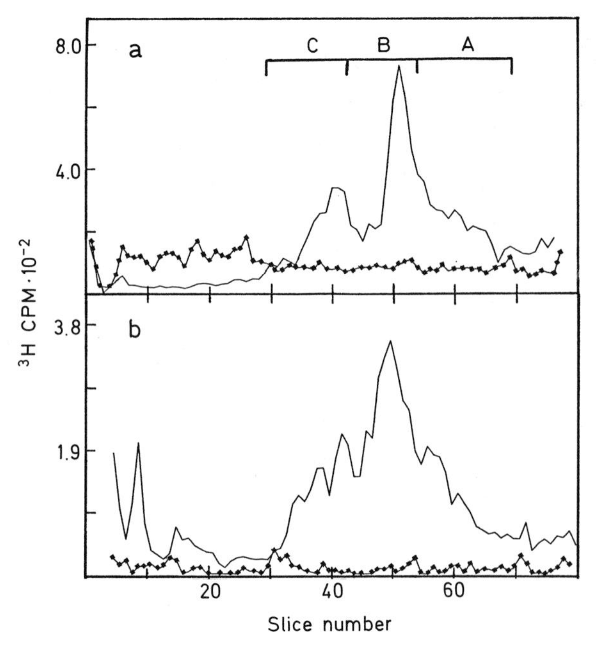

Fig. 45a and b. The proteins synthesized by wheat germ extract in response to equal amounts of total RNA from polyphemus follicles +3 (——) and 0 (◆——◆). The ^{3}H-labeled cell-free products from the two assays were run in parallel SDS-polyacrylamide gels and were aligned by comparison with internal ^{14}C-chorion standards *(brackets)*. The products were analyzed either directly (a) or after immunoprecipitation with chorion-specific antibody (b). RNA from follicle +3 leads to synthesis of chorion proteins of all three classes; no chorion synthesis is detectable with RNA from follicle 0. (From Gelinas and Kafatos, 1977)

can be recognized by the translation machinery. Messages that may be "masked" noncovalently will be detectable, since the RNA is deproteinized prior to cell-free assay. Nonadenylated message is probably also detectable, since the poly (A) does not appear to be an absolute requirement for translation (Williamson et al., 1974; Bard et al., 1974). It should also be noted that, although the antibody appears to precipitate a broad spectrum of chorion polypeptides, including C proteins (Fig. 45), it has not been shown to precipitate every single chorion species. Therefore, it is conceivable that some chorion polypeptides in the product of follicle 0 would escape detection by not being immunoprecipitable; the fact remains that no such polypeptides are detectable in the total product above background (Fig. 45a).

Figure 46 shows an autoradiogram of RNAs, synthesized in a developmental series of follicles and displayed so as to reveal the presence of labeled chorion mRNA. The follicles were labeled in vivo for 4 h with ^{32}P-phosphate. After dissection, the ribonucleoprotein complexes of the cytoplasm, including polysomes, were precipitated with high Mg^{2+}, the RNA was extracted with SDS-phenol-chloroform, and the poly (A) containing species were enriched by binding to oligo d(T) cellulose prior to electrophoresis on a slab gel. Clearly, chorion mRNA is synthesized throughout choriogenesis. Incorporation of ^{32}P is most active in early choriogenesis (positions 1 to 5); it declines gradually thereafter, but is still detectable

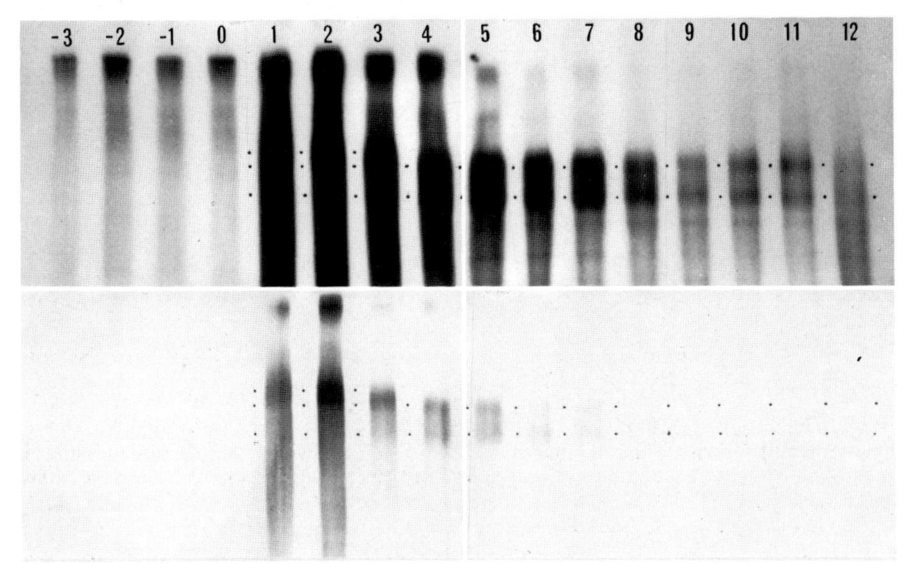

Fig. 46. Chorion mRNA synthesis in a developmental series of polyphemus follicles. A developing moth was labeled by injection of ^{32}P-phosphate. After 4 h, the eight ovarioles were removed and follicles of each position were pooled. One follicle from each position was labeled with ^{3}H-leucine for determination of synthetic stages; the rest were used to prepare RNA. Cytoplasmic, Mg^{2+} precipitable, oligo (dT) cellulose-bound RNA was collected, analyzed by electrophoresis on a 6% polyacrylamide gel, and detected by autoradiography. *Dots:* location of zones 1, 2, and 3 (*bottom* to *top*). *Top panel:* 48 h autoradiographic exposure; *bottom panel:* 24 h exposure, lightly printed. No chorion mRNA synthesis is detectable until position *1*; zone 3 is synthesized primarily at positions *1* and *2*, whereas synthesis of zones 1 and 2 continues to virtually the end of choriogenesis (position *12*). (From Gelinas and Kafatos, 1977)

by position 12. By contrast, no incorporation into chorion mRNA is detectable in vitellogenic follicles, up to and including follicle 0, the last prechorionating follicle in this ovariole.

It appears (Fig. 46) that the putative C protein message, zone 3, is labeled in early follicles which are known to synthesize predominantly C proteins, but not at later stages. Zone 3 synthesis is intense in follicles 1 and 2; by follicle 4, zones 1 and 2 predominate. These results suggest that specific message classes are produced in correlation with their translation in vivo.

The interpretation of this experiment should be tempered by consideration of the method of RNA purification prior to electrophoresis. This experiment would detect the synthesis of chorion mRNA or of a precursor which is cytoplasmically located, polyadenylated, and of message-size. In order to reduce the background sufficiently to permit rigorous mRNA detection and quantification, it was necessary to enrich for poly (A) containing species by binding to oligo d(T) cellulose. Thus, nonadenylated message precursors would have escaped detection. However, additional experiments failed to show (admittedly against a high background) RNA species of chorion message size even in the cytoplasmic "follicle 0" RNA which fails to bind to oligo d(T) cellulose. Precipitation with Mg^{2+} (a convenient step

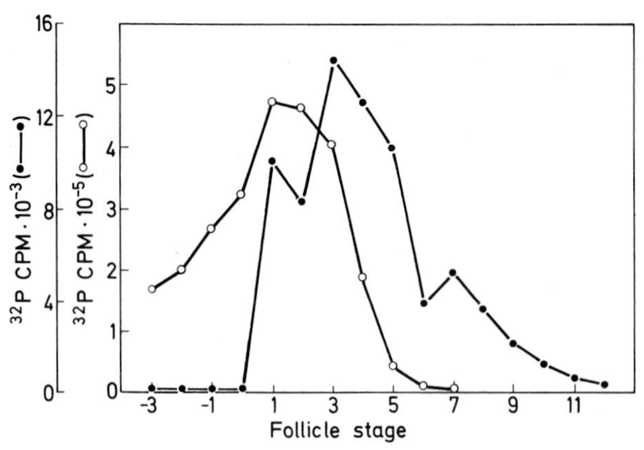

Fig. 47. The timing of synthesis of chorion mRNAs (●——●) and 18+28S rRNA (○——○) in polyphemus follicles. The radioactivity in the mRNA zones of Figure 46 was quantitated by excision of the zones, determination of Cerenkov radiation, and removal of background. The radioactivity of rRNA was similarly determined, after electrophoresis of the total Mg^{2+}-precipitable cytoplasmic RNA on 3% polyacrylamide—0.5% agarose slab gels. (From Gelinas and Kafatos, 1977)

in RNA preparation) is unlikely to lead to the overlooking of untraslated message: under our conditions, not only ribonucleoprotein complexes but even purified chorion mRNAs are partially precipitable, whether in buffer or mixed with a cell extract (Palmiter, 1974; Gelinas, 1974).

Synthesis of chorion mRNA shows an interesting contrast with synthesis of rRNA (Fig. 47). Quantification of electrophoretically fractionated labeled RNA shows that ^{32}P-incorporation into rRNA increases continuously in late vitellogenesis, reaching a peak at position +1 and remaining nearly constant in positions +1 to +3; thereafter, it declines precipitously. By contrast, synthesis of chorion mRNA is completely undetectable prior to position +1, increases rapidly to a peak later than rRNA does (positions +3 to +4), is high for a considerable period of time (positions +1 to +5), and then declines gradually. The absolute rates of synthesis cannot be evaluated without measurements of specific activities of the precursor pools. Nevertheless, it is striking that in positions +5 and later the radioactivity of cytoplasmic chorion mRNA is more than half the radioactivity of rRNA—implying that, in terms of numbers of molecules, mRNA is synthesized very much faster than rRNA.

In conclusion, the evidence indicates that chorion mRNA is synthesized throughout choriogenesis, and accumulates (at least in a translatable form) only during this period. There are indications that the parallelism between message production and message utilization holds even for the shift from early C proteins to the later A and B proteins. There is an uncertainty of 2 to 3 h in these correlations, because of the duration of ^{32}P labeling and the possible variation between ovarioles. No information is available about hypothetical message precursors; such information could be obtained by use of cDNA, synthesized on the mRNA templates by reverse transcription.

Within the above limits, it appears that chorion message synthesis and appearance in the cytoplasm play an important role in driving the developmental

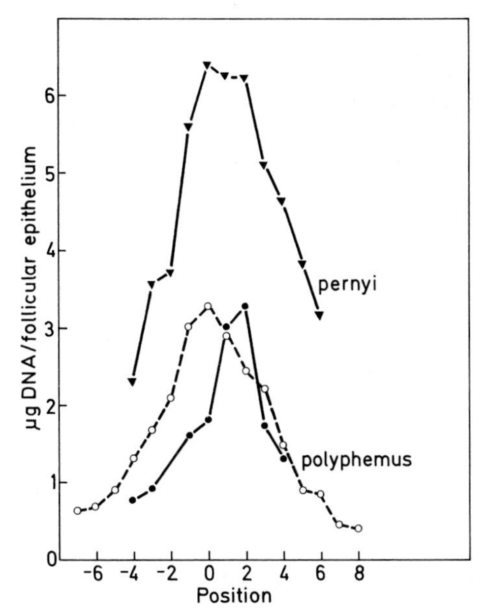

Fig. 48. DNA content of a developmental series of follicular epithelia. Pooled de-yolked follicles of the indicated positional stage (from single animals) were dropped into 20% perchloric acid, left for 0.5 h at room temperature, and pelleted. The supernatant was discarded and the pellet was resuspended in 200 μl 10% perchloric acid. After hydrolysis of DNA at 90° for 45 min, one volume of 4% diphenylamine in acetic acid was added and (separately) 10 μl acetaldehyde (1.6 mg/ml). The color was developed after 1–2 h at 56°, and was extracted into 1/4 volume of amyl acetate (Abraham et al., 1972). The sensitivity of the assay was 1 μg, and the standard curve (with calf thymus DNA) was linear between 1 and 25 μg. In several repeats the general shape of the curve was repeatable despite variations in the absolute amount of DNA accumulated in different animals

program of the follicular cells. Of course, with the limited resolution available to date we cannot exclude the possibility that the "fine tuning" of the protein synthetic program involves, in addition, some translational controls.

F. Follicular DNA

Quantification of chorion genes in whole body DNA and in the DNA of the follicular cells will be undertaken with the help of cDNA prepared with zone 1 and zone 2 mRNAs, separately, as templates. Even at a coarser level, however, the DNA of the follicular cells is of some interest.

The large follicular cells are highly polyploid; from measurements of DNA content, cell number and genome size, we estimate that in polyphemus the cells attain an average ploidy of 512n to 1024n by the beginning of choriogenesis. In a preliminary report on cecropia, Crippa and Telfer (1971) state that during this polyploidization there is detectable replication of only the unique DNA fraction (C_0t 10^2–10^4). We have found a surprisingly precipitous loss of DNA while choriogenesis (and chorion mRNA synthesis) are still proceeding rapidly. Figure 48 shows typical measurements of epithelial DNA content in developmental series of polyphemus and pernyi follicles.

Follicular cell polyploidization apparently occurs until the end of vitellogenesis. The maximum DNA content is attained at positions 0 to 2, depending on the animal. Soon thereafter DNA begins to disappear from the epithelium, usually decreasing by half in the span of four positions or less. Actively chorionating follicles (positions 7 to 8) have been observed with less than 20% of the maximum DNA level. It is not known whether DNA is lost in parallel from all nuclei, or whether DNA loss is random or selective.

XI. Chorion Mutants in Bombyx mori

A. Gross Description

The history of our understanding of prokaryotes amply demonstrates the importance of genetic analysis in the study of regulatory mechanisms. However, it is also clear that not all genetic studies are equally valuable in this respect. For understanding differentiation at the molecular level, genetics must be brought to bear in a system in which the ultimate gene products are biochemically defined and assayable. Only then can we hope to complement the genetic analysis with serious biochemical studies of gene expression at the protein and nucleic acid levels. Viewing choriogenesis as a eukaryotic analogue of T_4 development, we would like to study a broad collection of temperature-sensitive mutants which affect temporally or quantitatively the program of synthesis and accumulation of specific chorion proteins. We would then hope to gain insight into the regulatory circuits by the use of linkage and complementation analysis, temperature shifts, and biochemical measurements of the rates of synthesis and degradation of specific chorion messages and proteins. Drosophila may be the most appropriate organism for such studies (Sect. XII.), but the commercial silkworm, *Bombyx mori*, will also be of some use.

The study of Bombyx genetics is well developed, and a number of chorion mutants have been mapped to a few of the 28 chromosomes (Tazima, 1964; Chikushi, 1972). Four spontaneous "grey" egg mutants, Gr, Gr^{16}, Gr^{B}, and Gr^{col}, and at least three X-ray-induced mutants, Gr^{L}, Gr^{X1}, and Gr^{X2} produce eggshells of abnormal appearance. By determining recombination frequencies between various Gr mutants and the same outside markers, Takasaki (as quoted in Tazima, 1964) found that all the spontaneous mutations map to locus 6.9 of the second chromosome; the induced mutants are also probably located in the same locus. The genetic analysis is not fine enough to determine whether this locus corresponds to one or more cistrons (region of DNA coding for one polypeptide). Genetic analysis has also revealed an initially confusing but ultimately interesting pattern of gene interactions. These mutants exhibit incomplete dominance and a whole range of phenotypic expressions when crossed among themselves and with other strains. Particularly interesting are the effects of the nearby S allelic group (locus 6.1, chromosome 2) and the p^{S} mutation (locus 0.0, chromosome 2); Takasaki (as quoted in Tazima, 1964) concluded that particular Gr mutants are specifically modified in their expression by particular alleles in nearby loci, and that the effects are dependent on the location of the genes. An additional chorion mutant is White-side egg (Se, chromosome 15, locus 0.0). In this case, eggs laid by a homozygote have opaque eggshells, while those from a heterozygote show a whitish band in the ventral region (Tazima, 1964).

In terms of gross appearance (Takasaki, 1962; Tazima, 1964; Chikushi, 1972) the eggshells of Gr mutants (except Gr^{col}) have as a common denominator a milky white opacity, which gives hibernating eggs a grey hue in combination with the normal dark brown color of the underlying serosa. The opacity is probably due to abnormally numerous air bubbles, characteristically distributed in particular Gr mutants. In Gr^{B} heterozygotes, the opacity is confined primarily to the edge, leaving the central area of the two flat sides transparent (hence the name of the mutation, "Bird's eye"). The homozygotes of Gr^{B}, Gr, and Gr^{col} have eggshells thinner than normal. In the case of Gr^{col}/Gr^{col} the shell is so thin that the eggs dehydrate and collapse soon after oviposition. Finally, several of the mutations cause a macroscopically visible wrinkling of the chorion. None of the mutations has been reported to have effects other than on the chorion. Only the Gr^{col} and Gr^{B} homozygotes show egg lethality—the former presumably because of dehydration and the latter because of the inability of sperm to penetrate (Takasaki, 1962). Scanning electron microscopic studies of the mutant eggshells have been reported (Sakaguchi et al., 1973), but the morphological effects have not been documented by transmission electron microscopy.

B. The Protein Composition of Inbred Strains and Chorion Mutants in Bombyx

SDS electropherograms of chorion proteins from the mutants reveal characteristic protein defects. However, the biochemical effects of the mutations can be analyzed best by the use of isoelectric focusing (IF) slab gels.

Figure 49, left, shows the IF profiles of proteins from mature chorions of Gr strains: homozygous mutants, heterozygotes, and homozygous wild types. The profiles of each mutant must be compared to the respective wild type, since various wild type strains also show significant differences (Fig. 49, right). Although the homozygous wild type for Gr^{col} was not available for this study, Gr^{col}/Gr^{col} may be compared to the phenotypically normal $Gr^{col}/+$. In Figure 49, the 41 consistently recognized chorion proteins are identified by a numerical system.

Comparison of the wild type strains (Fig. 49, right) reveals that a substantial amount of protein polymorphism can be tolerated in the chorion without gross phenotypic effects. This would be expected if the chorion proteins are coded for by multigene families, in which individual members have partially overlapping functions (Hood et al., 1975; see Sect. V. B.). Some of the polymorphism may result from the resolution of additional components in particular strains, or, generally, from differences in protein mobilities (e.g., the component between proteins 7 and 8 in Ascoli, protein 24 in Ascoli, protein 23 in Sunrei, Ascoli and 703). It remains to be established rigorously that mobility differences are due to the evolution of structural genes. An additional striking feature of the IF profiles for wild type

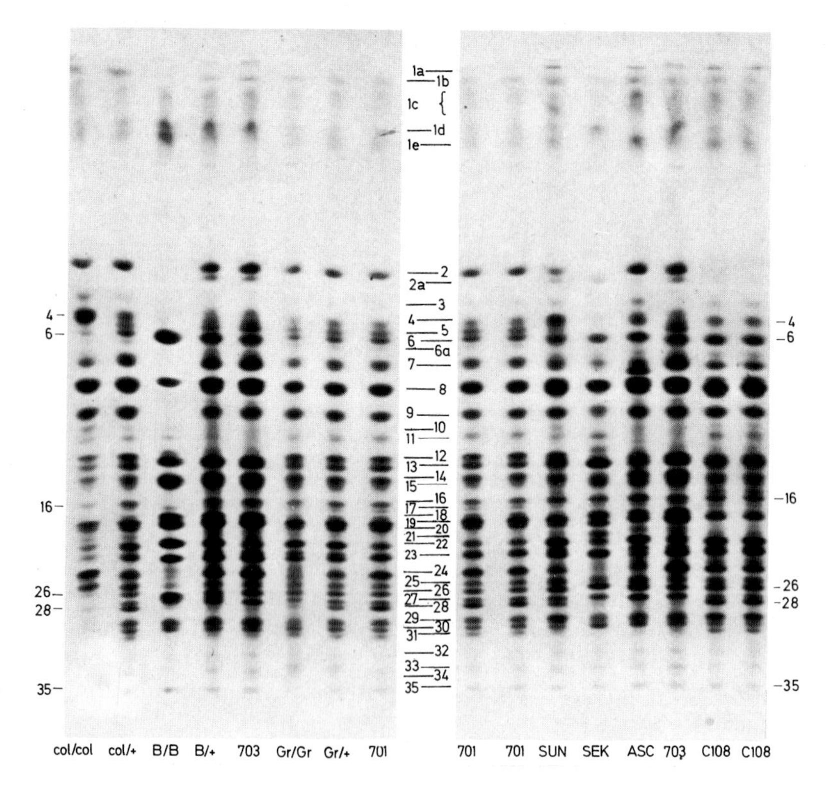

Fig. 49. Isoelectric focusing of Bombyx chorion proteins on a polyacrylamide slab gel. Mature eggshells were solubilized in 8 M urea, 70 mM dithiothreitol, 0.05 M Tris-HCl, pH 9.0, 1.5 mM lysine, 1.5 mM EDTA at room temperature, and carboxamidomethylated in the dark with an excess of iodoacetamide in one-half volume of 1.2 M Tris-HCl, pH 8.4. Reaction was terminated with an excess of β-mercaptoethanol. Fractionation was performed on a water-cooled slab gel containing 5% acrylamide, 0.02% bis, 6 M urea, 2% Ampholine (range pH 4–6; anode at *bottom*). Over a period of 1 h, voltage was gradually increased to a final value of 800 volts, and focusing was continued for 4.5 h. The gel was stained with Coomassie Brilliant Blue. *Right*: comparison of wild type strains. *Left*: comparison of Gr mutants with their heterozygote and homozygote wild types. The strains and their origins were (see Chikushi, 1972)

col/col, col/+: homozygote and heterozygote from strain dO21, Faculty of Agriculture, Kyushu University

B/B, B/+, 703: homozygote, heterozygote and wild-type from strain 703, Sericultural Experiment Station, Tokyo

Gr/Gr, Gr/+, 701: homozygote, heterozygote and wild-type from strain 701, Sericultural Experiment Station, Tokyo

Sun: Sunrei, National Institute of Genetics, Misima

Sek: Sekko, from Central China; National Institute of Genetics, Misima

Asc: Ascoli, from Italy; National Institute of Genetics, Misima

C108: National Institute of Genetics, Misima.

Numbers (1–35) are an arbitrary identification system for proteins separated under these conditions. Each track received proteins from 1.33 chorion, except for *col/col* which received proteins from 2.67 chorions

strains are marked quantitative changes in the abundance of particular bands (e.g., proteins 12 and 13 in 701 vs. Sekko, proteins 2 and 2a in all six strains examined). These differences are completely reproducible, and may suggest relatively rapid evolution of the regulatory circuits controlling production of individual chorion proteins.

Examination of the mutant IF profiles reveals that the mutations are pleiotropic with respect to the chorion (Fig. 49, left). A large number of chorion proteins are affected, almost exclusively quantitatively in terms of the amount accumulated by the end of choriogenesis. For example, in Gr^B/Gr^B several normally major proteins (2, 2a, 4, 7, 9, 16, 24) and a number of minor proteins are either present in very small amounts or completely absent; two proteins appear to be overrepresented (6, 22), and only one may show altered mobility (protein between 26 and 27; also compare the heterozygote and the homozygous wild type). Similarly, a large number of proteins are quantitavily affected in Gr^{col}/Gr^{col}; some are indicated in Figure 49 (4, 6, 16 and the entire complex of very acidic proteins, 26 to 35). In this case, no protein shows altered mobility. In Gr/Gr, the quantitative effects appear to be rather subtle, and three proteins show mobility differences (23 and 24, which split into doublets, and probably 27, which is slightly displaced toward the cathode; the difference in 23 is also expressed in the heterozygote).

C. The Phenotype of Gr^{col}/Gr^{col} and Its Development

The phenotype of Gr^{col}/Gr^{col} is particularly interesting. With the possible exception of 3, 4, and 24, every protein accumulates to subnormal amounts (the profile shown in Fig. 49 comes from twice the number of eggshells used for the adjacent $Gr^{col}/+$ profile). However, the degree of underaccumulation varies for different proteins, so that the protein profile shows a wide spectrum of quantitative differences when compared with the heterozygote.

The phenotype of Gr^{col}/Gr^{col} might suggest that the program of synthesis is altered in this strain. However, careful comparison of synthetic and accumulation profiles reveals instead that the post-translational fate of the proteins is involved. This is documented in Figure 50, which shows the IF profiles of accumulated proteins (stained preparation, upper part) and of newly synthesized proteins (autoradiogram, lower part) for a developmental series of follicles (staged by fractional position). The total proteins (epithelium plus chorion) are displayed from both a heterozygote and a homozygous mutant. In the heterozygote (Fig. 50, upper right) the chorion proteins accumulate continuously; recognizable chorion bands do not decrease in intensity over the course of choriogenesis. By contrast, in the homozygote (Fig. 50, upper left) the proteins which are severely underrepresented in the mature shell appear to accumulate in the early part of choriogenesis, and to be lost secondarily (e.g., proteins 6, 16, and 26 to 35). The progressive decline of these bands is not accompanied by an abnormal intensification of other components, suggesting that we are in fact dealing with protein disappearance, rather than modification. Comparison of the synthetic, i.e., labeling, profiles (Fig. 50, lower) reinforces the impression that the mutation acts posttranslationally: even the severely underrepresented proteins appear to be synthesized during the correct

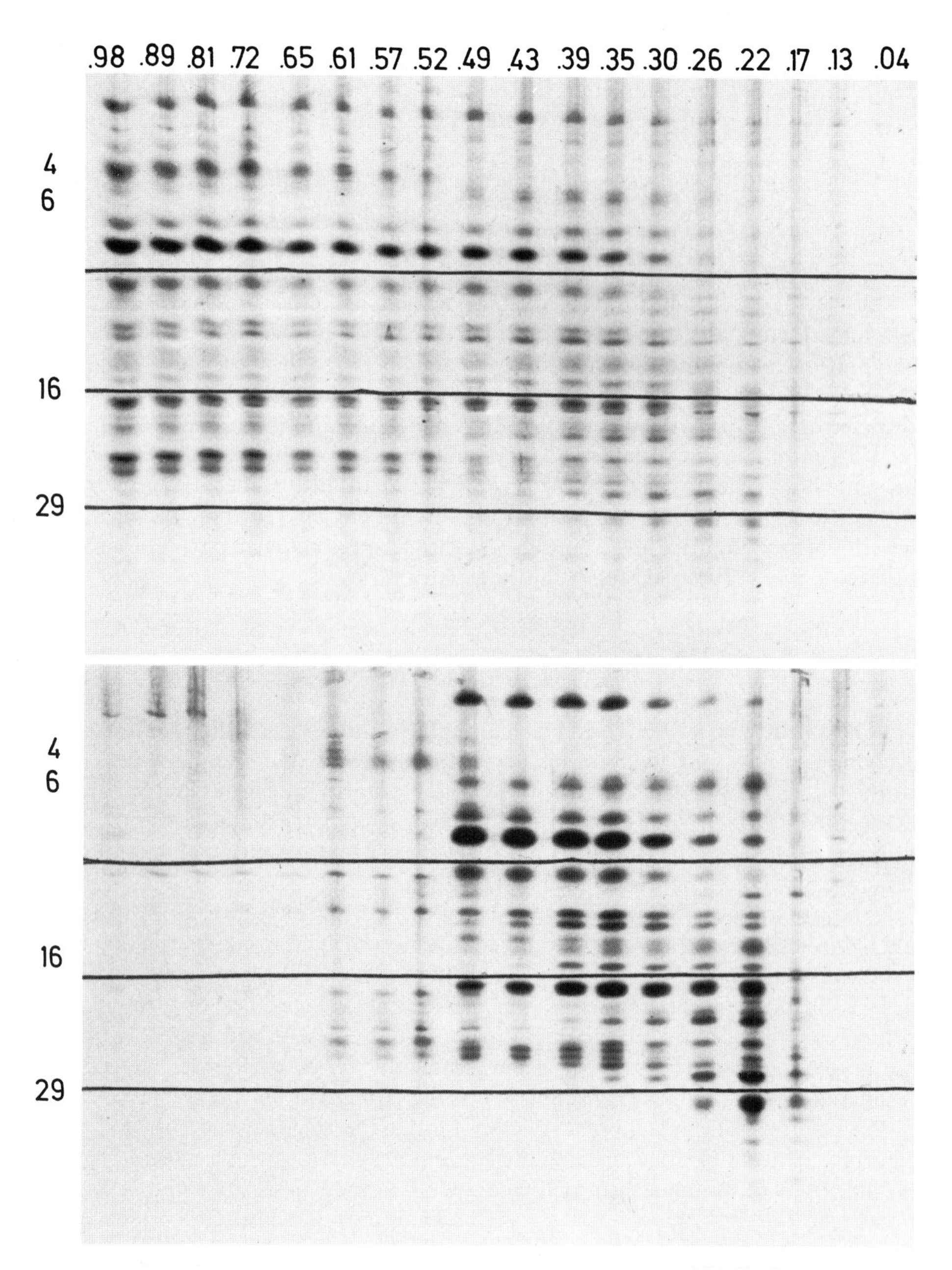

Fig. 50. Isoelectric focusing of developmental series of follicles from Gr^{col}/Gr^{col} *(left)* and $Gr^{col}/+$ *(right)*. Ovarioles were labeled for 1 h with ^{14}C-leucine (100 μCi/ml; 280 mCi/mmol) following a 10-min incubation in Grace's medium minus leucine. Follicles were then deyolked and the epithelium plus chorion solubilized and analyzed as in Figure 49. The numbers at the top represent the fractional position of the follicle between the end of vitellogenesis and the time of ovulation. For Gr^{col}/Gr^{col}, the equivalent of 2/3 of a single follicle was used in each track, except for positions 0.72 to 0.98 (2/3 of each of two adjacent

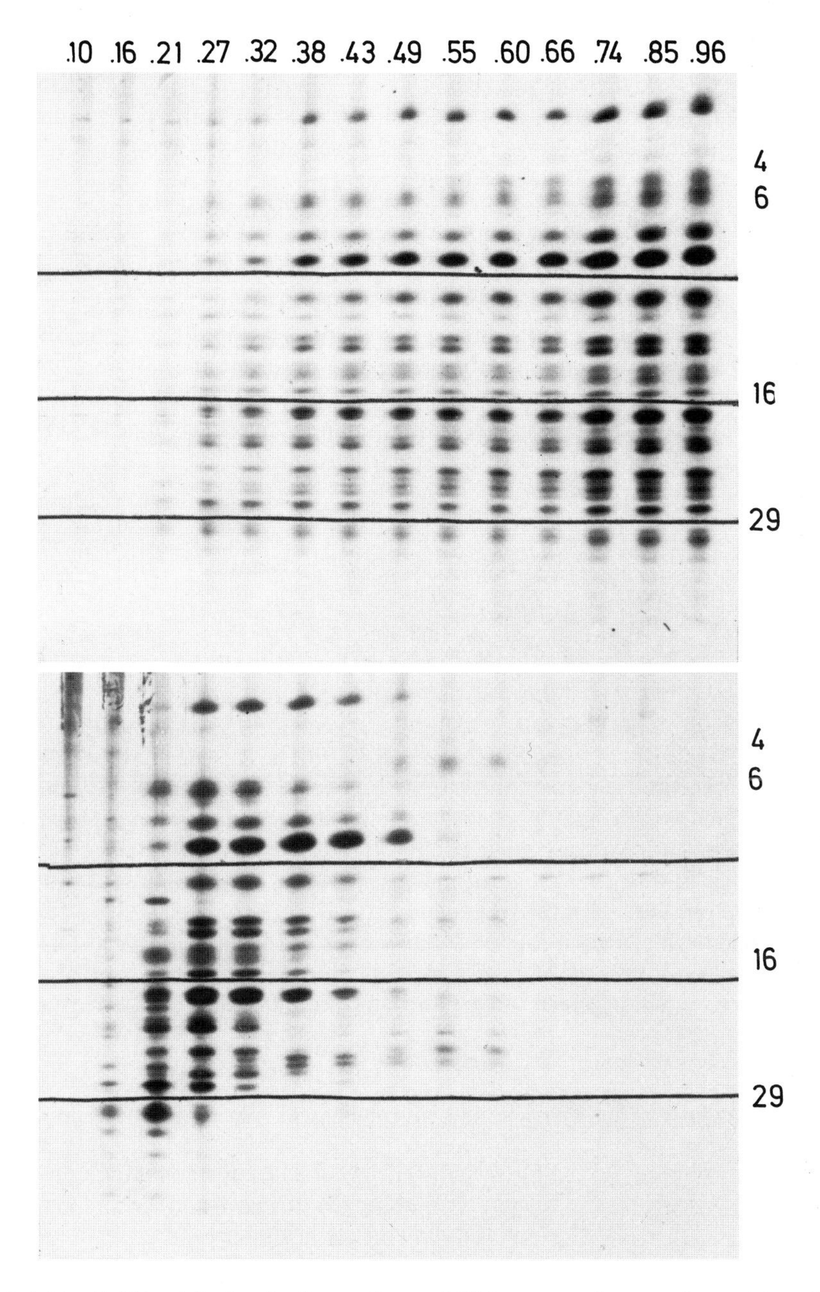

follicles). For $Gr^{col}/+$, 1/4 of each of two adjacent follicles was used, except for positions 0.74 to 0.96 (1/4 of each of four adjacent follicles). Numbers on *side* and *horizontal black lines* facilitate identification of proteins (see Fig. 49). The part of the gel above band 2 is not shown. *Top:* Protein accumulation profiles, obtained by staining the gel with Coomassie Brilliant Blue. *Bottom:* Protein synthetic profiles, obtained by autoradiography. The stained gel was dried and exposed for 4.5 days (Kodak RP Royal X-Omat)

interval and at normal rates relative to other proteins. Minor differences in the autoradiograms of the two genotypes can mostly be explained by imperfect matching of the developmental stages.

This unexpected result poses interesting—and answerable—questions. Is intracellular protein degradation occurring? Is secretion of some proteins perhaps abnormal, resulting secondarily in protein breakdown? Are the proteins eliminated after secretion? What is the role, if any, of turnover in normal choriogenesis? How is degradation directed to specific proteins? Is some turnover responsible for the generally depressed level of nearly all chorion proteins in Gr^{col}/Gr^{col}? Are other Gr mutants also acting posttranslationally? These questions can be answered by relatively simple experiments, involving quantification of accumulation and synthetic profiles, pulse-chase incubations, separation of chorion and cytoplasm, etc. Whether the mutation proves to affect primarily a step in secretion or a step in extracellular morphogenesis, it is a strong reminder that in eukaryotes differential gene expression does not end with translation.

XII. The Drosophila Chorion

A. Introduction

The silkworm chorion has proven to be a very fruitful system for biochemical studies; the Bombyx mutants are adding a genetic dimension to these studies. However, it is clear that fine genetic analysis will not be feasible in Bombyx for some time to come. Rather, we can hope to identify some regulatory or morphogenetic mutations, map them, and uncover their molecular mode of action, along the lines of recent work on human β-thalassaemia (Housman et al., 1973).

By contrast, with *Drosophila melanogaster* we can realistically set a much more ambitious, long-term goal: total genetic dissection of the mechanisms regulating programmed cell-specific protein synthesis over time. The richness of Drosophila genetics will undoubtedly offer new insights into developmental regulation, once applied to a system in which the products of differentiation can be biochemically defined.

B. Structure of the Chorion, and Morphological Aspects of Its Deposition

The morphology of the Drosophila chorion has been studied by King and his collaborators (reviewed in King, 1970) and by Quattropani and Anderson (1969). It is now agreed that the chorion is entirely an extracellular product of the follicular epithelium. The eggshell consists of three distinct layers (Fig. 51)—in Quattropani and Anderson's terminology Zones I, II, and III (counting from the oldest layer, next to the oocyte), or in the more traditional terminology vitelline membrane, endochorion, and exochorion. These are produced in distinct stages of oogenesis, which are defined by morphological criteria (King, 1970).

Deposition of Zone I between the follicle cells and the oocyte begins at stage 9, and is completed by the end of stage 10. Thus, yolk proteins from the blood (Gelti-

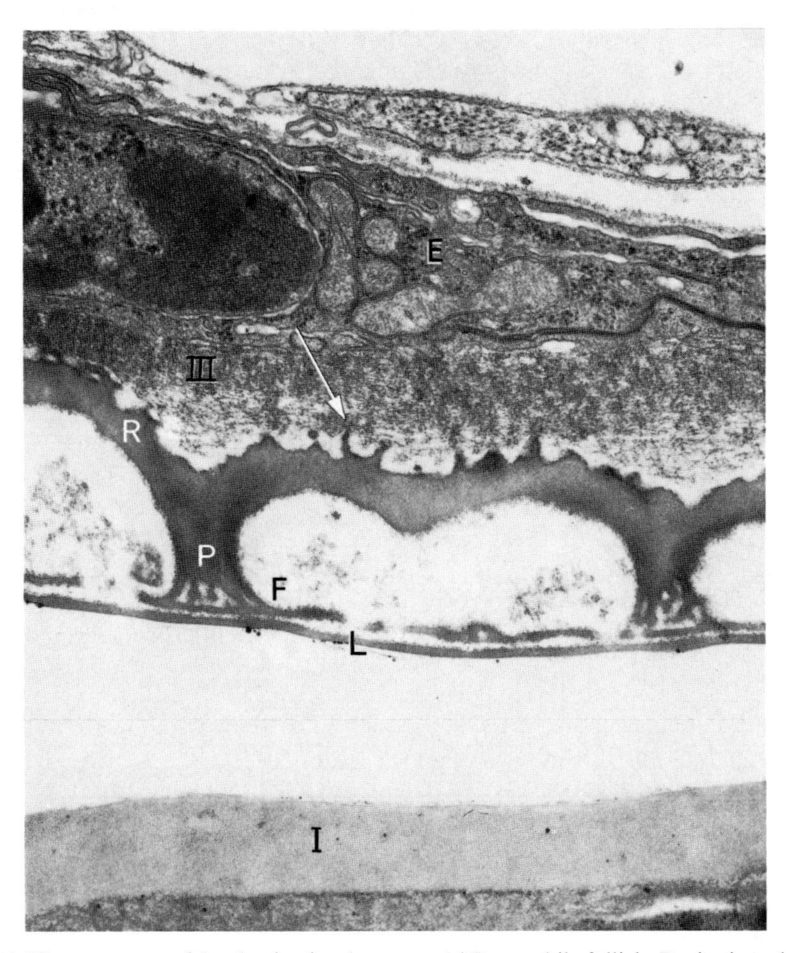

Fig. 51. Ultrastructure of the chorion in a late stage 14 Drosophila follicle. Beginning with the oocyte *(bottom)*, the layers encountered are Zone I (the vitelline membrane, *I*); the "innermost lamina" (Quattropani and Anderson, 1969, *L*); Zone II with its discontinuous floor (*F*), upright pillars (*P*), arched roof (*R*) and protruding honeycomb layer *(arrow)*; Zone III (*III*); and the attentuated follicular epithelial cell (*E*). Note the microvilli of the secretory surface. (× 32000). (Courtesy L. Margaritis)

Douka et al., 1974) are still being taken up by the oocyte while the vitelline membrane is forming (Mahowald, 1972). This apparent paradox is explained by the observation that Zone I is deposited as a discontinuous layer and is traversed by bundles of microvillar projections from both follicular cells and oocytes. At the end of stage 10 the microvilli are withdrawn, Zone I fuses into a continuous layer and septate desmosomes "zip together" adjacent follicular cells. The oocyte is thus sealed off from the blood (Mahowald, 1972) and becomes dependent on the nurse cells for its growth. The final thickness of Zone I is approximately 0.3 μm. Vitelline membrane isolated from mature eggshells is tough, transparent, and remarkably hydrophobic.

Zone II is deposited from stage 11 to the middle of stage 13 (King, 1970). During this period the oocyte attains its maximum size, at the expense of the nurse cells. Zone II, which is analogous to the trabecular layer occurring in Lepidoptera and other orders, has an elaborate architecture. It consists of a thin "floor" apposed to Zone I, an intermediate air layer traversed by upright pillars (perpendicular to the egg surface), and a thick "roof" with arches connecting the pillars (Fig. 51). The morphogenesis and structure of Zone II are intriguing (Quattropani and Anderson, 1969; Margaritis, 1974; Margaritis et al., 1976). The floor is laid down as a continuous thin layer by secretion of Golgi-derived electron dense vesicles across the entire microvillus-studded surface of the follicular cell. Thereafter, the microvilli retract except at regular intervals (of approximately 1 μ), where the microvilli remain attached while the pillars are built up under them by localized exocytosis of the dense vesicles. In between the pillars, a sparse, coarsely flocculent material is deposited. Finally, generalized exocytosis of dense vesicles resumes and produces the roof, initially in small patches which later become confluent. The last part of Zone II to be laid down is a honeycomb layer of hexagonal projections, which will bind Zones II and III together. Toward the end of Zone II formation, holes become visible in the floor and the roof appears to thin out (Quattropani and Anderson, 1969), probably as a result of stretching brought about by a final oocyte expansion. The final thickness of Zone II (including the pillars) is approximately 1 μm. The flocculent material in the space between the pillars eventually is replaced by air.

Zone III is deposited in stages 13 and 14 (King, 1970). It is a 0.5 μm layer containing oriented loose fibers, and is deposited both by Golgi-derived dense vesicles and via the tips of microvilli (Margaritis, 1974).

The chorion surface consists of polygons, each corresponding to a single follicle cell. The raised border of the polygon coincides with a local elaboration of the honeycomb outer layer of Zone II, at the junction between two cells. Under the scanning electron microscope, the chorion surface appears pitted. The surface details vary slightly among different Drosophila species, corresponding to the ecological niche of the egg (Kambysellis, 1973).

C. Composition of the Chorion

Until recently, the composition of the chorion in Drosophila had been studied primarily by histochemical techniques. As a result, no conclusions were possible about the quantitative composition and the number of macromolecular species in the chorion. The histochemical tests show that, in addition to protein, the chorion contains some non-proteinaceous material, including acid mucopolysaccharides or mucoproteins (assayed by Alcian Blue), carbohydrates (neutral polysaccharides and/or glycoproteins, assayed by Periodic Acid-Schiff and thiocarbohydrazide), and lipids (King, 1960a, 1970, King and Koch, 1963; Quattropani and Anderson, 1969). Lipids are reported to be present throughout the shell. It appears that Zone III is particularly rich in Alcian-Blue positive material. The flocculent material in the cavities of Zone II is Alcian-Blue negative but PAS and thiocarbohydrazide positive. A thin lamina between Zones I and II may be waxy (King, 1970), but also contains thiocarbohydrazide-positive material (Quattropani and

Table 10. Amino acid composition of the Drosophila eggshell (residues/100 residues)

	Zone I	Zones II and III
glu NH_2	N	0.6
gal NH_2	N	0.4
trp	N	1.0
lys	3.0	5.0
his	0.4	2.2
arg	0.0	3.8
cys	1.0	0.6
asx	4.2	5.9
thr	0.9	2.4
ser	16.9	8.4
glx	4.4	10.1
pro	18.3	10.9
gly	10.4	15.7
ala	28.6	14.7
val	3.0	6.1
met	0.0	0.5
ile	1.3	3.4
leu	3.4	4.9
tyr	3.5	3.8
phe	1.8	1.2

N: Not determined.

Anderson, 1969). Zone I includes both Alcian-Blue-positive and Alcian-Blue-negative carbohydrates.

The histochemical information might create the impression that a substantial proportion of the chorion is carbohydrate. Since we know from biochemical analysis that this is not the case in the silkmoth chorion (Sect. V.A.), we estimated the carbohydrate content of the *Drosophila* chorion directly by biochemical procedures. We found that only 5% of the dry weight of Zones II and III is neutral sugars, by the method of Dubois et al. (1956). Wilson (1960b) reported 0.6% glucosamine in whole eggshell, and failed to detect galactosamine. Our measurements of amino sugars in Zones II and III (using a Beckman amino acid analyser) showed an unmistakable amount of galactosamine (0.4%), as well as glucosamine (0.6%).

Protein is a major component of all three zones. Wilson (1960a) reported that protein constitutes more than 80% of the whole shell. We find that 73% of Zone I and 94% of Zones II and III is protein. Wilson (1960a) estimated the amino acid composition of the eggshell by the use of paper chromatography. We have determined the amino acid composition of Zone I, and of Zones II and III combined, using an amino acid analyser (Table 10). Vitelline membrane has an unusual composition: proline is 18 molar %, alanine 29% and glycine 10%. Cysteine amounts to only 1% of the residues, and arginine and methionine may be entirely absent. The amino acid composition of the vitelline membrane may account in part for its hydrophobic nature and for its physical strength, both of which are important in shielding the embryo from the external environment.

The amino acid composition of Zones II and III (Table 10) is quite different, but likewise remarkable for the preponderance of alanine (15%), glycine (16%) and proline (11%), and for the paucity of cysteine (less than 1%). Wilson's (1960a) observation that a large amount of tyrosine is found in the eggshells of Drosophila is not supported by our data. However, we have not eliminated the possibility that polyphenols and quinones may play a role in the hardening of the shell.

D. The Proteins of the Chorion, and Their Synthesis

The paucity of biochemical information on the structure and development of the Drosophila chorion may be attributed to two factors. The first is size, which has hindered most biochemical studies on Drosophila until recently; the other is the marked insolubility of the mature eggshell.

The Drosophila follicle is minuscule, having approximately one thousandth the volume of a polyphemus follicle during choriogenesis. To circumvent tedious and time consuming hand-dissection of large numbers of female flies, a mass isolation technique was developed, which in less than an hour will produce thousands of purified, individual follicles in stages 3 through 14 (Petri et al., 1976). The technique involves three basic steps. (1) To maximize the number of chorionating follicles per fly, adults are first conditioned for three days in a large population cage provided with fresh yeast suspension daily. (2) Conditioned flies are mass dissected by passage through a grinding mill in a Drosophila Ringer's solution, and the resulting brei is rapidly stirred to free follicles from the ovarioles. (3) Follicles are then purified by a combination of differential sieving and differential sedimentation. After the last step a good preparation will yield approximately 100000 mixed-stage follicles from 10000 females. Follicles of particular stages can then be easily picked for developmental studies. If stage 14 follicles are desired (for protein isolation), they can be easily purified by utilizing the fact that they stick to a clean glass surface more tenaciously than the younger stages. A good yield in this case would be about 30000–60000 stage 14 follicles from 10000 females.

In order to obtain chorion proteins free of epithelial cell and oocyte contamination, a procedure for preparing empty stage 14 shells was developed. When purified stage 14 follicles are shaken vigorously in distilled water, approximately 20% of the shells open, releasing the usually intact oocyte surrounded by the vitelline membrane. The empty stage 14 shells (Zones II and III) are purified en masse by differential sedimentation. Microscopic examination shows no evidence of cellular or yolk contamination.

The second problem of working with Drosophila chorion was to obtain in solution the individual protein components of the chorion. The mature, laid eggshell was found to be completely resistant to all treatments which normally disrupt protein structure without cleaving peptide bonds. Fortunately, eggshells from stage 14 follicles can be dissolved by incubation in strong denaturing reagents, such as 7 M guanidine hydrochloride, 8 M urea, or 1% SDS, buffered to pH 9.0. Reduction of disulfide bonds does not appear to promote solubility. The reactions which render the eggshell insoluble apparently occur prior to passage into the oviduct. When hydrolysates of the mature laid eggshell are subjected to thin-layer

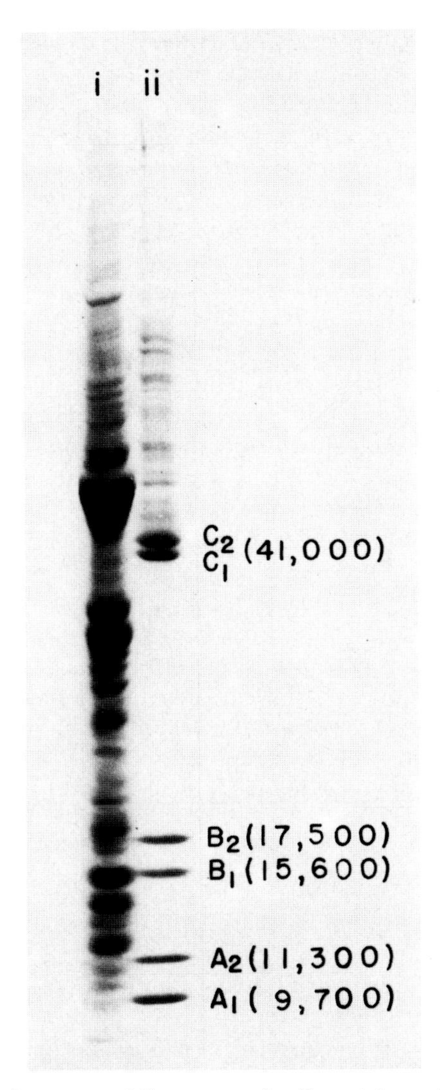

Fig. 52. Stained electropherogram of (*i*) oocyte and yolk proteins and (*ii*) proteins of purified eggshell from stage 14 follicles. The six major chorion proteins are identified. (Modified from Petri et al., 1976)

chromatography, two spots are observed with the same R_f values and blue fluorescence as found with dityrosine and trityrosine standards. In contrast, these spots are missing from hydrolysates of soluble stage 14 chorions. These results suggest that the insolubility of the eggshell is based, at least in part, on the formation of di- and trityrosine crosslinks.

These techniques for recovering soluble protein from stage 14 chorion produce sufficient amounts of material for biochemical studies. The proteins of chorion (Zones II and III) have been analyzed on polyacrylamide gels containing SDS and urea (Petri, 1976). The components vary widely in molecular weight; to display them with adequate resolution on a single gel we use a gel system with a gradient of acrylamide concentration. Figure 52 shows soluble proteins from approximately 200 shells. Six major components are highly enriched in the chorion preparations, and these have been assigned to three size classes A, B, and C. In addition, several minor bands are observed. The criteria for judging whether or not a particular component is a bona fide eggshell protein are discussed in detail by Petri et al. (1976).

Having defined the proteins in this gel system, we inquired into the timetable for their synthesis during choriogenesis. The morphological evidence for strictly sequential synthesis of the three zones and their sublayers encouraged us to expect sequential protein synthetic phases, as in the case of the silkmoth. This expectation was confirmed. About 25 mass-dissected follicles of each stage were labeled for 1 h in a defined tissue culture medium modified from those of Grace (1962) and Schneider (1964). ^{3}H-proline was used, because of the abundance of proline in the chorion. The follicles were then dissolved in toto and the proteins analyzed on SDS slab gels in the presence of carrier chorion protein. The incorporation profiles (Fig. 53) showed dramatic changes in the types of proteins synthesized as development proceeds. During stages 8 through 10, when the vitelline membrane is known to be deposited, four components, V_1 through V_4, are labeled which do not correspond to any of the stained bands seen in Figure 52. We believe these labeled bands to represent the proteins of Zone I. Later, during stages 11 through 14, when the endo- and exochorion are known to be deposited, an overlapping sequence of synthesis of the six major chorion components is observed. An analysis of many such experiments indicates the following general synthetic progression during eggshell development:

$$V_2 < V_1, V_3, V_4 < C_1, C_2 < B_2, A_2 < B_1 < A_1 .$$

E. Duration of Stages of Chorion Deposition

For further work on the biochemistry of chorion formation, it will be necessary to establish a timeable of synthetic stages. In this respect, the available detailed descriptions of the stages of oogenesis will be invaluable (reviewed in King, 1970). The duration of these stages has been estimated by the use of flies in steady state with respect to egg production (King, 1970). Briefly, if a "steady state" fly lays N eggs per day and contains n fully active ovarioles, each with v follicles at various stages of oogenesis, from X to Y, the total duration, T (in h), of stages from X through Y can be estimated as

$$T = \frac{24 \cdot n \cdot v}{N} \tag{3}$$

In flies with high fecundity, T has been estimated as 72–91 h for stages 2 through 14. If all ovarioles are not equally active, the product $n \cdot v$ in Equation (3) should best be replaced by N′, the total number of ovarian follicles (in stages X through Y).

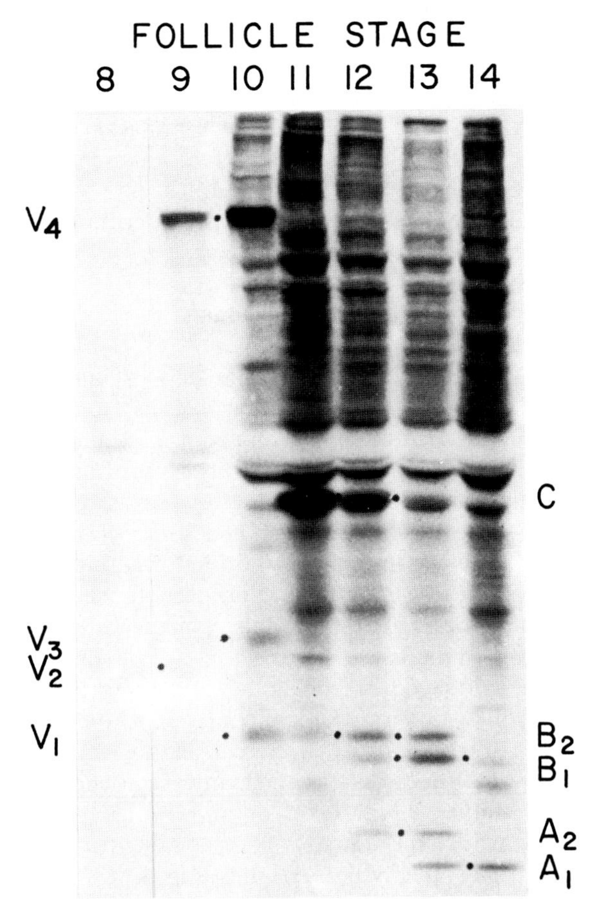

Fig. 53. Autofluorogram of ^{3}H-proline-labeled Drosophila follicles. Twenty follicles of each of the stages indicated above the gel tracks were incubated in tissue culture medium and in the presence of ^{3}H-proline for 1 h and the total labeled proteins were analyzed by electrophoresis and autofluorography. Purified unlabeled chorion was used as marker to locate precisely the major eggshell components. *Dots* placed between gel tracks indicate the location and major stages of synthesis of each component as evaluated in the original autofluorogram. (From Petri et al., 1976)

Now, if p is the proportion of follicles of a particular stage, within a steady state fly, the absolute duration of that stage, X, (in hours) will be

$$X = pT. \tag{4}$$

By this procedure, the duration of deposition has been estimated as 11 h for Zone I (stages 9 and 10), 2.7 h for Zone II (stages 11, 12 and half of stage 13) and 2.4 h for Zone III (remainder of stage 13 and stage 14; calculated from the data of David and Merle, 1968). Thus, chorion formation is quite rapid in Drosophila.

This method will give dependable estimates of the total duration of a stage, so long as the flies are truly in steady state and no resorption of follicles occurs. In some situations, retention of mature oocytes within the ovary may occur and block the maturation of follicles at certain stages (David and Merle, 1968). Any synthetic pause will be scored as increased duration of the stages involved.

Stage durations will be different under suboptimal physiological conditions and with different genotypes. Accurate comparative information on stage duration will be required for the study of regulatory mutants that may affect the timing mechanisms. Since it may be difficult to define equivalent physiological conditions for mutant and normal flies, the use of comparative stage duration studies in organ cultures may be necessary.

F. Chorion Mutants

Essential for a complete understanding of the steps in chorion formation will be the use of mutants. Obviously there are many aspects of chorion synthesis and deposition which, if altered, could lead to an abnormal chorion. For example, the chorion of the mutant *tiny* develops abnormally because of a precocious shift in the anterior follicular epithelial cells from a migratory to a secretory phase (Falk and King, 1964). The *ocelliless* mutant (Johnson and King, 1974) also shows eggshell defects. Particularly useful for analysis of chorion formation will be mutations in genes acting during or after the initiation of choriogenesis. Such genes should prove to be either structural or regulatory. While it may not be possible to select specifically for chorion mutants, we do expect many mutants having functionally altered eggshells to fall into the larger class of female sterile mutants. Since some regulatory mutants are expected to be dominant, temperature sensitive dominant female steriles will be of special interest. In recent years, a large number of recessive female sterile mutants have accumulated and many of these proved to have eggshell defects (King and Mohler, 1975; Gans et al., 1976; Mohler, personal communication). Thus, female sterile mutants will be a rich source of chorion mutations.

One such mutant has already been examined in detail. The mutation, which produces abnormally fragile chorion, exists in strain 8–854 which was isolated by Bakken (1973). In collaboration with Mohler, Margaritis (1974) showed that the chorion fragility and the ultrastructural defects causing it map in the same region of the second chromosome as the female sterility mutation, *fs(2)A9* (i.e., between *dp* and *b*). An ultrastructural analysis of choriogenesis in 8–854 (Margaritis, 1974) revealed that while Zone III is formed normally, Zone I is abnormally thin and Zone II is very incomplete, acquiring patches of material between the pillars, but never a complete roof. Surprisingly, the incomplete formation of these zones in mutant flies takes a disproportionately long period of time. It was determined by the method outlined in the preceding section that in homozygous mutant females oogenesis up to the beginning of vitelline membrane secretion requires 1.6 times as long and formation of Zone III 1.7 times as long as in heterozygous control females. In contrast, the time needed for formation of Zone I by mutant flies is 3.3 times normal, and for formation of Zone II, 9.4 times normal. It seems that in this mutant the program of choriogenesis may be incompletely implemented and temporally deranged, especially in the portion dealing with zone II formation. Electrophoretic comparison of the mutant and wildtype proteins has been hampered by the insolubility of mutant stage 14 eggshells. Precocious insolubility may be another aspect of temporal derangement.

The work on silkmoths established that the follicular epithelium of insects is favorable for studies on developmental regulation. This epithelium consists of highly differentiated, short time-constant cells, programmed to synthesize in sequence and autonomously several biochemically defined products. Among the insects, Drosophila is unique in facilitating detailed genetic investigations. It now appears that choriogenesis in Drosophila is also amenable to biochemical analysis. It is reasonable to expect that the biochemical basis of some temporal genetic lesions can be elucidated using existing methodologies. For example, identification of the chorion mRNAs will be facilitated by the small size of some of the polypeptides, by the developmental shifts among components varying in molecular weight, and by the G + C content of the mRNAs (which should be high, according to the amino acid composition). Once the mRNAs are identified, it should be possible to ascertain whether temporal mutations interfere with their production, accumulation or utilization.

The availability of structural variants of the polypeptides, and of purified mRNA, should permit mapping of the structural genes by genetic and in situ hybridization procedures. In turn, this should permit an exhaustive search for regulatory mutations possibly linked to the structural genes. A combined ultrastructural, biochemical, and genetic study of short time-constant cell differentiation in Drosophila now appears feasible and worthwhile.

XIII. Epilogue

The main part of this review was completed in October, 1974. To update the manuscript, minor changes were made in December, 1976. The purpose of this Epilogue is to identify relevant publications of the last two years, and to indicate some of the major directions of the current research on chorion.

Brief reviews on the insect chorion have appeared (Kafatos, 1975; Furneaux, 1976; Kafatos et al., 1977) dealing, respectively, with the chorion as a developmental system, with chorion biochemistry and with chorion as a model system for studying the structure and expression of eukaryotic genes. A summary of methods used in the study of the chorion system has also been published (Efstratiadis and Kafatos, 1976).

A major product of the follicular epithelium during vitellogenesis has been identified (Bast and Telfer, 1976). This product is unrelated to chorion, and appears to be involved in yolk formation.

Eleven A proteins from polyphemus and 1 A protein from pernyi have been partially sequenced from the N-terminus; in addition, 6 polyphemus B proteins, and 1 pernyi B protein have been partially sequenced following their single methionine residue (Kafatos et al., 1977; Regier, Rodakis and Moschonas, unpublished observations). All As show extensive homologies with the component described in Table 4, but also are distinguished by amino acid substitutions, eliminating the possibility that they are derived from each other by post-translational modifications. Similarly, the B proteins are both homologous and distinct from each other. No homologies have been detected thus far between A and B proteins. The results are in agreement with the hypothesis that the genes for A and B chorion proteins form two informational multigene families, in the sense of Hood et al., 1975. Further support comes from the results of genetic experiments, involving crosses between Bombyx strains which differ in the chorion protein patterns (see Fig. 49); these results suggest that the structural genes for chorion proteins are linked, and reside in chromosome 2, as do the Grey mutations (Goldsmith, unpublished observations).

Further work with Gr^{col} indicates that the lesion corresponds to an inability of the cells to secrete normally the proteins synthesized during specific developmental periods; the non-secreted proteins appear to degenerate intracellularly (Nadel and Goplerud, unpublished observations). The biochemistry of a second Grey mutation, Gr^{B}, has also been investigated (Nadel, unpublished observations). Quite in contrast to the situation in Gr^{col}, the proteins affected by Gr^{B} are always either undetectable or synthesized in only very minor amounts. Cell-free translation and electrophoretic analysis of chorion mRNA from Gr^{B}/Gr^{B}, $Gr^{B}/+$ and $+/+$ animals suggest that the mutation acts pre-translationally, by affecting production of specific chorion mRNAs (Nadel and Efstratiadis, unpublished observations). It is intriguing that a post-translational and a pre-translational mutation are so closely linked, and that they are apparently found in the same chromosome as the chorion structural genes.

Two types of post-translational modifications of chorion proteins have been revealed. One involves a shift of at least the major B proteins towards a slightly lower pI within the first half hour after synthesis; the A proteins are not similarly affected. The shift presumably corresponds to a modification of the N-terminal amino acid, since B but not A proteins have a blocked N-terminus, and since no significant change in molecular weight occurs in parallel with the shift in pI (Regier, unpublished observations). The second type of modification is suggested by the consistently higher MW distribution of chorion proteins synthesized in the wheat germ system, as compared with in vivo synthesized chorion proteins (Nadel, Thireos and Efstratiadis, unpublished observations). These data probably indicate that during translation an initiation peptide is synthesized as part of the chorion proteins, but is almost immediately cleaved by limited proteolysis in vivo, so that it is not detected in experiments such as those shown in Figure 21. Such initiation peptides appear to be characteristic of secretory proteins (e.g. Blobel and Dobberstein, 1975).

The silkmoth genome has been investigated by Cot analysis, and a pattern of extensive interspersion of unique and repetitive DNA has been detected (Efstratiadis et al., 1976a), comparable to that shown by most other metazoa, but unlike Drosophila (Davidson et al., 1975).

A major recent effort is the study of chorion genes at the DNA level (for a review, see Kafatos et al., 1977). The resolution of chorion mRNAs was substantially increased by deadenylation (Vournakis et al., 1975)—but it still remained far below the level necessary for isolation of individual mRNA species. Therefore, using rabbit globin mRNA, we developed methods for synthesizing full-length double-stranded DNA with mRNA as the template (Efstratiadis et al., 1975, 1976b). By molecular cloning of this synthetic DNA, it was possible to obtain hybrid plasmid clones, each of which originated from a single molecule of DNA, and hence corresponded to a single mRNA species; the mRNA sequence was represented with complete faithfulness in the cloned DNA (Maniatis et al., 1976). These procedures have now been applied to chorion mRNA, and a rich collection of individual pure chorion DNA sequences have been obtained; in parallel, chromosomal DNA has been cloned and clones containing chorion structural genes plus flanking sequences have been selected (Sim Gek Kee, Villa-Komaroff, Efstratiadis, Maniatis, and Kafatos, unpublished observations). Synthetic and chromosomal chorion DNA clones are being used to solve important questions about the structure and regulation of chorion genes: multiplicity, linkage relationships, similarities in primary structure, patterns of transcription, and the nature of the DNA flanking the chorion genes in chromosomal DNA. With the recent methodologies, the study of chorion as a model developmental system clearly enters a new exciting period.

Acknowledgements. The work presented reflects a major part of the team effort of our laboratory over the last five years. We want to acknowledge the participation of other members of the group, thanking L. Lawton and M. J. Randell for secretarial work, J. A. Jordan, B. Klumpar, V. Raidl and J. Goplerud for technical assistance, M. Tullis for help in the preparation of the figures, and Dr. L. Margaritis and N. Rosenthal for Figures 51 and 42a, respectively. The financial support for the work included research grants from the NIH (5-R01-HD04701), NSF (GB-35608X), and the Rockefeller Foundation (RF 73019), as well as student and facility support from the NIH (5-T01-HD00415, 2-T01-GM0036 and 5-R01-GM06637), individual fellowships from NIH, NSF and the Jane Coffin Childs Fund and Harvard University. We thank Drs. Y. Tazima, T. Ito, T. Ohtaki, A. Murakami, H. Doira and B. Sakaguchi for their kindness and help in obtaining the Bombyx mutants, Drs. R. Palmiter and L. Cherbas for helpful discussions and Dr. D. Mohler for collaboration in the study of Drosophila mutants. We thank Drs. L. Margaritis and D. Mohler for permitting us to discuss their unpublished results, and Drs. D. Smith, W. Telfer and A. Nevillle for permitting us to use Figure 12 from their publication. Part of the writing was done at the Marine Biological Laboratory, Woods Hole, while F.C.K. was an Instructor in the Embryology Course.

References

Abraham, G. N., Scaletta, C., Vaughan, J. H.: Modified diphenylamine reaction for increased sensitivity. Anal. Bioch. **49**, 547–549 (1972)

Anderson, L. M., Telfer, W. H.: A follicle cell contribution to the yolk spheres of moth oocytes. Tissue Cell **1** (4), 633–644 (1969)

Ashburner, M., Chihara, C., Meltzer, P., Richards, G.: Temporal control of puffing activity in polytene chromosomes. Cold Spring Harbor Symp. Quant. Biol. **39**, 655–662 (1974)

Bakken, A.: A cytological and genetic study of oogenesis in *Drosophila melanogaster*. Develop. Biol. **33**, 100–122 (1973)

Bard, E., Efrin, D., Marcus, A., Perry, R. P.: Translation capacity of deadenylated messenger RNA. Cell **1**, 101–106 (1974)

Bast, R. E., Telfer, W. H.: Follicle cell protein synthesis and its contribution to the yolk of the Cecropia moth oocyte. Develop. Biol. **52**, 83–97 (1976)

Beament, J. W. L.: The waterproofing process in eggs of *Rhodnius prolixus* Stahl. Proc. Roy. Soc. (London) Ser. B **133**, 407–418 (1946)

Beament, J. W. L.: The formation and structure of the micropylar complex in the eggshell of *Rhodnius prolixus* Stahl. J. Exptl. Biol. **23**, 213–233 (1947)

Beams,H.W., Kessel,R.G.: Synthesis and deposition of oocyte envelopes (vitelline membrane and chorion) and the uptake of yolk in the dragonfly. J. Cell. Sci. **4**, 241–264 (1969)

Birnstiel,M.L., Weinberg,E.S., Pardue,M.L.: Evolution of 9S mRNA Sequences. In: Molecular Cytogenetics. Hamkalo,B.A., Papaconstantinou,J. (eds.). New York: Plenum, 1974

Bishop,J.O., Freeman,K.B.: DNA sequences neighboring the duck hemoglobin genes. Cold Spring Harbor Symp. Quant. Biol. **38,** 707–716 (1974)

Blobbel,G., Dobberstein,B.: Transfer of proteins across membranes. I. Presence of proteolytically processed and unprocessed nascent immunoglobulin light chains on membrane-bound ribosomes of murine myeloma. J. Cell Biol. **67**, 835–851 (1975)

Bouligand,Y.: Sur une architecture torsadée répandue dans de nombreuses cuticles d'arthropodes. C. R. hebd. Acad. Sci., Paris **261,** 3365–3668 (1965)

Bouligand,Y.: Twisted fibrous arrangements in biological materials and cholesteric mesophases. Tissue Cell 4 (2), 189–217 (1972)

Chikushi,H.: Genes and Genetical Stocks of the Silkworm. Tokyo: Keigaku, 1972

Clayton,R.M.: Problems of differentiation in the vertebrate lens. In: Current Topics in Developmental Biology, New York: Academic Press, 1970, Vol. V, pp. 115–180

Counce,S.J., Waddington,C.H. (eds.): Developmental Systems: Insects. London–New York: Academic Press, 1973, Vol. II

Craig,S.P., Piatigorsky,J.: Protein synthesis and development in the absence of cytoplasmic RNA synthesis in non-nucleate egg fragments and embryos of sea urchins: Effect of ethidium bromide. Develop. Biol. **24**, 214–232 (1971)

Crippa,M., Telfer,W.: Hybridization analysis of DNA replication in nurse and follicle cells in the cecropia moth. Biol. Bull. **141**, 384 (1971)

Cummings,M.R.: Formation of the vitelline membrane and chorion in developing oocytes of *Ephestia kuhniella*. Z. Zellforsch. 127, 175–188 (1972)

Daneholt,B.: The giant RNA transcript in a balbiani ring of *Chironomus tentans*. In: Molecular Cytogenetics. Hamkalo,B.A., Papaconstantinou,J. (eds.) pp. 155–165, New York: Plenum Press 1973

Daneholt,B.: Transcription in polytene chromosomes. Cell **4**, 1–9 (1975)

David,J., Merle,J.: A reevaluation of the duration of egg chamber stages in oogenesis of *Drosophila melanogaster*. Drosophila Inf. Ser. **43**, 122–123 (1968)

Davidson,E.H., Galau,G.A., Angerer,R.C., Britten,R.J.: Comparative aspects of DNA organization in metazoa. Chromosoma **51**, 253–259 (1975)

Denny,P.C., Tyler,A.: Activation of protein biosynthesis in non-nucleate fragments of sea urchin eggs. Biochem. Biophys. Res. Commun. **14**, 245–249 (1964)

Dickerson,R.E.: The structure of cytochrome c and the rates of molecular evolution. J. Mol. Evol. **1**, 26–45 (1971)

Dubois,M., Gilles,K.A., Hamilton,J.K., Rebers,P.A., Smith,F.: Colorimetric method for determination of sugars and related substances. Anal. Chem. **28**, 350–356 (1956)

Efstratiadis,A., Crain,W.R., Britten,R.J., Davidson,E.H., Kafatos,F.C.: DNA sequence organization in the lepidopteran Antheraea pernyi. Proc. Natl. Acad. Sci. **73**, 2289—2293 (1976a)

Efstratiadis,A., Kafatos,F.C.: The chorion of insects: Techniques and perspectives. In: Methods in Molecular Biology **8**, 1—124, Last,J.,(ed.). New York: Marcel Dekker, 1976

Efstratiadis,A., Kafatos,F.C., Maxam,A.M., Maniatis,T.: Enzymatic in vitro synthesis of globin genes. Cell **7**, 279—288 (1976b)

Efstratiadis,A., Maniatis,T., Kafatos,F.C., Jeffrey,A., Vournakis,J.N.: Full length and discrete partial reverse transcripts of globin and chorion mRNAs. Cell **4**, 367—378 (1975)

Engelmann,F.: The Physiology of Insect Reproduction. Oxford: Pergamon, 1970

Falk,G.J., King,R.C.: Studies on the developmental genetics of the mutant *tiny* of *Drosophila melanogaster*. Growth **28**, 291—324 (1964)

Fan,H., Penman,S.: Regulation of protein synthesis in mammalian cells. II. Inhibition of protein synthesis at the level of initiation during mitosis. J. Mol. Biol. **50**, 655—670 (1970)

Fraser,R.D.B., Macrae,T.P., Rogers,G.E.: Keratins: Their Composition, Structure and Biosynthesis. Springfield, Illinois: Charles C. Thomas, 1972

Furneaux,P.J.S.: 0-Phosphoserine as a hydrolysis product and amino acid analysis of shells of newly laid eggs of the house cricket *Acheta domesticus* L. Biochem. Biophys. Acta **215**, 52—56 (1970)

Furneaux,P.J.S., Mackay,A.L.: The composition, structure and formation of the chorion and the vitelline membrane of the insect eggshell. In: The Insect Integument, Hepburn,H.R., (ed.). New York: Elsevier, 1976

Furneaux,P.J.S., James,C.R., Potter,S.A.: The eggshell of the house cricket *(Acheta domesticus)*: an electron microscope study. J. Cell. Sci. **5**, 227—249 (1969)

Furneaux,P.J.S., Mackay,A.L.: Crystalline protein in the chorion of insect eggshells. J. Ultrastruct. Res. **38**, 343—359 (1972)

Gage,L.P.: Polyploidization of the silk gland of *Bombyx mori*. J. Mol. Biol. **86**, 97—108 (1974)

Gans,M., Audit,C., Masson,M.: Isolation and characterization of sex-linked female sterile mutants in *Drosophila melanogaster*. Genetics **81**, 683—704 (1975)

Gaskill,P., Kabat,D.: Unexpectedly large size of globin messenger ribonucleic acid. Proc. Natl. Acad. Sci. **68**, 72—75 (1971)

Gelinas,R.E.: The chorion Messenger RNAs and Their Translation in the Silkmoth *Antheraea polyphemus*. Ph. D. thesis. Cambridge, Mass.: Harvard Univ., 1974

Gelinas,R.E., Kafatos,F.C.: Purification of a family of specific mRNAs from moth follicular cells. Proc. Natl. Acad. Sci. **70**, 3764—3768 (1973)

Gelinas,R.E., Kafatos,F.C.: The control of chorion protein synthesis in silkmoths: mRNA production parallels protein synthesis. Develop. Biol. **55**, 152—163 (1977)

Gelti-Douka,H., Gingeras,T.R., Kambysellis,M.P.: Yolk proteins in Drosophila: Identification and site of synthesis. J. Exptl. Zool. **187**, 167—172 (1974)

Gianni,A.M., Giglioni,B., Ottolenghi,S., Comi,P., Guidotti,G.: Globin α-chain synthesis directed by "supernatant" 10S RNA from rabbit reticulocytes. Nature New Biol. **240**, 183—185 (1972)

Gottschalk,A. (ed.): Glycoproteins, Vol. V A and B. Amsterdam: Elsevier, 1972

Grace,T.D.C.: Establishment of four strains of cells from insect tissues grown in vitro. Nature (London) **195**, 788—789 (1962)

Gross,P.R., Gross,K.W., Skoultchi,A.I., Ruderman,J.Y.: Maternal mRNA and protein synthesis in the embryo. In: Karolinska Symposia on Research Methods in Reproductive Endocrinology. No. 6: Protein Synthesis in Reproductive Tissue, Diczfalusy,E. (ed.), pp. 244—262. Stockholm: Karolinska Inst., 1973

Harris,S.E., Means,A.R., Mitchell,W.M., O'Malley,B.W.: Synthesis of the [^{3}H] DNA complementary to ovalbumin mRNA: Evidence for limited copies of the ovalbumin gene in chick oviduct. Proc. Natl. Acad. Sci. **70**, 3776—3780 (1973)

Hinton,H.E.: Respiratory systems of insect eggshells. Ann. Rev. Entomol. **14**, 343—369 (1969)

Hinton,H.E.: Insect eggshells. Sci. Am. **223** (2), 84—91 (1970)

Holland,J.J., Kiehn,E.D.: Specific cleavage of viral proteins as steps in the synthesis and maturation of enteroviruses. Proc. Natl. Acad. Sci. **60**, 1015—1022 (1968)

Hood,L.E., Campbell,J.H., Elgin,S.C.R.: The organization, expression and evolution of antibody genes and other multigene families. Ann. Rev. Gen. **9**, 305—353 (1975)

Hopkins,C.R., King,P.E.: An electron-microscopical and histochemical study of the oocyte periphery in *Bombus terrestris* during vitellogenesis. J. Cell. Sci. **1**, 201—216 (1966)

Housman,D., Forget,B.G., Skoultchi,A., Benz,E.J.: Quantitative deficiency of chain-specific globin messenger ribonucleic acid in the thalassemia syndromes. Proc. Natl. Acad. Sci. **70**, 1809—1813 (1973)

Humphreys,T.: Measurements of messenger RNA entering polysomes upon fertilization of sea urchin eggs. Develop. Biol. **26**, 201—208 (1971)

Jacobsen,M.F., Baltimore,D.: Polypeptide cleavages in the formation of poliovirus proteins. Proc. Natl. Acad. Sci. **61**, 77—84 (1968)

Jacobs-Lorena,M., Baglioni,C.: Messenger RNA for globin in the post-ribosomal supernatant of rabbit reticulocytes. Proc. Natl. Acad. Sci. **69**, 1425—1428 (1972)

Jamieson, J. D., Palade, G. E.: Intracellular transport of secretory proteins in the pancreatic exocrine cell. I. Role of the peripheral elements of the golgi complex. J. Cell Biol. **34**, 577—596 (1967a)

Jamieson, J. D., Palade, G. E.: Intracellular transport of secretory proteins in the pancreatic exocrine cell. I. Role of the peripheral elements of the Golgi complex. J. Cell Biol. **34**, 577—597—615 (1967b)

Jamieson, J. D., Palade, G. E.: Synthesis, intracellular transport, and discharge of secretory proteins in stimulated pancreatic exocrine cells. J. Cell. Biol. **50**, 135—158 (1971)

Johnson, C. C., King, R. C.: Oogenesis in the ocelliless mutant of *Drosophila melanogaster* Meigen (Diptera, Drosophilidae). Intern. J. Insect Morphol. Embryol. **3**, 385—395 (1974)

Kafatos, F. C.: The cocoonase zymogen cells of silkmoths: A model of terminal cell differentiation for specific protein synthesis. In: Current Topics in Developmental Biology, Vol. VII, pp. 125—191. New York: Academic Press 1972a

Kafatos, F. C.: mRNA stability and cellular differentiation. Acta Endocrin. Suppl. **168**, 319—345 (1972b)

Kafatos, F. C.: The insect chorion: programmed expression of specific genes during differentiation. In: Control Mechanisms in Development. Advances in Experimental Medicine and Biology, Meints, R. H., Davies, E. (eds.) **62**, 103—121. New York: Plenum 1975

Kafatos, F. C., Gelinas, R.: mRNA stability and the control of specific protein synthesis in highly differentiated cells. In: MTP International Review of Science—Biochemistry of Differentiation and Development, Vol. IX, Paul, J. (ed.). Oxford: Medical and Technical Publ., 1974

Kafatos, F. C., Maniatis, T., Efstratiadis, A., Sim Gek Kee, Regier, J. C., Nadel, M.: The moth chorion as a system for studying the structure of developmentally regulated gene sets. In: Organization and Expression of the Eukaryotic Genome. Bradbury, E. M., Javaherian, K. (eds.), pp. 393—420. London: Academic Press, 1977

Kambysellis, M. P.: Ultrastructure of the chorion in *Drosophila* species. Drosophila Info. Ser. **50**, 89—90 (1973)

Kambysellis, M. P., Williams, C. M.: In vitro development of insect tissues. I. A macromolecular factor prerequisite for silkworm spermatogenesis. Biol. Bull. **141**, 527—540 (1971)

Kawasaki, H., Sato, H., Suzuki, M.: Structural proteins in the silkworm eggshells. Insect Biochem. **1**, 130—148 (1971a)

Kawasaki, H., Sato, H., Suzuki, M.: Structural proteins in the eggshell of the oriental garden cricket *Gryllus mitratus*. Biochem. J. **125**, 495—505 (1971b)

Kawasaki, H., Sato, H., Suzuki, M.: Structural proteins in the eggshell of silkworms, *Bombyx mandarina* and *Antheraea mylitta*. Insect Biochem. **2**, 53—57 (1972)

Kawasaki, H., Sato, H., Suzuki, M.: Structural proteins in the egg envelopes of dragonflies, *Sympetrum infuscatum* and *S. frequens*. Insect Biochem. **4**, 99—111 (1974)

Kawasaki, H., Sato, H., Suzuki, M.: Structural proteins in the egg envelopes of the mealworm beetle, *Tenebrio molitor*. Insect Biochem. **5**, 25—34 (1975)

Kemp, D. J., Rogers, G. E.: Differentiation of avian keratinocytes. Characterization and relationships of the keratin proteins of adult and embryonic feathers and scales. Biochemistry **11**, 969—975 (1972)

King, P. E., Richards, J. G., Copland, M. J. W.: The structure of the chorion and its possible significance during oviposition in *Nasonia vitripennis* and other chalcids. Proc. Roy. Entomol. Soc. (London) Ser. A **43**, 13—20 (1968)

King, R. C.: Oogenesis in adult *Drosophila melanogaster*. IX. Studies on the cytochemistry and ultrastructure of developing oocytes. Growth **24**, 265—323 (1960)

King, R. C.: Ovarian Development in *Drosophila melanogaster*. New York: Academic Press, 1970

King, R. C., Aggarwal, S. K.: Oogenesis in *Hyalophora cecropia*. Growth **29**, 17—83 (1965)

King, R. C., Koch, E. A.: Studies on the ovarian follicle cells of *Drosophila*. Quart. J. Micr. Sci. **104** (3), 297—320 (1963)

King, R. C., Mohler, D.: The genetic analysis of oogenesis in *Drosophila melanogaster*. In: Handbook of Genetics. King, R. C. (ed.) **3**, 757—791 New York — London: Plenum Press, 1975

Koch, E. A., King, R. C.: The origin and early differentiation of the egg chamber of *Drosophila melanogaster*. J. Morph. **119**, 283—304 (1966)

Kuo, J. F., Wyatt, G. R., Greengard, P.: Cyclic nucleotide-dependent protein kinases. IX: Partial purification and some properties of guanosine 3′,5′-monophosphate-dependent and adenosine 3′,5′-monophosphate-dependent protein kinases from various tissues and species of Arthropoda. J. Biol. Chem. **246**, 7159—7167 (1971)

Kuwana, J., Takami, T.: Arthropoda, Insecta. In: Invertebrate Embryology. Kumé, M., Dan, K. (eds.). Belgrade, Yugoslavia: Nolit, 1968

Lucas, F., Shaw, J. T. B., Smith, S. G.: Comparative studies of fibroins: I. The amino acid composition of various fibroins and its significance in relation to their crystal structure and taxonomy. J. Mol. Biol. **2**, 339—349 (1960)

Mahowald, A. P.: Ultrastructural observations on oogenesis in *Drosophila*. J. Morph. **137**, 29—48 (1972)

Mandel, M., Schildkraut, C. L., Marmur, J.: Use of CsCl density gradient analysis for determining the guanine plus cytosine content of DNA. In: Methods in Enzymology Grossman, L., Moldave, K. (eds.). New York: Academic Press, 1968, Vol. XII B, pp. 184—195

Maniatis, T., Sim Gek Kee, Efstratiadis, A., Kafatos, F. C.: Amplification and characterization of a β-globin gene synthesized in vitro. Cell **8**, 163—182 (1976)

Margaritis, L.: Programmed Synthesis of Specific Proteins in Cellular Differentiation. A Contribution to the Study of Chorion Formation in the Follicles of *Drosophila melanogaster*. Ph. D. Thesis. Greece: Univ. Athens, 1974

Margaritis, L. H., Petri, W. H., Kafatos, F. C.: Three-dimensional structure of the endochorion in wild type *Drosophila melanogaster*. J. Exptl. Zool. **198**, 429—436 (1976)

Matsuzaki, M.: Electron microscopic observations on chorion formation of the silkworm *Bombyx mori*. J. Seric. Sci. Tokyo **37**, 483—490 (1968)

Matsuzaki, M.: Electron microscopic studies on the oogenesis of dragonfly and cricket with special reference to the panoistic ovaries. Develop. Growth Diff. **13**, 379—398 (1971)

Narita, K.: N-terminal group of ovalbumin. Biochem. Biophys. Res. Commun. **5**, 160—164 (1961)

Neurath, H.: Mechanism of zymogen activation. Federation Proc. **23**, 1—29 (1964)

Neville, A. C.: Cuticle ultrastructure in relation to the whole insect. In: Insect Ultrastructure; Symp. Roy. Entomol. Soc. London, Vol. V, pp. 17—39. Oxford: Blackwell 1970

Newell, P. C., Sussman, M.: Regulation of enzyme synthesis by slime mold cell assemblies embarked upon alternative developmental programs. J. Mol. Biol. **49**, 627—637 (1970)

Nijhout, M. M., Riddiford, L. M.: The control of egg maturation by juvenile hormone in the tobacco hornworm moth, *Manduca sexta*. Biol. Bull. **146**, 377—392 (1974)

O'Donnell, I. J., Thompson, E. O. P., Inglis, A. S.: N-acetyl groups in wool and extracted wool proteins. Australian J. Biol. Sci. **15**, 732—739 (1962)

O'Farrell, P. H.: High resolution two-dimensional electrophoresis of proteins. J. Biol. Chem. **250**, 4007—4021 (1975)

Paik, W. K., Kim, S.: Protein methylation. Science **174**, 114—119 (1971)

Palmiter, R. D.: Regulation of protein synthesis in chick oviduct. I: Independent regulation of ovalbumin, conalbumin, ovomucoid, and lysozyme induction. J. Biol. Chem. **247**, 6450—6459 (1972)

Palmiter, R. D.: Rate of ovalbumin messenger ribonucleic acid synthesis in the oviduct of estrogen-primed chicks. J. Biol. Chem. **248**, 8260—8270 (1973)

Palmiter, R. D.: Magnesium precipitation of ribonucleoprotein complexes. Expedient techniques for the isolation of undegraded polysomes and messenger ribonucleic acid. Biochemistry **13**, 3606—3615 (1974)

Palmiter, R. D., Schimke, R. T.: Regulation of protein synthesis in chick oviduct. III: Mechanism of ovalbumin "superinduction" by actinomycin D. J. Biol. Chem. **248**, 1502—1512 (1973)

Paul, M., Goldsmith, M. R., Hunsley, J. R., Kafatos, F. C.: Cellular differentiation and specific protein synthesis: Production of eggshell proteins by silkmoth follicular cells. J. Cell Biol. **55**, 653—680 (1972a)

Paul, M., Kafatos, F. C., Regier, J. C.: A comparative study of eggshell proteins in Lepidoptera. J. Supramolec. Struct. **1**, 60—65 (1972b)

Paul, M., Kafatos, F. C.: Specific protein synthesis in cellular differentiation. II: The program of protein synthetic changes during chorion formation by silkmoth follicles, and its implementation in organ culture. Develop. Biol. **42**, 141—159 (1975)

Peacock, W. J., Brutlag, D., Goldring, E., Appels, R., Hinton, C. W., Lindsley, D. L.: The organization of highly repeated DNA sequences in *Drosophila melanogaster* chromosomes. Cold Spring Harbor Symp. Quant. Biol. **38**, 405—416 (1974)

Petri, W. H., Wyman, A. R., Kafatos, F. C.: Specific protein synthesis in cellular differentiation. III. The eggshell proteins of *Drosophila melanogaster* and their program of synthesis. Develop. Biol. **49**, 185—199 (1976)

Pollack, S. B., Telfer, W. H.: RNA in Cecropia moth ovaries: Sites of synthesis, transport, and storage. J. Exptl. Zool. **170**, 1—23 (1969)

Pontz, B. F., Muller, P. K., Meigel, W. N.: A study on the conversion of procollagen. J. Biol. Chem. **248**, 7558—7564 (1973)

Quattropani, S. L., Anderson, E.: The origin and structure of the secondary coat of the egg of *Drosophila melanogaster* . Z. Zellforsch. **95**, 495—510 (1969)

Ray, P. M.: Radioautographic study of cell wall deposition in growing plant cells. J. Cell Biol. **35**, 659—684 (1967)

Regier, J. C., Kafatos, F. C.: Microtechnique for determining the specific activity of radioactive intracellular leucine and applications to in vivo studies of protein synthesis. J. Biol. Chem. **246**, 6480—6488 (1971)

Revel, J. P., Hay, E. D.: An autoradiographic and electron microscopic study of collagen synthesis in differentiating cartilage. Z. Zellforsch. **61**, 110—144 (1963)

Robinson, A. B., McKerrow, J. H., Cary, P.: Controlled deamidation of peptides and proteins: An experimental hazard and a possible biological timer. Proc. Natl. Acad. Sci. **66**, 753—757 (1970)

Rogers, G. E.: Electron microscopy of wool. J. Ultrastruct. Res. **2**, 309—330 (1959)

Rutter, W. J., Kemp, J. D., Bradshaw, W. S., Clark, W. R., Ronzio, R. A., Sanders, T. G.: Regulation of specific protein synthesis in cytodifferentiation. J. Cell. Physiol. **72**, (Suppl. 1) 1—18 (1968)

Rutter, W. J., Pictet, R. L., Morris, P. W.: Toward molecular mechanisms of developmental processes. Ann. Rev. Biochem. **42**, 601—646 (1973)

Sakaguchi, B., Chikushi, H., Doira, H.: Observations of the eggshell structures controlled by gene action in *Bombyx mori*. J. Fac. Agr., Kyushu Univ. **18**, 53—62 (1973)

Schmeckpeper, B. S., Cory, S., Adams, J. M.: Translation of immunoglobulin mRNA's in a wheat germ cell-free system. Mol. Biol. Reports **1**, 355—363 (1974)

Schneider, I.: Differentiation of larval Drosophila eye-antennal discs in vitro. J. Exptl. Zool. **156**, 91—104 (1964)

Slabaugh, R. C., Morris, A. J.: Purification of peptidyl transfer ribonucleic acid from rabbit reticulocyte ribosomes. J. Biol. Chem. **245**, 6182—6189 (1970)

Slayter, G.: Two phase materials. Sci. Am. **206**, 124—134 (1962)

Slifer, E. H., Sekhon, S. S.: The fine structure of the membranes which cover the egg of the grasshopper *Melanoplus differentialis*, with special reference to the hydropyle. Quart. J. Micr. Sci. **104**, 321—334 (1963)

Smith, D. S., Telfer, W. H., Neville, A. C.: Fine structure of the chorion of a moth, *Hyalophora cecropia*. Tissue Cell **3**, 477—498 (1971)

Spohr, G., Imaizumi, T., Stewart, A., Scherrer, K.: Indentification of free cytoplasmic globin mRNA of duck erythroblasts by hybridization to anti-messenger DNA and by cell-free protein synthesis. FEBS Lett. **28**, 165—168 (1972)

Steinberg, M., Levinson, B., Tomkins, G. M.: Kinetics of steroid induction and deinduction of tyrosine aminotransferase synthesis in cultured hepatoma cells. Proc. Natl. Acad. Sci. **72**, 2007—2011 (1975)

Steiner, A. L., Parker, C. W., Kipnis, D. M.: Radioimmune assay for cyclic nucleotides. I. Preparation of antibodies and iodinated cyclic nucleotides. J. Biol. Chem. **247**, 1106—1113 (1972)

Steiner, D. F., Oyer, P. O.: The biosynthesis of insulin and probable precursor of insulin by a human islet cell adenoma. Proc. Natl. Acad. Sci. **57**, 473—480 (1967)
Sullivan, D., Palacios, R., Stavnezer, J., Taylor, J. M., Faras, A. J., Kiely, M. L., Summers, N. M., Bishop, J. M., Schminke, R. T.: Synthesis of a DNA sequence complementary to ovalbumin mRNA and quantification of ovalbumin genes. J. Biol. Chem. **248**, 7530—7539 (1973)
Summers, D. F., Maizel, J. V., Jr.: Evidence for large precursor proteins in poliovirus synthesis. Proc. Natl. Acad. Sci. **59**, 966—971 (1968)
Takasaki, T.: Studies on the Second Linkage Group of the Silkworm. Ph. D. Thesis. Japan: Kyushu Univ., 1962, Chap. 3
Tazima, Y.: The Genetics of the Silkworm. London: Logos Press, and Englewood Cliffs, N. J.: Prentice-Hall, 1964
Telfer, W. H.: The mechanism and control of yolk formation. Ann. Rev. Entomol. **10**, 161—184 (1965)
Telfer, W. H.: Development and physiology of oocyte-nurse cell syncytium. Advan. Insect Physiol. **11**, 223—319 (1975)
Telfer, W. H., Anderson, L. M.: Functional transformations accompanying the initiation of a terminal growth phase in the cecropia moth oocyte. Develop. Biol. **17**, 512—535 (1968)
Telfer, W. H., Smith, D. S.: Aspects of egg formation. Proc. Roy. Entomol. Soc. (London) Symp. **5**, 117—134 (1970)
Vogel, S., Bretz, W. L.: Interfacial organisms: Passive ventilation in the velocity gradients near surfaces. Science **175**, 210—211 (1972)
Vournakis, J. N., Efstratiadis, A., Kafatos, F. C.: Electrophoretic patterns of deadenylated chorion and globin mRNAs. Proc. Natl. Acad. Sci. **72**, 2959—2963 (1975)
Vournakis, J. N., Gelinas, R. E., Kafatos, F. C.: Short polyadenylic acid sequences in insect messenger RNA. Cell **3**, 267—275 (1974)
Waller, J. P.: The NH_2-terminal residues of the proteins from cell-free extracts of *E. coli*. J. Mol. Biol. **7**, 483—496 (1963)
Weinberg, E. S., Birnstiel, M. L., Purdom, I. F., Williamson, R.: Genes coding for polysomal 9S RNA of sea urchins: Conservation and divergence. Nature (London) **240**, 225—228 (1972)
Wigglesworth, V. B., Salpeter, M. M.: The aeroscopic chorion of the egg of *Calliphora erythrocephala* Meig (Diptera), studied with the electron microscope. J. Insect Physiol. **8**, 635—641 (1962)
Williams, C. M.: The juvenile hormone. I. Endocrine activity of the corpora allata of the adult Cecropia silkworm. Biol. Bull. **116**, (2), 323—338 (1959)
Williamson, R., Grossly, J., Humphries, S.: Translation of mouse messenger ribonucleic acid from which the poly (adenylic acid) sequence has been removed. Biochemistry **13**, (1), 703—707 (1974)
Wilson, B. R.: Some chemical components of the eggshell of *Drosophila melanogaster*. I. Amino acids. Ann. Entomol. Soc. Am. **53**, 170—173 (1960a)
Wilson, B. R.: Some chemical components of *Drosophila melanogaster* eggshell. II. Amino sugars and elements. Ann. Entomol. Soc. Am. **53**, 732—735 (1960b)
Wilson, D. B., Dintzis, H. M.: Protein chain formation in rabbit reticulocytes. Proc. Natl. Acad. Sci. **66**, 1282—1289 (1970)
Wyatt, G. R., Linzen, B.: The metabolism of ribonucleic acid cecropia silkmoth pupae in diapause development and after injury. Biochem. Biophys. Acta **103**, 588—600 (1965)
Ycas, M.: De novo origin of periodic proteins. J. Mol. Evol. **2**, 17—27 (1972)

The Salivary Gland of *Chironomus* (Diptera): A Model System for the Study of Cell Differentiation*

U. Grossbach

Max-Planck-Institut für Biologie, Tübingen, and Max-Planck-Institut für Biochemie, Martinsried/München, FRG

I. Introduction

Among the various types of differentiated insect tissues, Chironomid salivary glands provide an especially fortunate combination of features. The glands consist of a limited number of very large cells that are committed to the continuous synthesis of specific proteins. Extensive polytenization renders their chromosomes visible in unparalleled detail during interphase so that units of genetic function can be studied at the level of the light-microscope. A gland comprises two or more related cell types that exhibit minor differences in terms of cell morphology, chromomere activity and protein synthesis and thus provide excellent tools for comparative studies of differentiated states in the same tissue.

Beermann (1952a, 1956a, 1961, 1962) and Mechelke (1953, 1958) have established the constancy of chromomere morphology as well as the cell-specificity of the functional puffing pattern in their investigations on the salivary glands of *Camptochironomus* and *Acricotopus*. The concept of chromomere puffing as a cytologically visible manifestation of genetic activity forwarded in the classical work of Beermann and Mechelke has opened a new approach to the problem of cell differentiation, and most subsequent research has, therefore, centered on the glands of these two organisms. In the present article, the biology of the *Camptochironomus* gland and its secretory proteins will be reviewed. For a survey of different aspects of transcription in this same tissue, the reader is referred to the recent reviews by Daneholt (1974, 1975). Cell differentiation and gene expression in the salivary gland of *Acricotopus* have recently been discussed by Baudisch (this vol., p. 137).

As is the case with many other cell types, the present state of our knowledge comprises not more than a rather incomplete descriptive view of the differentiated state in the salivary gland cells and some evidence on how this state may be

* Dedicated to Professor H. G. Callan, F.R.S., on occasion of his 60th birthday.

maintained and varied. The processes by which differentiation of these cells is initiated and established during development are almost completely unknown. The peculiar properties of these polytene cells, however, seem to render them ideally suited for future investigations of these basic processes.

II. The Subgenus Camptochironomus

The subgenus *Camptochironomus* of the genus *Chironomus* (*Chironomidae*, Diptera) comprises two species, *C. tentans* Fabr. and *C. pallidivittatus* Malloch which live in central and northern Europe as well as in North America. A third species of this subgenus, *C. flavofasciatus* Kieff., has been described from Asia minor (cf. Goetghebuer, 1937); however, the existence of this species is doubtful.

Adult *C. tentans* and *C. pallidivittatus* differ in several morphological characteristics. These include the relative lengths of the femora and tibiae, the form of the external sex organs, and the color of the thorax which is light brown in *C. tentans* and yellowish-green in *C. pallidivittatus* (Beermann, 1955). The color difference is most conspicuous in three prominent stripes on the prothoracal notum, dark brown in *tentans*- but yellow in *pallidivittatus*-imagines. In both species the head of the larvae exhibits a dark wedge-shaped spot which distinguishes them from other red *Chironomus* larvae (Beermann, 1955). The larval morphology is very similar within the subgenus, the species being most conveniently identified by the structure of their salivary gland chromosomes (Beermann, 1955, 1961).

C. tentans and *C. pallidivittatus* are frequently sympatric in the same habitat. The morphological differences of the external sex organs, however, provide a rather effective protection against interspecific hybridization (Beermann, 1955). Hybrid larvae have been found only very rarely under natural conditions (Beermann, 1955; Acton, 1959; Grossbach, unpubl.), but in the laboratory hybridization of *C. tentans* and *C. pallidivittatus* can be achieved rather easily. The F_1-hybrids are fertile and can be crossed back with both species (Beermann, 1955, 1961).

In contrast to other *Chironomus* species which perform a copulation flight, *C. tentans* and *C. pallidivittatus* copulate in small cultivation vials and can, therefore, be bred easily. The culture conditions have been described by Beermann (1952a). Briefly, the several hundred larvae which hatch from an eggmass are distributed among several plastic dishes of 20–30 cm diameter and about 10 cm height which contain a layer of 2 cm of water. Either tap water of good quality or distilled water with 0.06% NaCl is appropriate. Some sheets of cellulose-tissue and a small amount of decayed alder-leaves is added. The growing larvae are regularly fed with pulverized nettle *(Urtica maior)*.

The length of the generation period is to some extent variable with an average of 8 to 9 weeks, the first three larval stages lasting at least a week each and the fourth stage several weeks (Beermann, 1952a). The pupal and imaginal stages last a few days each. There are usually two generations per year in central Europe as well as in Scandinavia and Finland (Palmén and Aho, 1966).

The larval development is interrupted in autumn by a diapause period which is controlled by day length (Engelmann and Shappirio, 1965). Diapause may also occur in laboratory cultures, especially during the winter, but can be avoided by keeping the animals permanently under long-day conditions.

III. Biology of the Salivary Gland

A. Morphology and Development

The pair of salivary glands of the *Chironomus* larva is homologous to the labial glands of other insects, the ectodermal glands of the last head segment. In the adult larva the glands are flat bags of about 1.5 mm length, 0.75 mm breadth and a thickness of about 0.15 mm. They are composed of three larger and one smaller lobe (Fig. 1) and end in a narrow, chitinized duct which unites with the duct of the other gland before emptying at the hypopharynx. The glands extend through the three thoracal segments. Their position in relation to other organs in the thorax and head has been described in detail by Miall and Hammond (1900), Holmgren (1904), De Pozdniakoff (1932), and Churney (1940).

Each gland consists of 35–40 large cells which are arranged in a single-cell layer around the central gland lumen. The number of cells varies slightly from animal to animal and may even differ in the two glands of a pair. Most cells have a saddle-like shape (Fig. 2) while a few situated on the two sides of the gland are very flat. In the living tissue, the huge, more or less spherical nuclei with their giant chromosomes can easily be observed. The duct is formed by a layer of small epithelial cells. Electron microscope studies of the gland cells (Beermann and Meyer, unpubl.) have shown an ergastoplasm with numerous Golgi bodies, an endoplasmic reticulum with RNP-particles and a brush border typical for secretory cells (Fig. 3).

While no differences have been detected between the cells of the three larger lobes, the 3 or 4 cells forming the small lobe at the origin of the duct (Fig. 1) differ morphologically and physiologically from the other cells of the gland. Most conspicuous is the difference in the appearance of the secretion which fills the lumen of the *pallidivittatus*-gland; in this lobe, the secretion contains characteristic PAS-positive granules which are lacking in all other parts of the gland (Churney, 1940; Beermann, 1961). The homologous region of the *tentans*-gland produces another special type of secretion (see Sect. IV.C.1.). In both species this gland lobe has been designated "special lobe" and its cells "special cells" (Beermann, 1961). The appearance in the larval salivary gland of several different cell types has also been reported for other Chironomid species (Bauer, 1936; Beermann, 1952b; Mechelke, 1953).

Newly hatched larvae, about 0.8 mm long, possess fully functioning salivary glands with a length around 50 µm (Fig. 1). These grow within about two months to the dimensions mentioned above, which corresponds to a 30-fold increase in length. The increase in volume may be in the range of 10^4 times. This remarkable growth is brought about, not by cell divisions but exclusively, by the magnification of cells. It

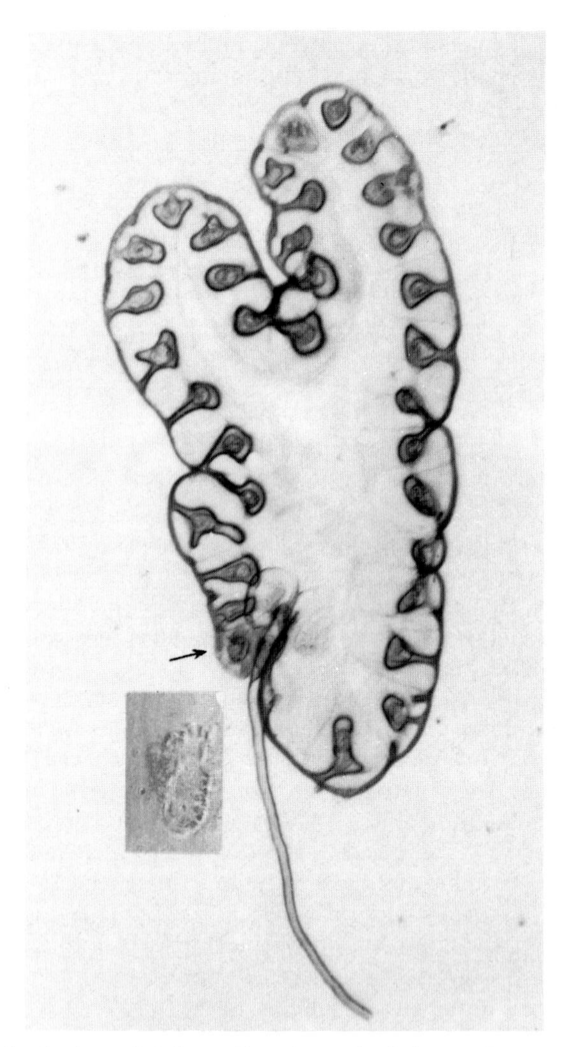

Fig. 1. Salivary gland of an adult larva (fourth stage) of *Camptochironomus pallidivittatus*. *Arrow:* special lobe. *Insert:* the salivary gland of the first larval stage about 48 h after hatching (×60, *insert* ×140, reduced to 80%)

is accompanied by an increasing polytenization of the nuclei, the different stages of gland development each being characterized by a certain structural type of giant chromosomes (Beermann, 1952a, 1962).

The larval salivary glands undergo histolysis shortly after the larval-pupal molt and degenerate completely. Tissue dissolution is preceded by a drastic increase in the activity of enzymes which probably are involved in autolysis. Acid phosphatase activity has been demonstrated to increase by several times in the gland concurrently with pupation (Schin and Laufer, 1973) when it appears freely distributed in the cytoplasm (Schin and Clever, 1965, 1968). A protease activity associated with lysosome-like structures, and exhibiting characteristics different

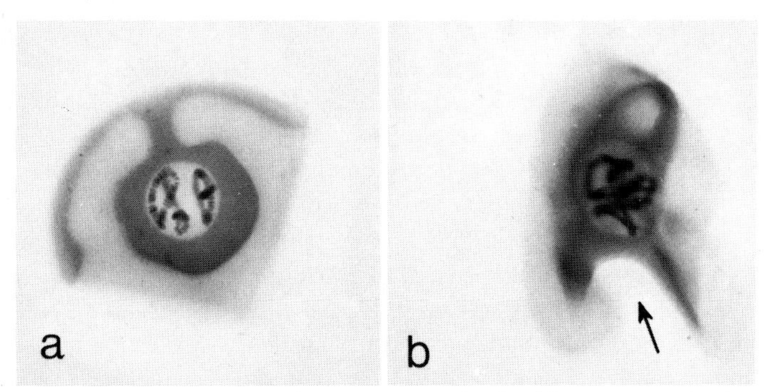

Fig. 2a and b. Salivary gland cells of *C. tentans* stained with aceto-orceine/aceto-carmine, phase contrast. (a) Apex cell from one of the posterior lobes. (b) Side-view of gland cell showing the typical saddle-shaped processes of the cell which surround the gland lumen *(arrow)*

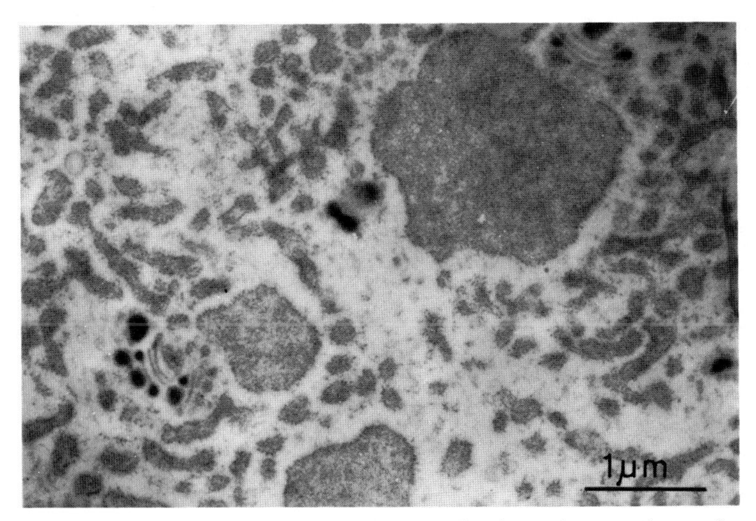

Fig. 3. Rough endoplasmic reticulum and Golgi bodies from the cytoplasm of a salivary gland cell of *C. tentans*. (Courtesy of Drs. W. Beermann and G. F. Meyer)

from the protease activity in earlier glandular stages, reaches its maximum shortly before cell degeneration begins (Rodems et al., 1969; Henrikson and Clever, 1972). These enzymes are synthesized at an earlier stage as inactive zymogens (Henrikson and Clever, 1972; Schin and Laufer, 1973).

B. The Function of the Gland

The function of the cells of the salivary gland is to synthesize specific structural proteins which are transported through the inner cell membrane and stored in the gland lumen. In addition to proteins, which will be described in Section IV, the secretion contains a small amount of polysaccharide. In typical analyses using the micromethod of François et al. (1962), 0.5–0.6 µg of neutral sugars were found in samples of 32 μg protein from *C. tentans* and 0.7–0.8 μg of neutral sugars in samples of 32 μg protein from *C. pallidivittatus* (Grossbach, 1969). This implies a polysaccharide content of about 2%. Using the micromethod of Parekh and Glick (1962), no hexosamine could be detected in similar samples of secretion (Grossbach, 1969), which should contain less than one percent hexosamine relative to the protein content, if any. Two-dimensional chromatography of acid hydrolysates of the total secretion (Grossbach, 1969) revealed no hexosamine spots. When the secretory proteins, after electrophoretic separation (see Sect. IV.A.), were stained for polysaccharides by the PAS-reaction, only one fraction was faintly colored and should, therefore, contain the polysaccharide moiety which may consist primarily of neutral sugars.

The mechanism of the secretory process has been described by Churney (1940). The orifice of the common gland duct is normally blocked by two broad elastic bands which insert laterally at the inner walls of the duct and together form a cone-shaped, tightly fitting plug. The proximal ends of these bands are attached to the hypopharynx. During the secretory process, retractor muscles inserting dorsally at the common duct of the glands contract rapidly and achieve quick forward-backward motions of the duct relative to the plug, by which a suction force is exerted on the secretion in the glands. Every retraction of the common duct stretches the two elastic bands so that they separate from each other and open a valve through which the secretion is sucked out. As the secretion is very viscous, a remarkable suction force must be exerted by this pumping mechanism, a force which should depend on the velocity of the contractions of the retractor muscles. Churney (1940) reports these muscles to be "the most rapidly contracting muscles in the head, if not in the entire body."

The secretory process can be stimulated by the administration of pilocarpine (Beermann, 1961). If the larvae are kept in 0.1% pilocarpine hydrochloride, the glands are completely emptied of secretion. Atropine, on the other hand, blocks the delivery of secretion (Grossbach, unpubl.). The drugs presumably act on synaptic transmission in the retractor innervation, or possibly directly on the muscles (cf. Pichon, 1974).

During or immediately after the secretory process the secretion molecules aggregate to form a paracrystalline thread of high mechanical strength and elasticity. The animal uses this thread to spin a cylindrical tube to protect itself. Long threads which run more or less in parallel and tend to form bundles can be isolated from the walls of tubes made by adult larvae (Fig. 4). In the electron microscope, these protein threads show a rather even diameter of 0.46–0.48 μ. Negative staining with phosphotungstic acid reveals their composition of smaller subelements (Fig. 5). Longitudinal subunits of the thread become separated to a certain extent upon treatment with uranylic acetate. A pretreatment with formic acid prior to staining with uranylic acetate unraveled numerous bundle-forming fibers which seem to be the subunits of the threads (Fig. 6). The thinnest and most frequent fibers had a diameter of about 45 Å. Thicker fibers may consist of more than one of these elements.

The aggregation of the protein molecules of the secretion in long fibers of 45 Å diameter and the arrangement of these fibers into a thread of specific diameter might be achieved by shearing forces which occur during the secretory process as

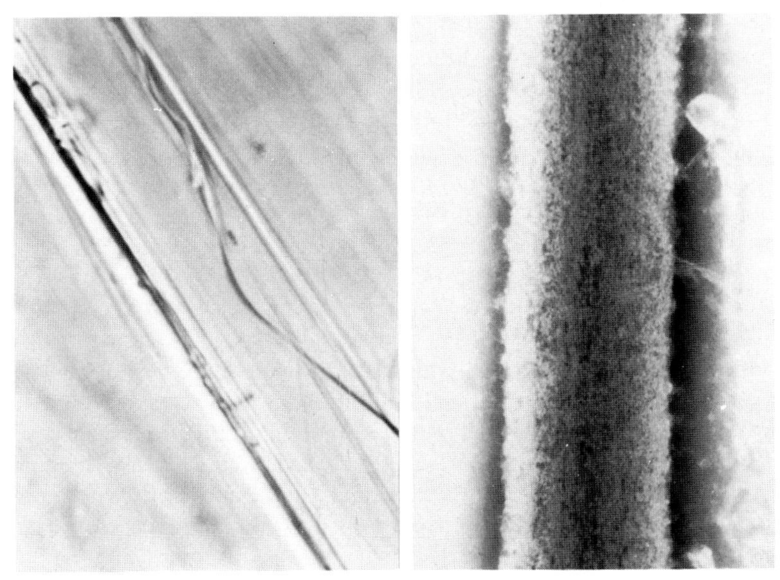

Fig. 4 Fig. 5

Fig. 4. Secretory threads from the tube spun by an adult larva of *C. tentans* ($\times 2800$, reduced to 80%). (From Grossbach, 1969)

Fig. 5. Secretory thread of *C. pallidivittatus*, negative staining with phosphotungstic acid ($\times 60000$, reduced to 80%)

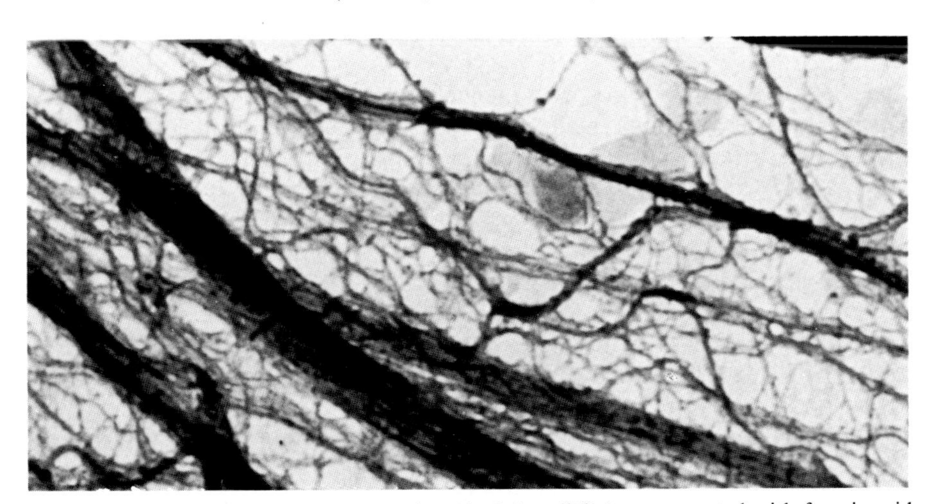

Fig. 6. Fibers composing the secretory thread of *C. pallidivittatus*, treated with formic acid and stained with uranylic acetate ($\times 20000$, reduced to 85%)

has been described for the formation of the silk thread in *Lepidoptera* (Lucas et al., 1958). It is also conceivable that the secretion of the "special lobe" which is mixed into the gland secretion immediately before it enters the duct, contains a component which induces aggregation.

The case which is spun from the secretory threads has a high mechanical stability and can last for weeks in the medium. It is usually fastened to the substrate and serves not only to protect the animal but also plays an important role in its feeding behavior. The case has the form of a cylindrical tube which is somewhat longer and some millimeters wider than the larva itself. Staying inside the tube, the larva performs rapid body undulations which obviously induce a stream of water to flow continuously through the tube in the anterior-posterior direction. This behavior probably facilitates the exchange of oxygen, the concentration of which in the bottom mud is certainly a critical point. However, the stream of water also provides the animal with food. As Walshe (1947) first observed, the larva spins a funnel-shaped net of secretion threads which is fastened to the internal walls of the tube near its anterior opening. Keeping a thread from the center of the web in its mouthparts, the larva stretches the net by withdrawing a short distance into the tube in a posterior direction. Detritus particles and planktonic organisms are thus filtered from the stream of water and caught in the net. Walshe (1947) reports that after a short period of undulations the larva stops and eats the whole net along with the filtrate, immediately after which a new net is spun and the cycle repeated. These observations have been confirmed by other authors.

According to Walshe, a new net is formed every two minutes for an hour or more, implying a very high rate of secretory thread production. Larvae also construct new cases frequently, for which even more material is needed. Their behavior, therefore, indicates that a high level of secretory protein synthesis in the salivary gland is continuously maintained. This assumption has been substantiated by quantitative methods.

Tubular cases are spun throughout all four larval stages. Newly hatched larvae of the first stage possess fully functioning salivary glands (Beermann, 1952a) and are immediately able to spin, perhaps after a short initial period of planktonic life (Lellák, 1968). This ability is maintained until pupation, the gland cells, therefore, synthesizing secretory proteins more or less continuously during several weeks. In quantitative terms, this synthesis of specific proteins comprises many times the protein content of the cells themselves. Table 1 lists the protein content of five pairs of fully grown glands from whose lumina the secretion had been withdrawn for separate determination. As can be seen, emptied glands from larvae in the late fourth instar contain about 7–12 μg protein or 0.175–0.3 μg protein per cell. It is obvious from these values that the secretion in the gland lumen was equivalent to 65–94% of the cells' protein content and that about 40–48% of a gland's total protein is accounted for by the secretion in the lumen. The numbers for the glands devoid of secretion may suggest amounts which are slightly too high, as it is difficult to completely withdraw the secretion from the lumen. Furthermore, it must be assumed that these values include growing peptide chains as well as newly synthesized secretory protein molecules not yet transported through the cellular membrane. Therefore, more than 50% of the protein content of a gland should be accounted for by secretory proteins.

This remarkable amount of secretory protein can be newly synthesized by the gland cells within 24 h. As was mentioned earlier, the larvae empty their glands completely under the influence of pilocarpine. When the animals are brought into fresh culture medium subsequent to this treatment, newly synthesized secretion is

Table 1. Protein content of salivary glands from *C. tentans* as determined by a micromodification of the Lowry technique

Gland pair	Gland	Protein content after withdrawal of secretion µg	Secretory protein µg	Percent of secretory protein in total gland protein
I	1	11.9	7.9	39.9
	2	10.8	7.3	39.8
II	3	7.7	7.2	48.3
	4	7.3	6.6	47.5
III	5	7.2	6.1	45.9
	6	7.8	6.8	46.6
IV	7	7.7	5.0	39.4
	8	7.1	5.2	42.3
V	9	14.0	3.6	20.4
	10	10.4	7.7	42.5

again accumulated in the gland lumen, a normal degree of filling being reached after about 24 h. Taking into account the values given in Table 1, the rate of secretory protein synthesis can thus be estimated to be in the range of 0.2–0.3 µg per cell in 24 h.

C. The Polytene Chromosomes of the Gland Cells

The salivary glands of *Chironomus* have aroused the interest of cytologists for almost a hundred years because of their giant chromosomes. The "cordons cylindriques" which Balbiani (1881) detected in the cell nuclei have since been studied and portrayed by many authors (see Beermann, 1962, for a survey of the older literature). It was Rambousek (1912) who first recognized their chromosome nature, but his conclusions, published in a remote journal and language, remained unknown to the scientific world. Heitz and Bauer (1933), working on the midge *Bibio*, and Painter (1933), on *Drosophila*, provided convincing evidence that these nuclear structures represent paired homologous chromosomes in the interphase state. This detection made it possible to correlate in detail the genetic maps of *Drosophila* to the sequence of chromomeres (bands) in its chromosomes, an analysis which has recently been refined to a precise cytological localization of adjacent complementation groups (Judd et al., 1972; Shannon et al., 1972). From these and other studies which include several different regions of the *Drosophila* genome, it seems inevitable to assume that one complementation group on the genetic map usually corresponds to one chromomere on the cytologic map and that the structural counterpart of a functional genetic unit is not substantially smaller than a chromomere (Hochman, 1973; Judd and Young, 1973; Lefevre, 1973; Sorsa et al., 1973).

The morphological organization of the polytene chromosome may, therefore, reflect its underlying genetic organization in a rather direct way (for a detailed discussion of this subject, the reader is referred to Beermann, 1972).

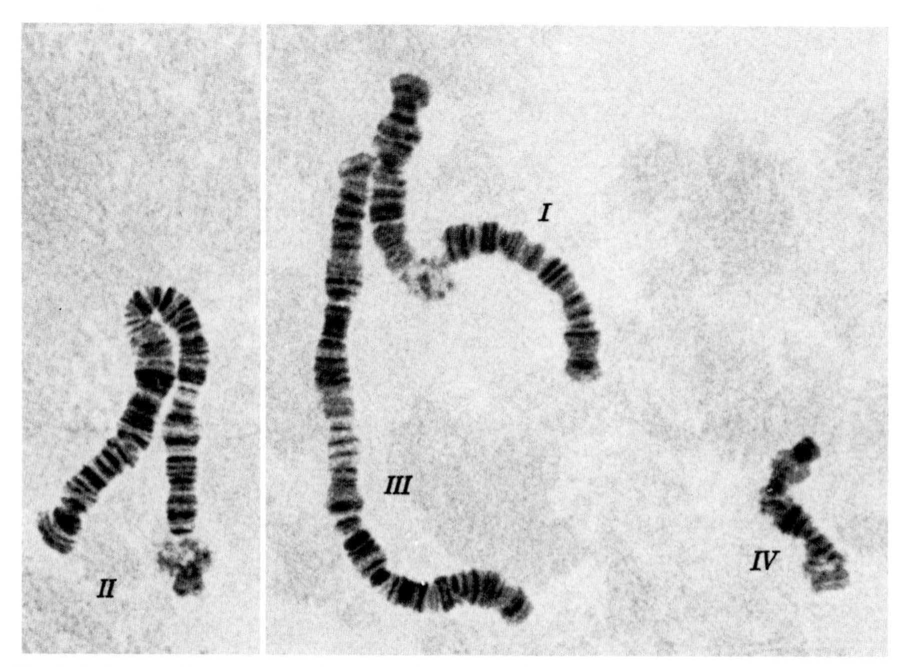

Fig. 7. Polytene chromosomes from a salivary gland cell of *C. tentans.* stained with aceto-orceine/aceto-carmine, phase contrast

Cells with polytene chromosomes provide substantial advantages for studies on cell differentiation. This holds true especially for the *Chironomus* and *Camptochironomus* salivary glands which reach the highest stages of polyteny known and have, therefore, been the favored material for investigations on the fine structure of giant chromosomes (Bauer, 1935; Bauer and Beermann, 1952; Beermann, 1952a, 1962; Beermann and Bahr, 1954). From these morphological studies originated the concept of differential gene activation as a basis for cell differentiation (Beermann, 1956a, 1959) which has greatly influenced the current view of development and has also stimulated experimental work on gene expression in the *Chironomus* gland and comparable systems.

Both *C. tentans* and *C. pallidivittatus* possess haploid sets of three long metacentric and one short telocentric chromosome. In the salivary glands these chromosomes perform up to 13 replication rounds and reach polyteny values of 8192 C (Beermann, 1952a; Daneholt and Edström, 1967). As the homologous chromosomes are somatically paired the nuclei contain only one set of polytene chromosomes (Fig. 7) which are thus each composed of 16384 chromatids. Not all nuclei of a gland, however, reach this high polyteny level, and in many larvae one or two replication rounds less are performed even by the largest nuclei. According to Beermann (1952a), the maximal length of the largest chromosome (no. I) is 270 μ, the maximal diameter 20 μ, and the total number of chromomeres or bands in all chromosomes about 1900. The size of the genome of *Chironomus* is 0.2 μμg (Daneholt and Edström, 1967), from which value a haploid DNA amount of 10^{-16} g is derived for an average chromomere. Beermann (1972) estimates the heaviest bands to contain about 10^5 base pairs and the finest bands below 5×10^3 base pairs of DNA per chromatid.

Besides the salivary glands, most of the other cell types of the larva also contain polytene chromosomes (Beermann, 1952a), but only in a few tissues do these reach a degree of polyteny sufficient for a detailed analysis of their structure. Where this is the case, the shape, arrangement and relative distances of the chromomeres in all chromosomes strictly coincide with the homologous regions in the salivary gland (Beermann, 1952a). The morphological organization of the gland chromosomes is thus not specific to this one tissue but rather an extraordinarily clear manifestation of a structure common to all cells of the organism.

C. tentans and *C. pallidivittatus* can be easily identified and distinguished by the morphology of their chromosomes. Though all chromosome segments except a few short sections can be homologized in both species, their arrangement is specific, all chromosome arms being characterized by one or several species-specific inversions (Beermann, 1955). The positions of the nucleolus organizers are also species-specific and provide prominent "landmarks:" chromosome II carries a nucleolus on its left arm in *C. tentans* and one on its right arm in *C. pallidivittatus*, while chromosome III carries a second nucleolus in *C. tentans*. Somatic pairing of the homologous segments usually occurs in hybrids of the two species. Beermann (1955) assumes that the differentiation of the caryotypes in the two species is a result of paracentric inversions also in the nonhomologized sections, though other mechanisms of chromosome evolution, e.g. duplications (cf. Keyl, 1964, 1965) or "minute rearrangements" (cf. Metz and Lawrence, 1938), cannot be excluded. All populations of *C. tentans* and *C. pallidivittatus* which have been investigated also showed intraspecific chromosome polymorphism due to paracentric inversions (Beermann, 1955, and pers. comm.).

Local modifications of the chromosome structure, the puffs (Poulson and Metz, 1938), have been observed in *Chironomus* by Balbiani as early as 1881. Beermann's studies (1952a) revealed that these structures arise from single chromomeres in which the homologous chromatids lose their tight lateral contact and expand to many times their original length, thus forming a local dispersal and swelling of the chromosome's shape. These morphological processes are concomitants of RNA synthesis, the degree of puffing and the transcription rate in a chromomere being directly and positively correlated (Pelling, 1959, 1964). Whether the puffing of a chromomere is a prerequisite or rather a consequence of RNA synthesis remains to be elucidated. It is interesting to note that the first noticeable event in newly induced puffs is the accumulation of nonhistone proteins, a process taking place even if RNA synthesis is experimentally inhibited (Berendes and Beermann, 1969). Nothing is known about these proteins, but it is tempting to speculate that some of them may be involved in regulating transcription of the sites at which they accumulate. Before this hypothesis can be tested it will be necessary to isolate and characterize such proteins. It has been reported that new fractions appear in the electrophoretic pattern of nuclear proteins from *Drosophila* salivary glands after the induction of specific puffs by hormone treatment or change of temperature (Helmsing and Berendes, 1971; Berendes, 1972; Helmsing, 1972). The quantity of these protein fractions, however, which form a major band in the total pattern of nuclear proteins (cf. Berendes, 1972), reders them unlikely for very specific

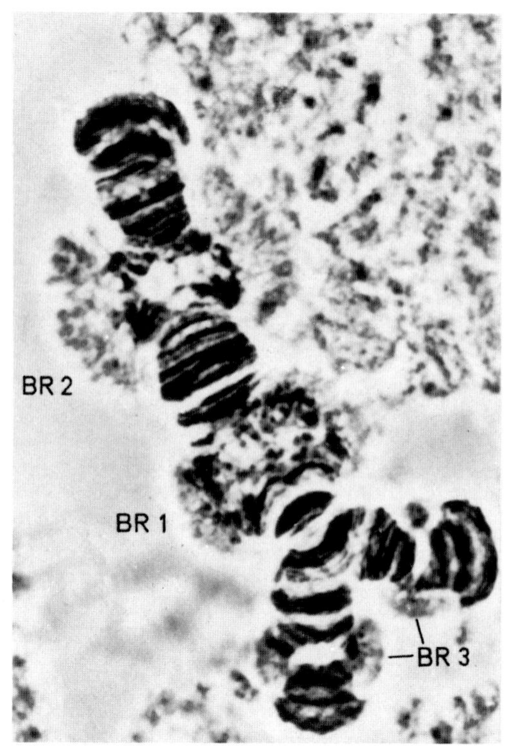

Fig. 8. Chromosome IV from the salivary gland of *C. pallidivittatus* with the predominant puffs (Balbiani rings) BR 1, BR 2, and BR 3. Aceto-orceine/aceto-carmine, phase contrast

functions at a few chromosome sites. Furthermore, recent analyses suggest that the newly arising protein fractions described by Helmsing and Berendes may not be genuine constituents of native *Drosophila* chromatin (Elgin and Boyd, 1975).

The pattern of puffed chromomeres, which includes only a small part of the total number of bands, has been shown to be cell-specific in *Chironomus* (Beermann, 1952a, 1961) as well as in other species (Mechelke, 1953, 1963; Becker, 1959; Berendes, 1967). In a given tissue certain puffs are restricted to a specific developmental stage. As development proceeds, several puffs in the salivary gland regress and their RNA synthesis ceases; on the other hand, new chromomeres become active under the influence of the molting hormone, ecdysone (Clever and Karlson, 1960; Panitz, 1960, 1964; Clever, 1961, 1962, 1963a, b). In the chromosomes of the *tentans*-gland, Pelling (1964) has mapped a total number of 275 puffs up to two-thirds of which can be active at the same time in a cell, corresponding to 10% of all chromomeres. Most of these puffs are, however, also present in other cell-types of the larva. The number includes chromomeres at all levels of puffing, from a slight disorder of the chromatids to most conspicuous structures, and a correspondingly wide range of transcription rates.

The most prominent puffs of the salivary gland nuclei are the three Balbiani rings in the short telocentric chromosome IV (Fig. 8). These structures arise from a maximal local unraveling of the chromatids (Bauer and Beermann, 1952;

Beermann, 1952a; Beermann and Bahr, 1954) and exhibit levels of RNA synthesis two or three orders of magnitude above those in average puffs (Pelling, 1964). They also accumulate remarkable amounts of nonhistone proteins. Balbiani rings are thus certainly a special class of puffs, but the difference appears to be a quantitative rather than a qualitative one. They originate from individual chromomeres via stages resembling puffs (Beermann, 1952a, 1962) and in certain cases can also regress to small puffs (Mechelke, 1953, 1963; Beermann, 1973).

By a morphological comparison of Balbiani ring loci in different states of activity, Beermann (1952a, 1955, 1962, 1973) has been able to identify the chromomeres from which the Balbiani rings 1 and 2 in the salivary gland of *Camptochironomus* originate. Balbiani ring 1 arises from a region homologous in both species which comprises five bands. The origin can be in both ends of the region, the first indication of puffing occurring in either of the bordering rather faint bands. The fully developed Balbiani ring then includes the three heavy medial chromomeres, though possibly only for mechanical reasons.

The loci of Balbiani rings 2 in *C. tentans* and *C. pallidivittatus* are probably homologous (Beermann, 1955; Lambert and Beermann, 1975). The formation of Balbiani ring 2 has been ascribed to a faint band in region 3 B of chromosome IV (Beermann, 1952a, 1973; Lambert and Beermann, 1975); recent studies on chromosomes with experimentally repressed Balbiani rings, however, have revealed a neighboring prominent double-band as the origin of Balbiani ring 2 (Beermann, pers. comm.). It seems, therefore, that the large Balbiani rings 1 and 2 both originate from more than one chromomere and that these chromomeres have a DNA content above that of average bands. A similar situation has been found to exist in a Balbiani ring of the Chironomid, *Smittia* (Bauer, 1957). In the salivary gland of the Chironomid, *Acricotopus* (see Baudisch, this vol., p. 137), all Balbiani rings formed probably result from the joint puffing of groups of two to three adjacent chromomeres that exhibit extraordinarily high DNA contents (Mechelke, 1953, 1959).

On the other hand, a Balbiani ring which is formed in the special cells of the *pallidivittatus*-gland (Balbiani ring 4) obviously originates from a single chromomere of average appearance (Beermann, 1961). Three other Balbiani rings, one regularly present in the Malpighian tubules and two others occasionally occurring in the salivary gland of *Camptochironomus* (Balbiani rings 5 and 6), arise from single chromomeres of not more than average size (Beermann, 1952a, 1959, 1973), though in the latter case two neighboring chromomeres of higher DNA content may become included as puffing proceeds (Beermann, 1973). Balbiani ring 3 which appears in a region homologous in both species (Beermann, 1955) has not yet been localized in a specific chromomere.

It is important to note that the Balbiani rings in the salivary gland are strictly cell-specific and never develop in any other cell type of *Camptochironomus* which can be analyzed, although Balbiani rings are actually formed in the Malpighian tubules by other chromosome loci. Their cell-specificity as well as their other properties led Beermann (1952a, 1956a, 1959) to presume that puffing indicates a genetic activity, and he also suggested that the Balbiani rings might be especially active gene loci, a view substantiated by the quantitative difference in RNA synthesis rates found between Balbiani rings and average puffs (Pelling, 1959, 1964).

This hypothesis implies a number of assumptions which can be tested experimentally. Cells developing one or several Balbiani rings along with ordinary puffs should synthesize specific proteins at a rate remarkably exceeding that of the majority of their proteins. The genes coding for this class of proteins should be located within the Balbiani rings. In particular, if the genetic correlations of chromomeres and complementation groups in *Drosophila* can be generalized, the number of these proteins in a given cell should be in the range of the number of chromomeres giving rise to the Balbiani rings. Evidence in support of these assumptions is provided by biochemical and cytogenetic work on the salivary gland of *Camptochironomus*.

IV. The Secretory Proteins of the Salivary Gland

A. General Properties

The protein secretion, the continuous synthesis of which accounts for more than 50% of the total glandular protein synthesis (see Sect. III.B.), is stored in the gland lumen as a viscous, ropy substance insoluble in aqueous buffer solutions even in the presence of high concentrations of urea or nonionic detergents. Its viscosity is considerably decreased when calcium ions are omitted from the medium. The secretion can be dissolved in formic acid and in the ionic detergent, sodium dodecyl sulfate (SDS). 50% formic acid or 1% SDS in water have been used to dissolve the secretion which flows out of the lumen of punctured glands in a calcium-free medium (Grossbach, 1969). Solubility is relatively low even under these conditions, the protein dissolved in 1% SDS precipitating when concentrated to about 1% protein.

In the secretory threads spun by the larvae the protein has become even less accessible to solvents. It is now insoluble in SDS solutions and formic acid, conceivably because of newly formed intermolecular crosslinks.

The denaturing conditions used to dissolve the secretion will, of course, destroy noncovalent bonds. Covalent linkages, however, are not broken by SDS (Hofmann and Harrison, 1963; Reynolds et al., 1967). Prolonged treatment with 50% formic acid can be expected to hydrolyse peptide bonds. Samples were, therefore, rapidly evaporated prior to further procedures in order to remove the acid.

Electrophoretic analysis of the secretion dissolved in 1% SDS has revealed several protein fractions, though the quality of the separation was seriously impeded by the low solubility of the samples. On layers of cellulose acetate, two heavy and two faint, faster migrating zones were obtained (Fig. 9), while electrophoresis on agarose-polyacrylamide resulted in separation into 6 bands (Fig. 10), the two fastest of which may correspond to the two fast bands on cellulose acetate. Only the most slowly migrating zone showed a slightly positive reaction when stained for polysaccharide by the periodic acid-Schiff reagent. While the PAS-reaction is not quantitative, this result certainly coincides with the low contents in the secretion of neutral sugars and hexosamine as determined by quantitative methods (see Sect. III.B).

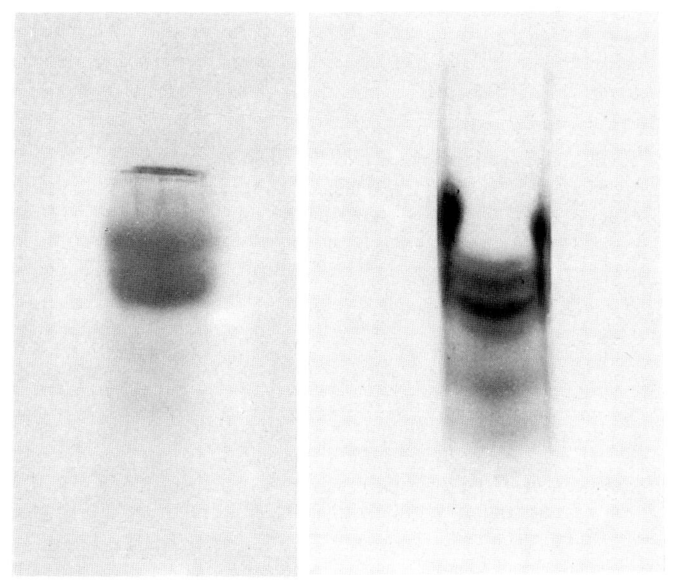

Fig. 9 Fig. 10

Fig. 9. Electrophoretic separation of the secretory proteins of *C. pallidivittatus* on cellulose acetate in Tris-EDTA-disodium-boric acid (pH 9.2) containing 1% sodium dodecyl sulfate. The protein zones were stained with amido black; cathode above. (From Grossbach, 1969)

Fig. 10. Electrophoretic separation of the secretory proteins of *C. pallidivittatus* on a gel of 1% acrylamide and 0.5% agarose in Tris-EDTA-disodium-boric acid (pH 9.2) with 1% sodium dodecyl sulfate. The gel column was sliced longitudinally and the center piece stained with amido black. The slowest protein zone formed a ring-shaped band and is visible only at the sides of the gel

The secretory proteins do not migrate into gels of polyacrylamide. They are even excluded from gels polymerized from solutions of 2.4% acrylamide and 0.1% methylenebisacrylamide that provide a rather open molecular sieve. This behavior suggests an unusually large molecular size. An estimation of the molecular weights of the secretory proteins has been attempted by thin-layer chromatography on cross-linked dextran (Sephadex G 200 superfine, Pharmacia). The migratory velocity of the secretory proteins in 1% SDS was compared to that of apoferritin (molecular weight 480000 daltons; Harrison, 1963) which, in contrast to most other proteins, does not dissociate into monomers during prolonged treatment in 1% SDS (Hofmann and Harrison, 1963). Apoferritin was run in parallel with other proteins in buffer devoid of the detergent. Dextran Blue 2000 (Pharmacia, molecular weight 2×10^6 daltons) was used as a reference substance. The migration of apoferritin relative to Dextran Blue was not changed in the presence of SDS, and the relative distances reached by the proteins in both series of experiments could, therefore, be directly compared.

The secretory proteins migrated as one fraction with a velocity well above that of apoferritin and equal to that of the reference substance (Fig. 11). Their sizes should, therefore, exceed that of apoferritin (i.e. 4.8×10^5 daltons) and should be

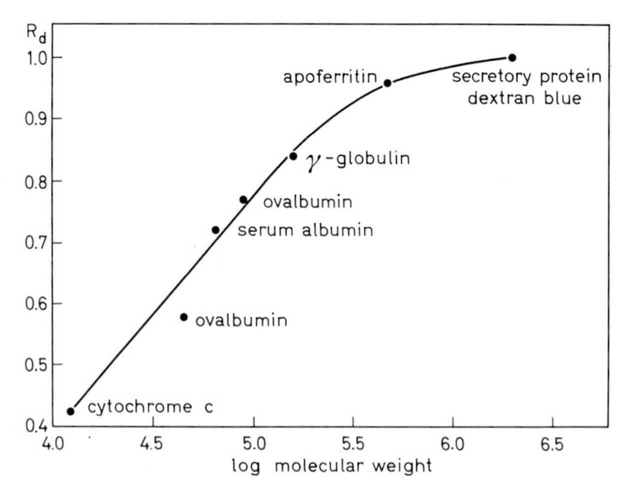

Fig. 11. Chromatography of *C. pallidivittatus* secretory protein on a thin-layer of Sephadex G 200 superfine, as compared to standard proteins. (Redrawn from Grossbach, 1969)

comparable to those of molecules which are excluded from the pore system of the Sephadex particles, i.e. about 8×10^5 daltons. Exact estimations are, however, impossible, as the relation of relative migration and logarithm of molecular weight is not linear in the upper range of the curve, a phenomenon also reported earlier by other authors (cf. Andrews, 1964, 1965).

An evaluation of the molecular weights of the secretory proteins by chromatography on materials with a lower degree of crosslinking and a correspondingly decreased molecular sieving action has not been undertaken because of the limited amounts of material that have been available. Molecular weight determinations by ultracentrifugation seem difficult to perform since the proteins in solution bind to a great number of dodecylsulfate ions (Reynolds et al., 1967).

The electrophoretic separation of the secretory proteins into a number of reproducible fractions, despite their large size, excludes the possibility that we are dealing with randomly formed aggregates of smaller molecules. The proteins are, on the contrary, assembled in a regular way from smaller subunits covalently linked by intermolecular disulfide bonds (see Sect. IV.C).

B. Amino Acid Composition

The salivary secretion, isolated from the gland lumina after discarding the lobe of special cells in each gland, has been subjected, without prior fractionation, to complete acid hydrolysis in order to determine its amino acid composition (Grossbach, 1969). This was found to be very similar in *C. tentans* and *C. pallidivittatus* (Tables 2 and 3). The high content of the basic amino acids lysine and arginine which together account for about 25% of the amino acid residues is the most conspicuous property of this group of proteins. There seem to be no similarities to the amino acid compositions of the silk proteins of other insects, which usually exhibit a quantitative preponderance of a few amino acids and low

Table 2. Amino acid composition of the secretion from the main part of salivary glands from *C. tentans* in percent of amino acid residues detected. Each value represents the mean of two analyses of an acid hydrolysate (from Grossbach, 1969)

Amino acid	Sample no.				
	I	II	III	IV	V
Gly	4.51	3.85	4.19	3.97	4.88
Ala	10.54	8.62	8,93	9.53	9.50
Val	2.84	3.01	3.44	3.20	3.12
Leu	1.88	2.80	2.44	1.42	2.01
Ile	1.96	2.24	2.36	1.42	2.11
Pro	2.49	2.69	4.34	1.16	5.91
Phe	—	—	—	—	—
Tyr	—	—	—	—	—
$MetO_2$	7.42	1.61	4.47	3.68	2.08
Ser	9.39	16.79	11.88	14.67	11.73
Thr	2.96	2.92	1.96	3.27	3.51
$CySO_3H$	12.88	13.58	12.99	13.43	9.03
Asp	6.18	4.78	4.93	4.25	6.62
Glu	12.79	11.62	12.82	10.63	14.08
Lys	11.09	12.58	11.84	15.57	13.87
Arg	13.02	12.54	13.36	12.26	11.48
His	—	0.23	—	1.49	—

Table 3. Amino acid composition of the secretion from the main part of salivary glands from *C. pallidivittatus* in percent of amino acid residues detected. Each value represents the mean of two analyses of an acid hydrolysate (from Grossbach, 1969)

Amino acid	Sample no.				
	I	II	III	IV	V
Gly	3.62	5.14	4.99	5.55	6.04
Ala	9.20	9.49	10.13	9.72	7.58
Val	2.22	4.07	2.88	2.72	3.66
Leu	1.52	1.79	1.78	1.85	1.71
Ile	1.32	1.54	1.97	1.36	1.82
Pro	3.20	4.10	2.33	0.89	3.62
Phe	—	—	—	—	—
Tyr	—	—	—	—	—
$MetO_2$	3.67	5.43	{ 19.13	{ 18.14	2.84
Ser	14.76	10.89			14.58
Thr	3.29	2.70	2.92	2.69	3.73
$CySO_3H$	9.95	10.08	11.78	13.13	10.67
Asp	5.68	5.57	4.49	7.73	6.15
Glu	11.48	12.56	10.61	12.68	11.52
Lys	17.59	15.41	15.84	13.78	14.11
Arg	10.48	10.72	11.10	11.35	10.90
His	1.68	0.53	—	—	—

percentages of the others (cf. Lucas et al., 1958, 1960; Seifter and Gallop, 1966; Suzuki, this vol., p. 9).

The deviations between different samples of secretion (Tables 2 and 3) are well above the inherent error of the method (see Grossbach, 1969, for technical details) and should, therefore, reflect actual variations in the relative quantities of the different proteins in the sample. A biological variation between different animal cultures and stocks of the relative quantities of the subunits composing these proteins (see below) has frequently been observed.

C. The Subunits of the Secretory Proteins

1. The Secretory Subunits in Different Cell Types of the Salivary Gland

The large molecular sizes which the secretory proteins exhibit under denaturating conditions suggest their being composed of subunits that are assembled by intermolecular covalent bonds. It has actually been found that interchain disulfide linkages play an important role in establishing the structure of the secretory proteins. Reduction of these linkages by mercaptoethanol or Cleland's reagent resulted in a remarkable decrease in size of the components of the secretion (Grossbach, 1969). At the same time, solubility was enhanced.

Prior to further treatment the SH-groups formed by reduction were alkylated by recrystallized iodine acetamide to avoid random formation of new disulfide bonds. The reduced and alkylated subunits of the gland secretion were separated into a number of fractions by polyacrylamide gel electrophoresis in the presence of SDS or high concentrations of urea. Best separations were achieved in a discontinuous buffer-system at low pH in the presence of 8 M-urea. Under these conditions, the *pallidivittatus*-secretion revealed six major fractions of subunits and a few additional faint bands (Fig. 12). Electrophoresis for longer periods of time and in gels of different acrylamide concentration did not change the number of bands. The slowly migrating fraction no. 1, in particular, could not be subdivided into several bands in gels of lower acrylamide concentration. Because of the preceding extensive reduction and alkylation and the electrophoretic conditions used for separation, aggregation of molecules is very unlikely to occur, and the fractions obtained probably represent single polypeptides that become separated because they differ in their primary structure.

The gland secretion of *C. pallidivittatus* is thus apparently composed of six quantitatively predominant and some less abundant constituents, presumably polypeptide chains, assembled by disulfide bonds to form the complex proteins described above. A certain protein thereof could be composed of several polypeptides of one type or, alternatively, of different types of polypeptides. The occurrence of two groups of fractions, one slowly migrating and heavily stained and one faster, less intensely stained, in electropherograms of the protein complexes as well as in electropherograms of the subunits, seems to support the former view, but direct evidence is not available at present.

A comparative analysis of the secretion produced by the cells in different regions of the salivary gland has revealed the presence of two different cell types

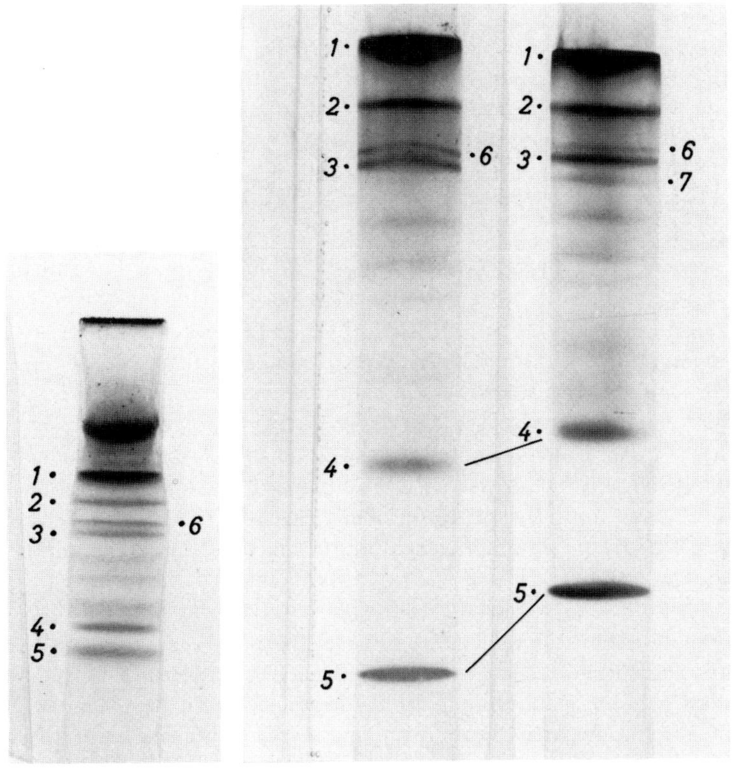

Fig. 12 Fig. 13

Fig. 12. Electrophoretic separation of the secretory polypeptides from *C. pallidivittatus* salivary glands. The secretion from 20 glands was reduced with 1 M mercaptoethanol in 8 M urea and alkylated with iodine acetamide. Disc-electrophoresis at pH 2.3 was performed in 7% polyacrylamide in the presence of 8 M urea, and the protein zones were stained with amido black; gel diameter 5 mm

Fig. 13. Separation by disc-electrophoresis of the reduced and alkylated secretion obtained from the main part *(left)* and the special lobes *(right)* of *C. pallidivittatus* salivary glands. Gels were overloaded with protein sample to prove the absence *(left)* and the presence *(right)* of secretory fraction no. 7 in the secretion of the respective gland cells. Electrophoretic conditions as described for Figure 12. (From Grossbach, 1969)

which synthesize similar, but not identical patterns of secretion subunits (Grossbach, 1966, 1968). Beermann (1961) has described the small group of "special" cells in the *pallidivittatus*-gland that differ from the other cells by morphological features and by the production of microscopically visible secretory granules. A biochemical differentiation underlying these morphological differences was suggested by Beermann (1961); it has been substantiated in terms of protein synthesis by the finding that the special cells synthesize another component, fraction no. 7, in addition to the six secretory constituents (Fig. 13).

For technical reasons it cannot be excluded that one or more of the normally present polypeptides is lacking in the secretion of the special cells, as it is impossible to avoid contamination by the secretion from the other gland cells when collecting the product of this region. The majority of components should, however, also be synthesized in the special cells.

The two types of cells constituting the salivary gland may hence be envisaged as having very similar functions and representing two very similar, but not identical differentiated states. Comparative studies of such related cell types may be expected to prove especially useful for a better understanding of the mechanisms by which cells differentiate.

In *C. pallidivittatus* stocks of different geographical origin, the same pattern of secretory polypeptides has been found. Mutants showing slight changes in the electrophoretic mobility of certain fractions might, however, have escaped attention, as the migration distances in disc electrophoresis are not strictly reproducible from gel to gel. Small variations in the relative distance of fractions 3 and 6 from each other have been repeatedly observed in electropherograms from single glands as well as in those from many pooled glands. Multiple bands, which would be expected if different genetic types of fractions 3 and 6 were present in the material, have never been found. Therefore, these variations are considered due to uncontrollable parameters of the electrophoretic techniques, and the upper fraction was designated no. 6, the lower one no. 3 in all cases.

A certain degree of variation in the relative amounts of the secretory components between different cultures of animals has been observed even after continued inbreeding for many generations. Such variations are considerably less accentuated between sister animals from the same culture vial. The two glands of one larva, when analyzed separately by microelectrophoresis, regularly exhibited patterns that were very similar also in quantitative terms. The observed variations should, therefore, reflect actual differences in secretory protein synthesis, rather than being due to technical errors.

The qualitative composition of the salivary secretion of *C. tentans* is more variable than that of *C. pallidivittatus*. While the pattern is in general very similar to that of the neighbor species, qualitative differences have been detected that have been proven useful for the cytogenetic localization of secretory genes. Inbred stocks of *C. tentans* derived from material collected at Plön (Germany) have been found to be devoid of secretory component no. 6, while fractions no. 1–5 were indistinguishable from those of the neighbor species (Fig. 14). The group of cells homologous to the special cells of *C. pallidivittatus* produces a secretion that exhibits the same complement of fractions no. 1–5. While the possibility must be considered that one or more of these polypeptides originate from an adjacent gland region rather than from the special cells themselves, there is no doubt that both cell types in the *tentans*-gland have at least some of the five fractions in common. Electrophoretically separable components specific for the special cells have not been detected in *C. tentans*; in particular, no secretion subunit corresponding to the fraction no. 7 of *C. pallidivittatus* has been found. There are, however, morphological features characterizing the secretion of the special cells of *C. tentans*. These will be described below.

The production of the secretory polypeptides, no. 6 and 7, in the salivary gland of *C. pallidivittatus*, has provided the possibility to locate in the polytene chromosomes genes of gland-specific activity. Localization of the genes for polypeptides no. 6 and 7 was achieved by interspecific hybridization with *C. tentans*

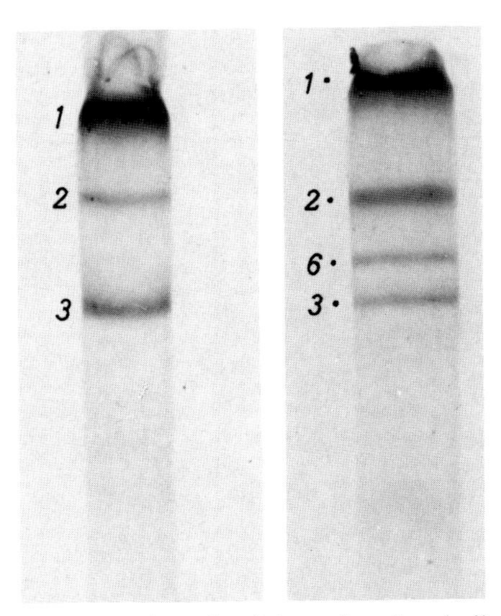

Fig. 14. Separation by disc-electrophoresis of the reduced and alkylated secretion from a single gland of *C. tentans (left)* and *C. pallidivittatus (right)*. While fractions no. 1–3 and no.4–5 (not shown) run in parallel in the two species, the *tentans* strain is devoid of fraction no. 6. Electrophoretic conditions as described for Figure 12; gel diameters 0.7 mm. (From Grossbach, 1969)

stocks that lack these secretory components. The results substantiate the interpretation of chromosome puffing as a morphological manifestation of genetic activity. They will be discussed in Section VI together with other evidence supporting the same view.

Stocks of *C. tentans* collected at different times from several geographically distinct populations exhibited a secretion pattern corresponding to fractions no. 1–5 of *C. pallidivittatus*. This pattern and, in particular, the absence of fraction no. 6 was, therefore, considered to be typical of *C. tentans* (Grossbach, 1968, 1969, 1973). More recently, a stock has been isolated from material collected at Plön which in addition to no. 1–5 produces a secretory component that upon electrophoresis migrates very similarly to the *pallidivittatus*-fraction no. 6 and may be identical to it (Fig. 15). The absence of this polypeptide in many *tentans*-populations probably, therefore, does not qualify as a species character. A polymorphism with regard to the presence or absence of a secretory subunit that may correspond to fraction no. 6 in our material has also been found by Pankow (1973) in mass cultures of *C. tentans*.

From a population of *C. tentans* in southern Finland, a stock carrying two different alleles for fraction no. 5 has been isolated. Heterozygous animals synthesize equal amounts of the two electrophoretically different variants (Fig. 16). In a stock of *C. tentans* from Canada that was kindly provided by Dr. H. Tichy, a pattern of secretory subunits has been found in which fraction no. 3 exhibits an electrophoretic mobility different from that in the material from several locations in Europe (Fig. 17). It seems that the qualitative composition of the *tentans*-secretion is more variable than is the case with *C. pallidivittatus*. Quantitative variations in the relative amounts of the secretory subunits occur in cultures of *C. tentans* to a similar

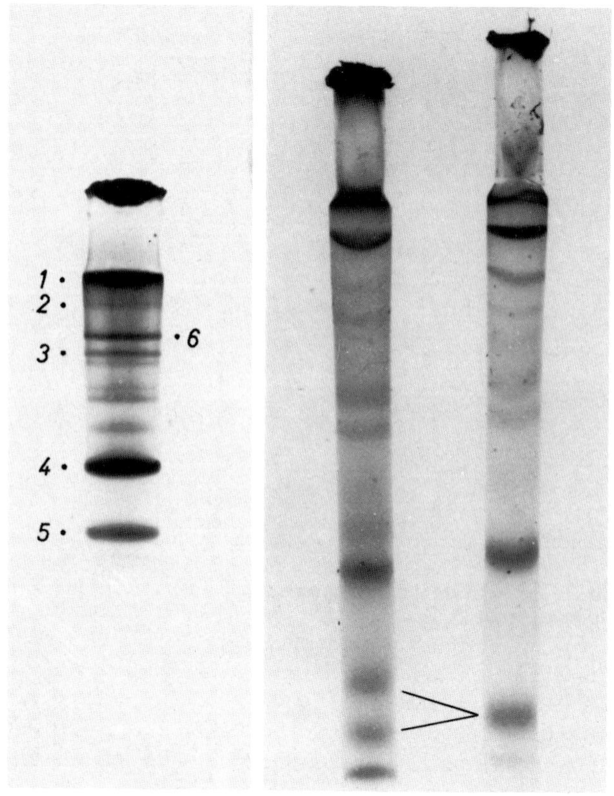

Fig. 15 Fig. 16

Fig. 15. Separation by disc-electrophoresis of the reduced and alkylated secretion from a *C. tentans* strain exhibiting a polypeptide fraction in the position of no. 6. Electrophoretic conditions as described for Figure 12; gel diameter 5 mm

Fig. 16. Separation by disc-electrophoresis of the reduced and alkylated secretion from single glands of *C. tentans*. *Left:* larva from a strain collected in Finland exhibiting two electrophoretically different variants of fraction no. 5 *(arrow)*. *Right:* larva from a strain collected at Plön (Germany) exhibiting standard fraction no. 5. Electrophoretic conditions as described for Figure 12; gel diameters 0.7 mm

extent as in *C. pallidivittatus*. Systematic studies have, however, not been undertaken.

It is interesting to note that the special cells of the *tentans*-gland, besides exhibiting morphological peculiarities (Beermann, 1961), also produce a secretory component that is found neither in the other gland regions nor in the homologous cells of *C. pallidivittatus*. This component has not yet been characterized by physical methods. Its presence is evident from the structure of the secretion of the special cells upon several hours of fixation in ethanol-acetic acid (3:1). Under these conditions, the secretion is denatured as long fibers, clearly separated from the amorphous, glassy secretion of the other cells and easily identified microscopically

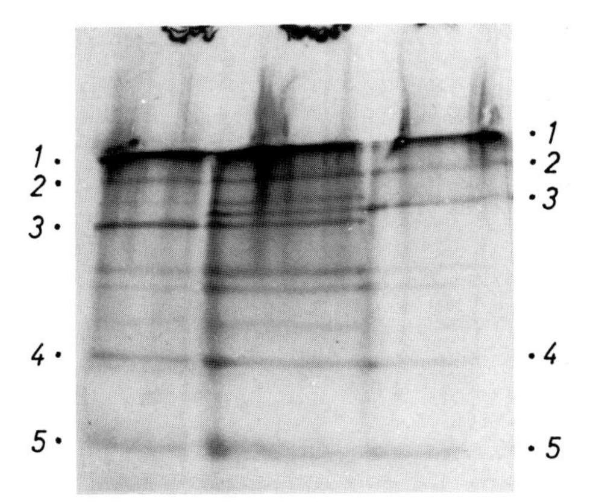

Fig. 17. Separation by disc-electrophoresis of the reduced and alkylated secretion of *C. tentans*. *Left*: secretion from 4 glands of a strain from Canada. *Right*: secretion from 4 glands of a strain from Germany. *Middle*: secretions from two glands of each strain mixed and separated together to demonstrate different electrophoretic mobilities of fractions no. 3 in the two strains. Electrophoresis was performed as described for Figure 12 on a slab gel of 0.3 mm diameter

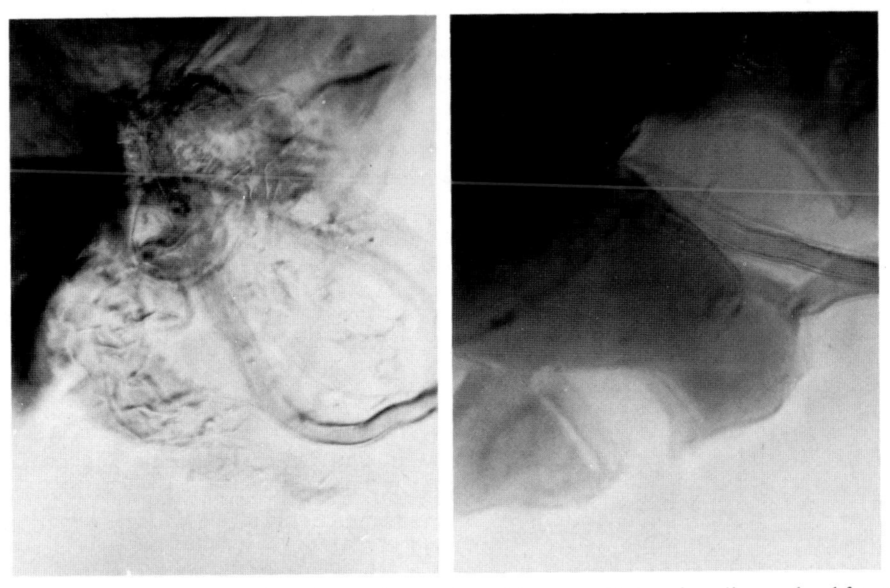

Fig. 18. Acetic acid-ethanol precipitated secretion of the special lobe of a salivary gland from *C. tentans (left)* and *C. pallidivittatus (right)*. The cells have been removed. For further explanation see text. *Left*: ×210; *right*: ×180, reduced to 80%. (From Grossbach, 1969)

(Fig. 18). In the special cells of *C. pallidivittatus*, a similar component is obviously lacking, their secretion showing the same precipitation behaviour as that of the adjacent cells.

Nothing is known at present about the nature of a secretory component which seems to induce this deviating behavior of the entire secretion of these cells. The absence from the electropherograms of protein fraction(s) specific for the special cells intimates that it is not a major component of protein nature. While it seems difficult to establish the chemical nature and actual role of this secretory constituent, the morphologically manifest denaturating behavior of the secretion as a whole can be considered as a species-specific character. In crosses with *C. pallidivittatus*, the inheritance of this character has been followed by cytogenetic techniques and the gene for the underlying secretory component was found to be situated in a section of the small chromosome IV (Grossbach, 1969).

2. *Molecular Size of the Secretory Subunits*

The molecular size of fractions no. 2–6 has been approximately estimated from their retardation during electrophoretic migration through a molecular sieve of polyacrylamide in the presence of sodium dodecyl sulfate (SDS). The secretion was reduced with Cleland's reagent, alkylated with iodine acetamide and subsequently separated by disc electrophoresis at pH 2.3 in 8 M urea. The fractions were identified by staining with coomassie brilliant blue and isolated as gel slices which were then placed in a small sample of 1% sodium dodecyl sulfate on top of a column of 7% polyacrylamide and subjected to electrophoresis in a buffer containing 0.1% of the detergent. Under these conditions, polypeptides bind 1.4 g of sodium dodecyl sulfate per gram of protein (Reynolds and Tanford, 1970a, b) and, therefore, exhibit a constant ratio of negative charge to unit mass. Their relative electrophoretic mobilities within a gel consequently depend on mass rather than charge differences and can be used to estimate molecular weights (Shapiro et al., 1967; Weber and Osborn, 1969, 1975; Laemmli, 1970). A calibration curve was obtained by means of a number of reduced and alkylated polypeptides of known molecular weights migrating under identical conditions.

From the group of fractions no. 2, 3, and 6, three protein bands were obtained in dodecyl sulfate electrophoresis. Two of them exhibited relative mobilities of polypeptides in the range of 140000–160000 daltons and the third in the range of 55000–60000 daltons. Fraction no. 2 seems to represent the largest of these molecules. For fraction no. 4 a molecular weight of 30000–40000 daltons was obtained in repeated experiments, while fraction no. 5 appears to be still smaller and similar in size to myoglobin from sperm whale (17800) which was used as a reference in all experiments. The quantitatively less prominent constituents of the secretion that band between fractions no. 3 and 4 upon electrophoresis at pH 2.3 (Fig. 12), were isolated in one gel slice and in SDS-electrophoresis formed three major zones with apparent molecular weights of about 40000, 50000, and 55000–60000 daltons, respectively.

These values have to be considered with some care. The secretory proteins contain about 25% of lysine plus arginine and thus have extraordinarily high net charges which may influence their relative mobilities even if the normal amount of sodium dodecyl sulfate is bound. Histones which contain similarly high amounts of positive charge are known to exhibit low mobilities in dodecyl sulfate electrophoresis (Panyim and Chalkley, 1971), and their apparent molecular weight determined by this method is therefore higher than the actual one. For histone Fl (molecular weight 21000 daltons) an apparent molecular weight of 35000 daltons has been obtained by this method (Weber and Osborn, 1975).

The influence of intrinsic positive net charge on the mobility of a polypeptide in SDS-electrophoresis can be expected to be relatively high in an open network of polyacrylamide. It is less important in a tight gel of high acrylamide concentration where the electrophoretic mobility is primarily determined by the molecular weight of the polypeptide. While the apparent molecular weights in SDS-electrophoresis of typical polypeptides without intrinsic net charge are not affected by the acrylamide concentration, the relative mobilities of protein-SDS complexes carrying high intrinsic net charges may vary when gels of different porosities are used. The apparent molecular weight obtained for such a molecule can be expected to be too high in gels of low acrylamide concentration, and to approach the actual value as the gel concentration is increased. For fractions no. 2–6 of the *Chironomus* secretion, apparent molecular weights considerably higher than in 7% polyacrylamide have been obtained in gels of 4% and 5% concentration (Grossbach, 1971, and unpubl.), suggesting a migration behavior influenced by intrinsic positive net charge. While the values obtained in 7% polyacrylamide are certainly more reliable than those observed in gels of higher porosity, it cannot be excluded that the actual molecular weights of fractions no. 2–6 are in fact somewhat lower.

The size range of the secretory constituents is thus comparable to the molecular weights of the subunits of other proteins (cf. Klotz and Darnall, 1969), and it seems reasonable to assume that fractions no. 2–6 as well as the less prominent fractions represent single polypeptide chains which upon synthesis become assembled by interchain disulfide bonds in an ordered way to form a number of different proteins of higher molecular weight.

The nature and molecular weight of secretory fraction no. 1, on the other hand, is still questionable. This subunit hardly migrates into gels of 5% or more polyacrylamide and shows a very low electrophoretic mobility even in gels of high porosity (Fig. 19). Its apparent molecular weight as determined by SDS-electrophoresis in 3.3% polyacrylamide is in the range of 5×10^5 daltons (Fig. 20). This value must be considered an approximation because of the low mobility and because no reference proteins of similar molecular weight are available. A high intrinsic net charge, if present in this molecule, should have little influence on the electrophoretic mobility under the conditions used, where the molecular sieving action of the gel is the prevailing factor.

The remarkably low mobility of fraction no. 1 is probably not due to a carbohydrate moiety. Glycoproteins are known to exhibit atypical mobilities in SDS-electrophoresis (cf. Weber and Osborn, 1975), but the carbohydrate content of the secretion has been shown to be below 2% (see Sect. IV.A) and can thus be expected not to impair the relative mobility of the molecule. It seems, therefore, reasonable to assume that the low mobility observed for fraction no. 1 actually indicates a molecular weight in the range of 5×10^5 daltons.

Attempts to diminish the size of this secretory component by boiling in urea-mercaptoethanol and SDS-mercaptoethanol solutions have not been successful. It has, therefore, been suggested that fraction no. 1 may either be a single polypeptide chain with a molecular weight of approximately 5×10^5 daltons or, alternatively,

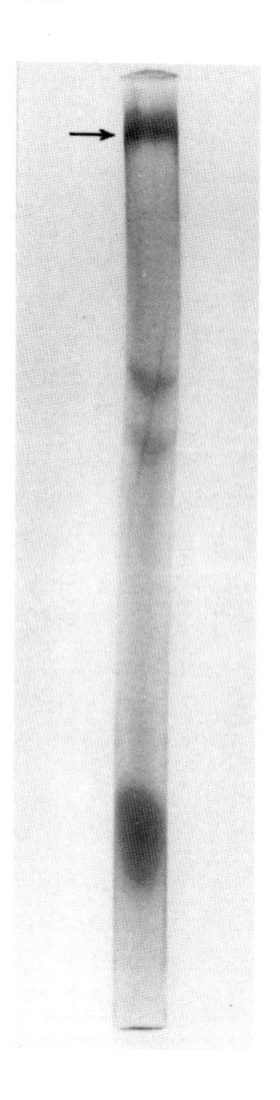

Fig. 19. Electrophoresis of secretory fraction no. 1 *(arrow)* in 3.3% polyacrylamide in the presence of 0.1% sodium dodecyl sulfate. Chymotrypsinogen was used as a reference protein and migrated near the front

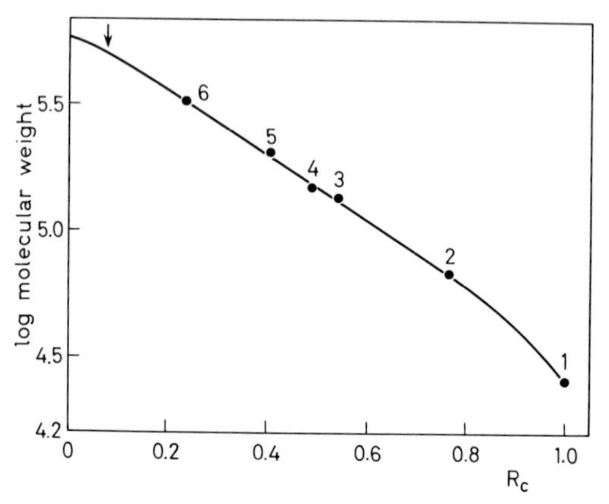

Fig. 20. Determination of the molecular weight of secretory fraction no. 1 from *C. tentans* by sodium dodecyl sulfate electrophoresis. The relative electrophoretic mobility in phosphate buffer—0.1% SDS (Weber and Osborn, 1969) on 3.3% polyacrylamide was compared to the relative electrophoretic mobilities of proteins with known molecular weights: Chymotrypsinogen (1), bovine serum albumin (2), dimer of bovine serum albumin (3), γ-globulin (4), trimer of bovine serum albumin (5), fibrinogen (6). The relative electrophoretic mobility of secretory fraction no. 1 *(arrow)* suggested a molecular weight of about 5×10^5 daltons. (From Grossbach, 1973)

may consist of several subunits bound by crosslinks of less common type such as ε-amino peptide or ester bonds (Grossbach, 1971, 1973).

A single polypeptide chain of 5×10^5 daltons does not appear inconceivable. Several structural proteins are known to comprise subunits of remarkable molecular weight (Seifter and Gallop, 1966; Klotz and Darnall, 1969). The major polypeptide of silk worm fibroin, a protein with a function comparable to that of the *Chironomus* secretion, has a molecular weight of 3×10^5 daltons (Tashiro et al., 1972). Subunits of 5×10^5 daltons could be imagined to form the core structure of the protein thread spun by the *Chironomus* larva.

The presence or absence of interchain cross-links, on the other hand, does not seem easy to establish. If present in small numbers per molecule, such bonds may escape detection unless high amounts of secretion can be analyzed by specific methods. Attempts to demonstrate the presence or absence of interchain cross-links in indirect ways have so far not yielded unequivocal results. Sasse (unpubl.) exposed *Chironomus* larvae for three weeks to the lathyrogen β-aminopropionitrile which is known to irreversibly inhibit lysyl oxidase, the enzyme converting ε-amino groups of lysine to aldehydes in collagen cross-link formation (Siegel et al., 1970; Deshmukh et al., 1971; Layman et al., 1972; Piez, 1968). The pattern of newly synthesized secretory components was not altered after this treatment and the apparent size of fraction no. 1, in particular, remained unchanged. However, it has not been established that the agent actually entered the glands. Furthermore, it can be imagined that insects may possess a lysyl oxidase not sensitive to β-aminopropionitrile.

Rydlander and Edström (1975) have reported that the size of the high molecular weight fraction as determined by agarose gel filtration is diminished to about 40000–60000 daltons when the secretion is isolated from the glands into a medium containing glycine ethyl ester, an inhibitor of interchain cross-link formation between glutamine side chains and lysine ε-amino groups in fibrin (Lorand and Jacobsen, 1964; Pisano et al., 1969). This result suggests that covalent cross-links of the type occurring in fibrin polymerization are not native constituents of the secretory proteins in the gland lumen, but artifacts formed during the isolation procedure. An enzyme analogous to the transglutamidase involved in fibrin cross-linking would be expected to mediate the reaction which may normally occur outside the gland between parallely arranged polypeptides in the silk thread.

The secretory components isolated from the gland lumen in the absence of a cross-link inhibitor would thus appear to be randomly formed artifacts. It is difficult to see how such randomly arising artifacts could yield the qualitatively and quantitatively reproducible patterns of discrete fractions which are actually achieved in electrophoresis of both reduced and unreduced secretion samples. That the formation of some of these artifacts could be species-specific and be inherited as single Mendelian factors (see Sect. VI.B) appears very unlikely.

While the evidence provided by Rydlander and Edström (1975) seems indisputable, we have been unable to achieve corresponding results. Secretion isolated from *pallidivittatus*-glands in 0.1 M glycine ethyl ester was subsequently reduced, alkylated and separated by gel electrophoresis at pH 2.3 in the presence of 8 M urea. The pattern of fractions obtained exactly resembled the pattern previously described (cf. Fig. 21), and the position of fraction no. 1, in particular,

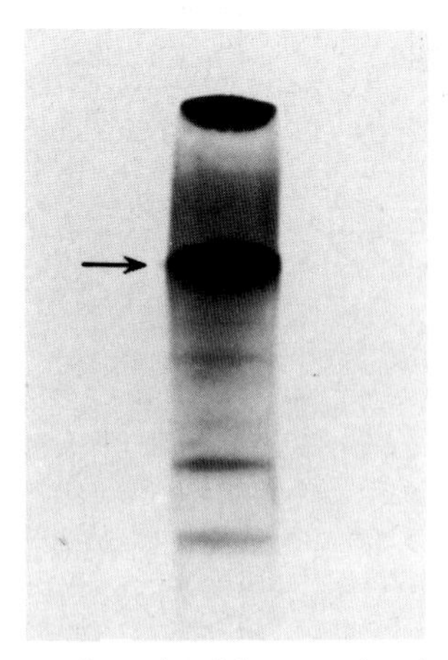

Fig. 21. Separation by disc-electrophoresis of the secretion from *C. tentans* isolated in 0.1 M glycine ethyl ester and subsequently reduced and alkylated. Electrophoretic conditions as described for Figure 12; note unchanged position of fraction no. 1 *(arrow)*

was not changed. A direct demonstration of native or artifact cross-links in the reduced secretion must thus be awaited before the actual molecular weights of its constituting polypeptides can be established.

The molecular weight of the polypeptide(s) of fraction no. 1 is of particular interest. Should it turn out to be in the size range apparent from SDS-electrophoresis (i.e. 5×10^5 daltons), then this would imply interesting consequences for the organization of Balbiani ring 2 and its transcript (Daneholt and Hosick, 1973a, b; Grossbach, 1973; Daneholt et al., 1975; Rydlander and Edström, 1975). These aspects will be discussed in Section VII.

D. Protein Synthesis in the Salivary Gland

It is evident from the abundant production of protein threads at all stages of larval development that the gland cells synthesize large amounts of secretory proteins. A *Chironomus* larva can produce a new net for food filtration every two minutes over a long period of time (Walshe, 1947), and it can also spin a new case frequently. The amount of secretion stored in the gland lumen was found to be equivalent to 65–94% of the gland cells' protein content (Table 1), and the rate of secretory protein synthesis in fully grown glands of the late fourth larval instar was estimated to be about 0.2–0.3 μg per cell in 24 h (see Sect. III.B). This amount is equivalent to $1–2 \times 10^{11}$ molecules of fraction no. 1 (or more, if the molecular weight is actually less than 5×10^5 daltons) and about 10^{11}–10^{12} molecules of each of the other polypeptides.

The synthesis of the secretory polypeptides as compared to other gland proteins has been studied in glands kept in an in vitro medium supplied with a protein hydrolysate (Grossbach, 1972, 1973). Lysine and arginine were labeled with ^{14}C, and the incorporation of ^{14}C into gland proteins was documented in autoradiographs prepared of the gels upon electrophoresis. After 15 min of incorporation,

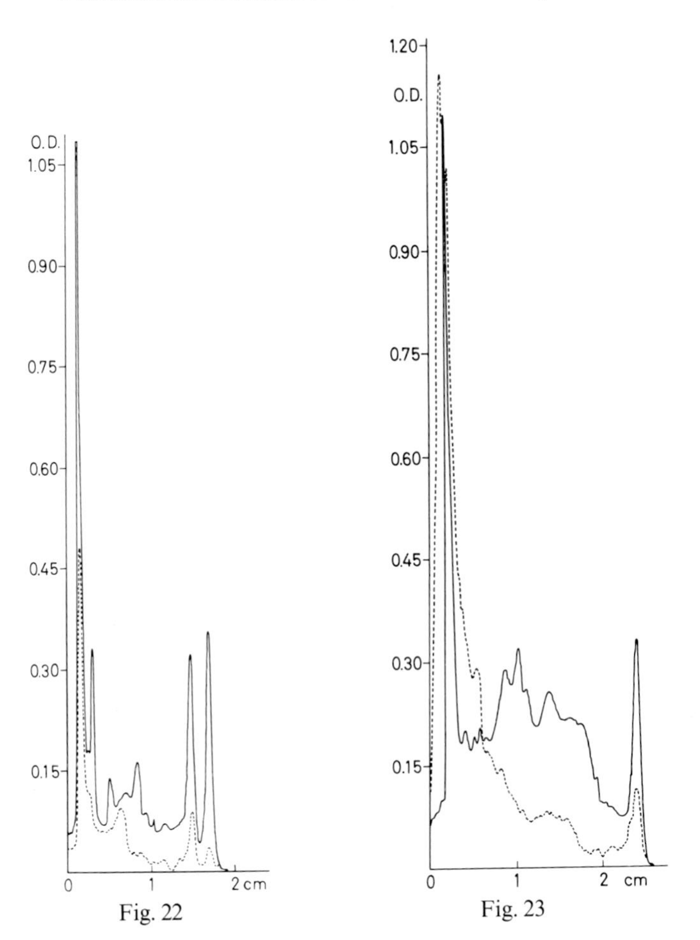

Fig. 22

Fig. 23

Fig. 22. Electrophoresis of the secretory polypeptides synthesized by salivary glands in organ culture. 16 salivary glands of *C. pallidivittatus* were isolated, washed in Ringer solution and incubated in a medium containing ^{14}C-arginine (15.6 μCi/ml, spec. act. 0.324 Ci/mM) and ^{14}C-lysine (15.6 μCi/ml, spec. act. 0.318 Ci/mM). After 30 min the glands were carefully washed in Ringer solution and the secretion was isolated from the gland lumina. It was then reduced, alkylated and subjected to electrophoresis as described for Figure 12. The optical density of the protein zones was recorded with a densitometer *(solid line)*, and the gel was then cut lengthwise, dried to a thin film, and exposed for 5 days to an X-ray film. The density of the radioautogram was recorded densitometrically *(dashed line)* (From Grossbach, 1973)

Fig. 23. Electrophoresis of the proteins from whole glands. 16 glands of *C. pallidivittatus* were isolated and incubated in organ culture as described for Figure 22. After 30 min they were carefully washed in Ringer solution and homogenized in 8 M urea and 1 M mercaptoethanol. The following procedures were performed as described for Figure 22. *Solid line:* optical density of protein zones stained with amido black; *Dashed line:* density of the gel autoradiogram. (From Grossbach, 1973)

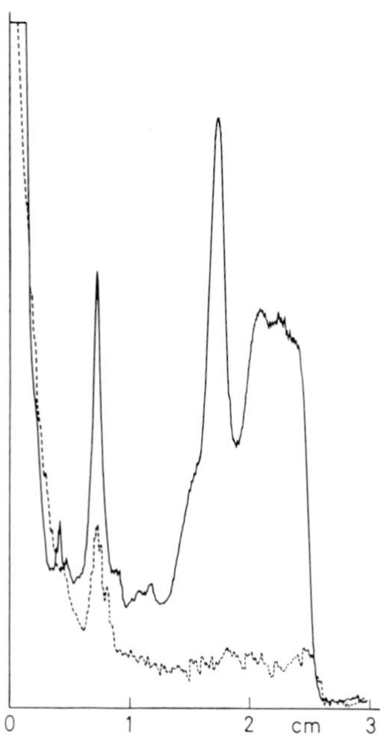

Fig. 24. Electrophoresis of the proteins from whole salivary glands. 15 glands of *C. pallidivittatus* were incubated for 2 h in ^{14}C-arginine and ^{14}C-lysine, homogenized and reduced as described for Figures 22 and 23. The proteins were then separated on 7% polyacrylamide in phosphate buffer containing 1% sodium dodecyl sulfate. *Solid line:* optical density of protein zones stained with amido black; *Dashed line:* density of the gel autoradiogram. (From Grossbach, 1973)

labeled secretory proteins were detected in the secretion collected from the gland lumen. This implies that the posttranslational processes including transport from the site of translation into the extracellular gland lumen take place rather rapidly. After 30 min of incorporation in vitro, all secretory fractions in the gland lumen including the minor components were labeled (Fig. 22).

The role which the synthesis of the secretory proteins plays within the total protein synthesis of the cells has been determined by comparing the ^{14}C-incorporation in secretory proteins to that of the other gland proteins. When the proteins from homogenized salivary glands are separated electrophoretically, a complex pattern is obtained in which the peaks of the secretory fractions can be clearly discriminated. After the glands had been incubated for 30 min in the presence of ^{14}C-lysine and ^{14}C-arginine, these peaks were preferentially labeled while incorporation into the other cellular proteins was comparatively low (Fig. 23).

The different levels of amino acid incorporation into secretory and nonsecretory proteins, respectively, became even more obvious when the reduced proteins from homogenized glands were separated according to size differences in SDS-gels. Under suitable conditions, the large polypeptides of the secretion migrated rather slowly as compared to many nonsecretory polypeptides of the gland, and a reasonable separation of the two groups of polypeptides was achieved, while the dissolution of fractions was incomplete. As is illustrated in Figure 24, the slowly migrating polypeptides in the upper part of the column were labeled to a much higher extent than the preceding smaller molecules. In autoradiograms of fixed salivary glands, Clever et al. (1969) found that 80% of the silver grains were above the gland lumen 4 h after a short pulse of ^{3}H-lysine. Protein synthesis in the salivary gland of the closely related species, *Chironomus thummi*, has been carefully studied by Wobus et al. (1972). ^{14}C-labeled amino acids from protein hydrolysates and ^{14}C-lysine were readily incorporated in vivo and in vitro into all secretory components with the exception of one fraction that migrated with the buffer front and probably represents degradation products. The most prominent polypeptides were already labeled after 10–15 min of incubation with ^{14}C amino-acids. On the other hand, electropherograms of the proteins extracted from gland cells revealed very low incorporation of ^{14}C-amino acids into cellular proteins. These results thus agree well with those reported for *Camptochironomus* (Grossbach, 1972, 1973). Wobus et al. (1972) also compared the ^{14}C-amino acid incorporation into the secretory fractions during different stages of larval development. By means of densitometry of gel autoradiograms, no striking changes in the labeling pattern of secretory proteins were detected from early 4th larval instar to late prepupa.

It thus appears that the salivary gland cells for weeks continuously synthesize secretory proteins at a rate much above that of the other proteins. The amounts of protein produced per cell in 24 h are comparable to those in other particularly specialized cell types committed to the preferential synthesis of one or a few specific proteins. In the huge galea cells of metamorphosing silk moths, the synthesis of the cocoonase zymogen reaches a level of 0.75 ng/cell/24 h (Kafatos, 1972); silk glands of *Bombyx mori* synthesize fibroin at a maximum rate of 70 μg/cell/24 h (Tashiro et al., 1968). These values are equivalent to maximum levels of 1.4×10^{10} zymogen molecules and 1.4×10^{14} fibroin molecules synthesized per cell per day (Tashiro et al., 1968; Kafatos, 1972), as compared to $1\text{–}2 \times 10^{11}$ molecules of fraction no. 1 and about $10^{11}\text{–}10^{12}$ molecules of the other secretory fractions in *Camptochironomus*. These translation rates can be more readily compared when the cells' different levels of ploidy are taken into consideration, which are 128 in the galea (Kafatos, 1969), 8192 and sometimes 16384 in *Chironomus* glands (Beermann, 1962; Edström and Daneholt, 1967), and about 400000 in the posterior silk gland (Suzuki, this vol., p. 11). 10^{8} zymogen molecules per genome and 3.5×10^{8} fibroin molecules per genome are synthesized in 24 h in the galea and silk gland, respectively, while the corresponding values in the *Chironomus* salivary gland are about 2×10^{7} molecules of fraction 1 per genome and about $10^{7}\text{–}10^{8}$ molecules per genome of the other fractions, if a ploidy of 8000 is assumed. The translation efficiency per genome in the salivary gland is thus somewhat lower than in the two other insect systems. It is also about one order of magnitude lower than in highly specialized vertebrate cells such as chick oviduct, rat embryo pancreas cells or erythroid cells, which synthesize ovalbumin, chymotrypsinogen, and hemoglobin at maximum rates of 3×10^{8} molecules per genome (Palmiter and Wrenn, 1971), 4.5×10^{8} molecules per genome (Rutter et al., 1968), and 7×10^{8} molecules per genome (Fantoni et al., 1975) per day, respectively. It must, however, be remembered that the values quoted are maximum synthesis rates that are reached only in a certain, often short, period of the cells' life history. It is not known whether there are such periods in the *Chironomus* salivary gland and whether its protein synthesizing machinery works at its maximum efficiency when the gland is replenished upon pilocarpine induced emptying, the process for which the number of 10^7–10^8 molecules per genome per day was estimated.

There is another aspect with respect to which the *Chironomus* gland cells differ from most other similarly specialized cell types. While the specific product of those cells is usually synthesized with increasing rate and accumulated exponentially, the *Chironomus* secretion is synthesized and secreted continuously for about six weeks during which period the ploidy is increased by several steps of polytenization and the translation efficiency per genome may remain rather constant. The total amount of fibroin synthesized by a cell of the posterior silk gland of *Bombyx mori* during its lifetime is about 10^9 molecules per genome (Suzuki, this vol., p. 36); the total amount of secretory molecules produced by the *Chironomus* gland must be at least as high and may considerably exceed that value.

Hybridization saturation experiments have excluded the possibility of specific gene amplification in highly specialized cells such as erythroid cells (Bishop et al., 1972) or the posterior silk gland (Suzuki et al., 1972). In the latter tissue, not more than three fibroin genes per haploid complement of DNA can be present (Suzuki et al., 1972). With regard to the translation rate observed in the *Chironomus* gland, there is thus no need to postulate the presence of more than one gene per genome for each of the secretory polypeptides. This would hold even if a molecular weight of 30000–60000 daltons for fraction no. 1 (Rydlander and Edström, 1975) would turn out to be correct, though the efficiency of fraction no. 1 translation would in this case be similar to the maximum efficiency for fibroin, which has been shown to be the result of a transcription rate and mRNA stabilization comparable to the transcription and stabilization of rRNA (see Suzuki, this vol., p. 26).

Cells that are committed to the synthesis of "luxury molecules" (Holtzer, 1968) such as hemoglobin, myosin, chymotrypsinogen or fibroin, exhibit an extraordinarily high stability of the specific messenger RNAs. The high messenger RNA half life appears to be a prerequisite for the high total amount of "luxury molecules" synthesized per cell (Kafatos, 1972; Suzuki and Brown, 1972; Holtzer et al., 1972; Fantoni et al., 1975; Suzuki, this vol.). It may thus not be surprising that different lines of evidence indicate very stable templates of secretory protein translation in the salivary gland of *Camptochironomus*. Clever (1969) observed that the granules of the special cells are still produced in appearingly normal amounts upon prolonged treatment of the glands with actinomycin, and the incorporation of ^{14}C-amino acids into total gland protein under these conditions remained at a level similar to controls (Clever et al., 1969). Wobus et al. (1972) found little decrease in total protein synthesis of *C. thummi* salivary glands upon actinomycin D treatment for up to five days, and the ^{14}C-amino acid incorporation into most of the secretory protein fractions was not significantly reduced under these conditions. The complex effects actinomycin exerts on cells exclude, however, as Wobus and his associates are aware, more detailed conclusions being drawn from these experiments. The time lag of several days which is observed before an experimentally induced shift of specific genetic activities (Beermann, 1973) becomes manifest on the translationary level (Grossbach, 1973, 1975; cf. Sect. VI.B.) is also most satisfactorily interpreted as a consequence of mRNA stability.

V. The Salivary Gland of *Chironomus*, an Excretory Organ?

From a series of studies on the salivary glands of *C. pallidivittatus* and *C. thummi*, it has been concluded that a remarkable fraction of the proteins in the salivary gland lumen originate from the hemolymph and are merely excreted rather than synthesized by the gland cells (Laufer et al., 1964; Laufer, 1965, 1968; Laufer and Nakase, 1965; Doyle and Laufer, 1968, 1969a; Kloetzel and Laufer, 1968, 1970; Schin and Laufer, 1974). While it is evident that the vast majority of proteins in the gland lumen, including those designated "minor fractions" in the present article, are synthesized de novo in the gland cells and thus establish their secretory nature (Baudisch, 1961, 1964, 1967; Beermann, 1961; Grossbach, 1966, 1968, 1969, 1972, 1973; Baudisch and Panitz, 1968; Doyle and Laufer, 1969b; Kloetzel and Laufer, 1970; Wobus et al., 1972), it cannot be excluded that a minor amount of hemolymph proteins is transferred to the gland lumen. The evidence hitherto presented in favor of such an excretory gland function does, however, not seem to be unequivocal.

Laufer and his associates (Laufer, 1965; Laufer and Nakase, 1965; Doyle and Laufer, 1969a) dissolved proteins secreted by living larvae in NaCl-Tris buffers, thus leaving aside the bulk of the glands' secretion which is insoluble under these conditions (Grossbach, 1969). Among the proteins dissolved a number of unspecific enzymes were found (Laufer and Nakase, 1965). Antisera prepared against the proteins dissolved from the secreted material were very similar to hemolymph antisera and did not reveal antigens that were not present in hemolymph (Laufer and Nakase, 1965). When interpreting these results it must be considered that it seems very difficult to exclude contamination by other proteins when the secretion is collected from larvae creeping around on sand (Laufer and Nakase, 1965; Doyle and Laufer, 1969a); that proteins immunologically related to hemolymph proteins actually occur in the gland lumen has not yet been demonstrated. Antisera prepared against the total complement of proteins from the gland lumen did not react with hemolymph in complement fixation experiments (Jacoby and Grossbach, unpubl.). A positive reaction was observed after the hemolymph had been concentrated about 3–5 times. This result might indicate the presence of a small amount of hemolymph protein(s) in the gland lumen, but is equally well interpreted to suggest the presence of gland produced proteins in the hemolymph.

An excretory function of the *Chironomus* salivary gland has also been deduced from studies in which larvae were injected with heterologous or hemolymph proteins that had previously been labeled with ^{131}I or ^{3}H (Laufer and Nakase, 1965; Schin and Laufer, 1974). Gland cells and secretion were found to become increasingly labeled upon such injections, but the label has not been shown to be actually due to an accumulation of the injected proteins in the gland lumen. Secretory proteins are synthesized at high rates and may rapidly incorporate labeled amino acids upon breakdown of injected proteins. The preferential uptake into glands of ^{131}I from the hemolymph fraction comprising proteins between 25000 and 45000 daltons, which was observed by Schin and Laufer (1974), does not necessarily exclude this possibility. In gel filtration more than 80% of all hemolymph proteins are obtained in this fraction which would thus provide a more abundant source of labeled amino acids for de novo protein synthesis than the other hemolymph fractions tested by Schin and Laufer (1974).

VI. Correlations Between Chromosome Puffs and the Synthesis of Cell-Specific Proteins

In the salivary glands of *C. pallidivittatus*, the cells of the special lobe synthesize a specific type of PAS-positive secretory granules (Churney, 1940; Beermann, 1961). This character is inherited as a single Mendelian factor (Beermann, 1961). In crosses with *C. tentans*, the glands of which lack the granules, Beermann (1961) has localized the site of this factor in a short section of chromosome IV where a large puff is found. This puff (Balbiani ring 4, cf. Sect. III.C) is restricted to the special cells of *C. pallidivittatus*, whereas an ordinary, unpuffed chromomere is present at the homologous chromosome site in the other cells and also in the special cells of *C. tentans*. Furthermore, cells in which the manifestation of Balbiani ring 4 is suppressed due to the influence of modifier genes, do not synthesize secretory granules (Beermann, 1961). The strict correlation of Balbiani ring 4 with a specific secretory component is the best evidence available at present in favor of the hypothesis that puffed chromomeres represent active gene loci.

The most conspicuous feature of the chromosomal puffing pattern in the salivary gland cells is the appearance of a small number of extraordinarily large puffs with high transcription rates, the Balbiani rings 1–4. These arise from 6 (and possibly a few more) chromomeres (cf. Sect. III.C), a number that is surprisingly close to the number of secretory polypeptide species which are the predominant protein products of the gland cells. This relatively simple system, in which only a few polypeptides are synthesized at exceedingly high rates, therefore, lends itself to an examination of the general validity of the hypothesis. If the puffing of chromomeres is actually a visible manifestation of genetic activity, then the genes of the secretory polypeptides would be expected to be located in gland-specific puffs.

By using Beermann's chromosome maps of the two species (Beermann, 1952a, 1955) and his cytogenetic techniques (Beermann, 1961), the gene loci of polypeptides no. 6 and 7 have been localized. Both are situated in the short chromosome IV in sections where specific Balbiani rings are developed in the gland cells.

It has been possible in some cases to experimentally change the activity of single puffs. Concomitant shifts within the pattern of secretory polypeptides were observed in such glands, intimating a correlation between the activity level of certain puffs and the translation rate of some polypeptides.

A. Localization of Genes of Secretory Components

The species *C. pallidivittatus* and *C. tentans* can be interbred and produce fertile hybrids. All chromosome arms are characterized by species-specific patterns of chromomeres from which the genetic constitution of F_2-animals and backcrosses can easily be determined (Beermann, 1955, 1961). By comparing the inheritance of the chromosome arms, that are protected against crossing-over by species-specific inversions, with the inheritance of species- or stock-specific polypeptides, the gene loci were localized in one of the chromosome arms. Recombinations within this arm made possible more precise localizations in spite of no genetic markers being available.

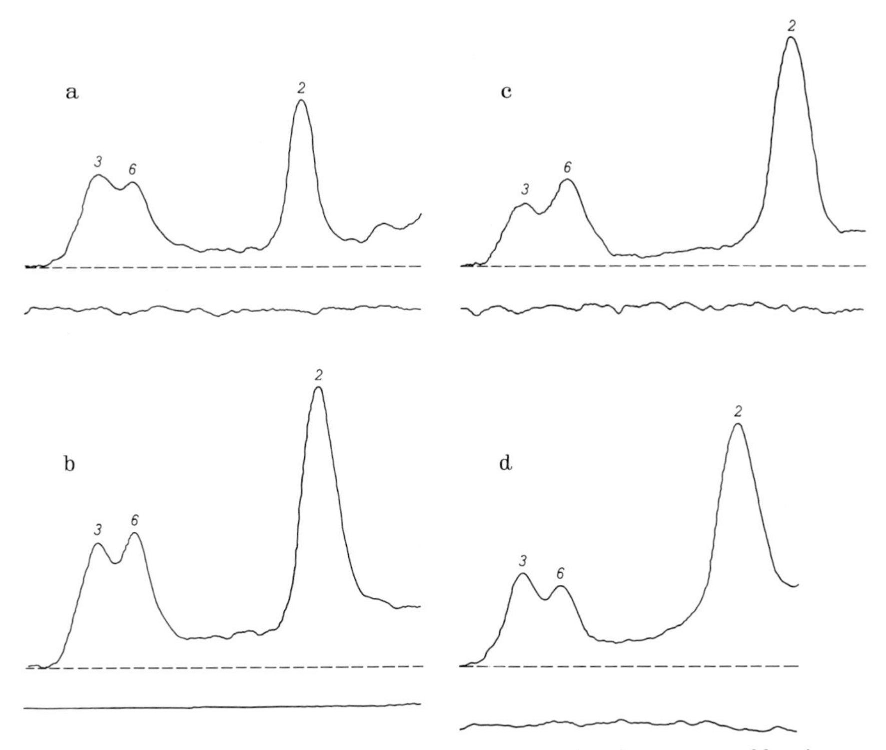

Fig. 25a—d. Amount of secretory fraction no. 6 as compared to the amounts of fractions no. 2 and 3 in single glands of *C. pallidivittatus* (a) and heterozygotes (b–d). Densitometer tracings of proteins stained with amido black. (From Grossbach, 1969)

1. The Gene Locus of Polypeptide no. 6

C. pallidivittatus was crossed with a *tentans*-stock devoid of polypeptide no. 6. Larvae of the F_1-generation were found to produce, within the limits of the usual interindividual variation, the same amount of polypeptide no. 6 as do larvae of *C. pallidivittatus* (Fig. 25).

The inheritance of polypeptide no. 6 was examined in animals derived from repeated backcrossing of hybrids with *C. tentans*. One gland of each larva was used for a squash preparation of the chromosomes; the secretion from the second gland was reduced, alkylated and analyzed by microelectrophoresis on a capillary gel. Table 4 lists the chromosome constitutions and secretion phenotypes of 50 such animals. It is obvious that the capacity to produce polypeptide no. 6 is inherited along with the *pallidivittatus*-chromosome IV while other chromosome arms are not correlated with this character.

Some of the animals listed in Table 4 produced polypeptide no. 6 in spite of carrying two homozygous *tentans*-chromosomes IV (no. 8, 13, 16); on the other hand, this polypeptide was not detected in some others (no. 2, 4, 28, 36) that possessed a heterozygous pair of chromosomes IV. These cases indicate crossover events by which the gene locus in question has become separated from the species-

Table 4. Chromosome constitution and secretory type in larvae obtained by repeated backcrossing of interspecific hybrids with *C. tentans* (from Grossbach, 1969)

No.	backcross	1L	1R	2L	2R	3L	3R	4	Polypeptide 6
1	D1 F_1	t/p	t/p	t/t	t/t	t/t	t/t	t/p	+
2	D1	t/t	t/t	t/t	t/p	t/t	t/t	t/p	–
3	D1	t/t	t/p	t/t	t/t	t/t	t/t	t/t	–
4	D1	t/p	t/p	t/t	t/p	t/t	t/t	t/p	–
5	D1	t/p	t/p	t/t	t/p	t/t	t/t	t/t	?
6	D1	t/p	t/p	t/p	t/p	t/t	t/t	t/p	+
7	D1	t/p	t/t	t/p	t/p	t/t	t/t	t/p	+
8	D1	t/p	t/p	t/t	t/t	t/t	t/t	t/t	+
9	D1	t/p	t/p	t/p	t/p	t/t	t/t	t/p	+
10	D1	t/p	t/p	t/p	t/p	t/t	t/t	t/p	+
11	D1	t/t	t/t	t/t	t/p	t/t	t/t	t/p	(+)
12	D1	t/p	t/p	t/p	t/p	t/t	t/p	t/p	+
13	D1	t/t	t/p	t/t	t/t	t/t	t/t	t/t	+
14	D1	t/p	t/t	t/t	t/p	t/t	t/t	t/p	+
15	D1	t/t	t/p	t/t	t/t	t/p	t/t	t/p	(+)
16	D1	t/p	t/p	t/t	t/p	t/t	t/t	t/t	+
17	D1 × t	t/p	t/p	t/t	t/t	t/t	t/t	t/t	–
18	D1 × t	t/p	t/p	t/t	t/t	t/t	t/t	t/t	?
19	D1 × t	t/p	t/p	t/t	t/t	t/t	t/t	t/t	–
20	D	t/t	t/t	t/t	t/p	t/p	t/t	t/p	+
21	D	t/p	t/t	t/t	t/p	t/p	t/t	t/p	+
22	D	t/t	t/t	t/t	t/p	t/p	t/t	t/p	+
23	D	t/t	t/t	t/t	t/p	t/p	t/t	t/p	+
24	D	t/t	t/t	t/t	t/p	t/p	t/t	t/p	+
25	D	t/t	t/t	t/t	t/p	t/p	t/t	t/p	+
26	D	t/p	t/t	t/t	t/t	t/t	t/t	t/p	+
27	D6	t/p	t/t	t/p	t/p	t/p	t/t	t/p	+
28	D6	t/p	t/t	t/t	t/p	t/p	t/t	t/p	–
29	D5	t/p	t/p	t/t	t/p	t/p	t/t	t/t	–
30	D5	t/p	t/t	t/t	t/p	t/p	t/t	t/t	–
31	D5	t/t	t/t	t/t	p/p	t/p	t/t	t/p	+
32	D5	t/p	t/t	t/t	t/t	t/p	t/t	t/t	–
33	D5 F_1	t/p	t/t	t/t	t/p	t/t	t/t	t/p	+
34	D5 F_1	t/p	t/t	t/t	t/p	t/p	t/t	p/p	+
35	D5 F_1	t/t	t/t	t/t	t/p	t/p	t/t	t/t–p	–
36	D5 F_1	t/p	t/p	t/t	t/p	t/t	t/t	t/p	–
37	D5 F_1	t/p	t/t	t/t	t/p	t/t	t/t	t/p	+
38	D5 F_1	t/p	t/t	t/t	p/p	t/t	t/t	p/p	+
39	D5 F_1	t/p	t/t	t/t	t/t	t/p	t/t	t/p	+
40	D5 F_1	t/p	t/t	t/p	t/t	t/p	t/t	t/p	+
41	D5 F_1	t/p	t/t	t/t	t/p	t/p	t/t	t/p	(+)
42	D F_1	t/t	t/t	t/t	t/p	t/p	t/t	t/p	+
43	D F_1	t/t	t/t	t/t	t/p	t/t	t/t	t/p	+
44	D F_1	t/t	t/t	t/t	t/p	t/t	t/t	t/p	+
45	D F_1	t/t	t/t	t/t	t/t	t/t	t/t	t/p	+
46	D1 × t	t/t	t/p	t/t	t/t	t/t	t/t	t/p	+
47	D1 × t	t/p	t/p	t/t	t/t	t/t	t/t	t/p	(+)
48	D1 × t	t/p	t/t	t/t	t/t	t/t	t/t	t/t	–
49	D1 × t	t/t	t/p	t/t	t/t	t/t	t/t	t/p	+
50	D F_1	t/t	t/t	t/t	t/t	t/t	t/t	t/p	+

D, D1, D5, and D6 are sister animals from different cultivation vials. t = *tentans*, p = *pallidivittatus*

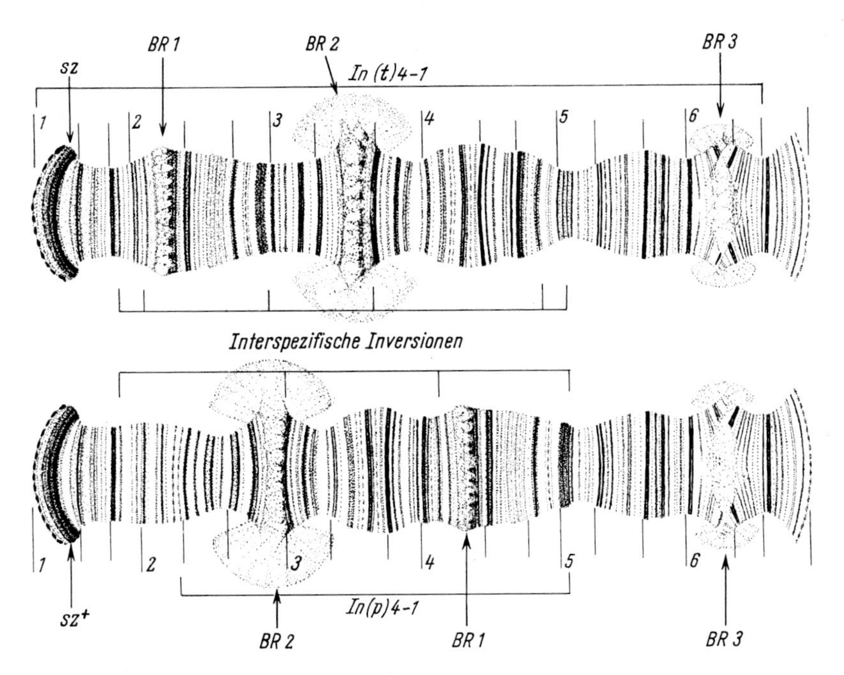

Fig. 26. Map of chromosomes IV from *C. tentans (left)* and *C. pallidivittatus (right)* showing the interspecific and intraspecific inversions. (From Beermann, 1955)

specific inversions that characterize the chromosomes of the two species; they render possible a more precise localization by proving that the gene site must in fact be situated outside these inversions, i.e. either close to the centromere or in the distal segment of *pallidivittatus*-chromosome IV (section 1A–1B or segment 5A–6C of Beermann's chromosome map; see Fig. 26).

Recombinations within the species-specific inversions of chromosome IV can be expected to occur very rarely. In a large amount of material from interspecific hybrids, one such case has been detected (Grossbach, 1969). In a very extensive amount of crosses examined, Beermann (1955, 1956b, and pers. comm.) did not find crossovers in this region. The one case detected is included in Table 4 (no. 35) and shown in Figure 27. One of the chromosomes IV exhibited the standard chromomere pattern of *C. tentans*. The other chromosome IV had the chromomere sequence 1A–3B of *C. pallidivittatus* attached to the sequence 4B–6C of *C. tentans*, i.e. it consisted of a proximal third of a *pallidivittatus*- and a distal half of a *tentans*-chromosome and had lost a sequence including Balbiani ring 1. Homologous regions of the two chromosomes were somatically paired (Fig. 27). This chromomere arrangement must have resulted from crossing-over between the homologous regions 4A of *C. tentans* and 3B of *C. pallidivittatus* inside the species-specific inversions and provided a decision concerning the site of the gene locus of polypeptide no. 6. The glands of animal no. 35 produced secretory granules in their special cells. The chromosome constitution in region 1A–1C of chromosome IV must, therefore, be t/p, as the gene determining this character lies in this region of the

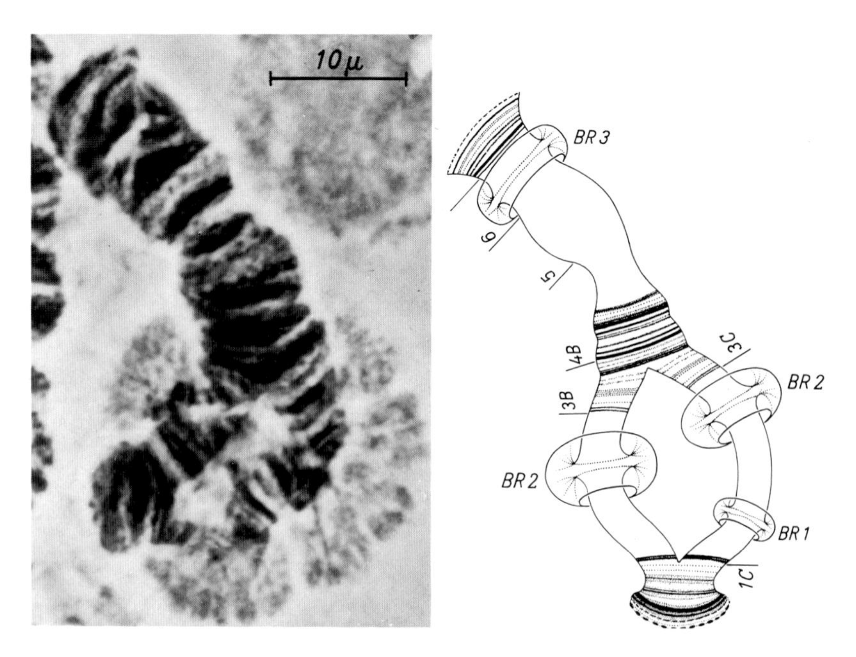

Fig. 27. Chromosome IV from the salivary gland of larva no. 35 from Table 4. For explanation see text. (From Grossbach, 1969)

pallidivittatus-genome. As larva no. 35 did not synthesize polypeptide no. 6 the gene of this secretory component must be located not in region 1A–1C but on the other side of the species-specific inversions, i.e. in segment 5A–6C of *pallidivittatus*-chromosome IV. A somewhat more precise localization within this region may be deduced from the chromosome constitution in two larvae (no. 13 and 16 of Table 4) which produced polypeptide no. 6 in spite of possessing the homologous constitution t/t in chromosome IV. However, the distal regions 5C–6C of the two homologous chromosomes were not somatically paired in spite of their apparently identical chromomere patterns. As failure of polytene chromosomes in *Camptochironomus* to pair somatically usually indicates heterozygosity (Beermann, pers. comm.), it may be concluded that the regions 5C–6C of chromosome IV are possibly heterozygous in these animals, and that the crossover which separated the gene of polypeptide no. 6 from the *pallidivittatus*-specific inversions probably occurred in 5C. The gene locus in question would thus be located within 5C–6C, a region which comprises about 20 chromomeres (see Fig. 26).

A more precise localization will depend on the detection of cytologically visible recombinations in this region that can be expected to occur only at very low frequencies. However, even the present result deserves interest since one of the preponderant puffs of the gland cell, Balbiani ring 3, is found within section 5C–6C. The localization of a cell-specifically expressed gene within a region that comprises about 1% of the genome and carries one of the major cell-specific puffs strongly suggests a genetic correlation between the puff and the expressed character.

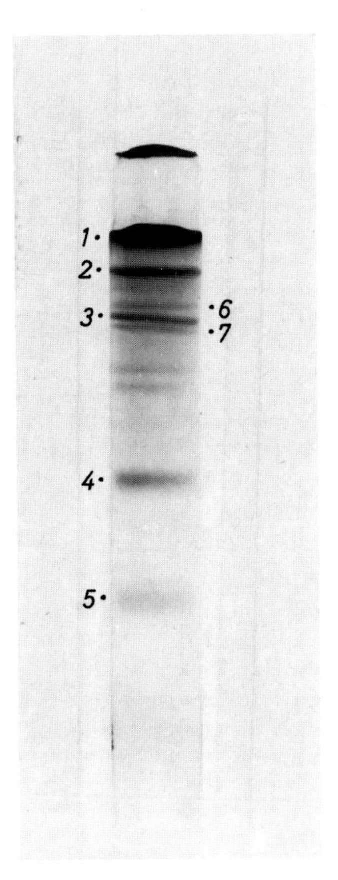

Fig. 28. Separation by disc-electrophoresis of the reduced and alkylated secretion from the special lobes of 40 salivary glands from backcross animals which carried a heterozygous chromosome IV while all other constituents of the genome had a homozygous *tentans/tentans* constitution. Electrophoretic conditions as described for Figure 12; note presence of fraction no. 7. (From Grossbach, 1969)

2. The Gene Locus of Polypeptide no. 7

The gene of polypeptide no. 7 that is expressed only in the special cells has also been found, by means of the cytogenetic techniques described above, to be localized in chromosome IV of *C. pallidivittatus* (Grossbach, 1969). Factors in other parts of the genome have no influence on the manifestation of this secretion character, as is evident from backcross stocks that carried a homozygous *tentans*-chromosome complement with the exception of a t/p constitution in chromosomes IV. Such larvae regularly produced polypeptide no. 7 (Fig. 28).

The gene locus may be located in either of the two chromosome segments outside the species-specific inversions, i.e. within 1A–1B or within 5A–6C (Fig. 26). The material hitherto available does not permit a more precise localization. It is, however, worth mentioning that a puff specific for the special cells of the *pallidivittatus*-gland is located in one of the two regions in question, namely Balbiani ring 4 in section B1. It is at the site of this Balbiani ring that Beermann (1961) has localized the gene for the production of characteristic and cell-specific secretory granules. Secretory fraction no. 7 that is likewise restricted to the special cells may possibly contribute to or represent these granules which account for a considerable amount of the secretion's mass and are dissolved under the same conditions.

B. Correlations Between Puff Activities and the Synthesis Rate of Certain Proteins

According to the hypothesis of puffing, each particular pattern of puffed chromomeres is characteristic of a specific differentiated state. Every pronounced change of puff activities would thus be expected to initiate an alteration and to be succeeded, after a certain delay, in the cell's acquiring a new differentiated state, noticeable e.g. from its synthesizing another set of proteins. In especially simple cases the dependency of certain proteins upon the activity of a specific puff may thus become directly evident.

Programmed changes of the pattern occur in the salivary gland of *Camptochironomus* and other insects when metamorphosis is initiated under the influence of the hormone, ecdyson (Clever and Karlson, 1960; Panitz, 1960, 1964; Clever, 1961; Berendes, 1967). In *Chironomus* as well as in *Drosophila*, a few puffs are activated within minutes upon administration of the hormone, probably by reacting directly with a hormone-protein complex, while a series of other stage-specific puffs come into play later, following an ordered schedule (Clever, 1961; Berendes, 1967).

The primary products of these hormone-induced puffs, however, are completely unknown and seem difficult to detect. Balbiani rings, on the other hand, are very stable structures throughout larval development, as would be expected for puffs that direct the synthesis of continuously produced proteins. For the investigation of possible relationships between chromomere activity and cell-specific protein synthesis, it is, therefore, advantageous that the activity of certain Balbiani rings can be selectively impeded and augmented experimentally. Such experiments may be expected to unravel influences of puff activities on the differentiated state of the cell.

It has been mentioned above that the synthesis of secretory granules in the special cells of the *pallidivittatus*-gland depends on the activity of Balbiani ring 4. Beermann (1961) found that the formation of this puff is repressed in animals derived after several generations of backcrossing of interspecific hybrids with *C. tentans*. The specific genetic conditions to which this repression of a *pallidivittatus*-gene can be ascribed include a general preponderance of the *tentans*-genome and the presence of modifier genes situated in the left arm of chromosome III (Beermann, 1961). Cells with repressed Balbiani rings 4 were found by Beermann not to produce secretory granules.

The selective repression of another salivary gland puff, Balbiani ring 1, has been found to occur under reproducible genetic conditions that can be established by interspecific hybridization (Grossbach, 1969). *Camptochironomus* larvae with a general preponderance of the *tentans*-genome, a heterozygous constitution in

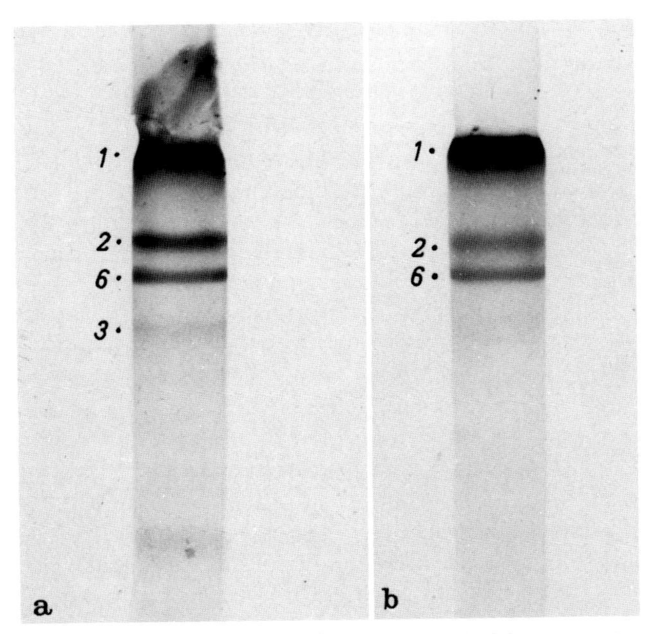

Fig. 29a and b. Reduced amount (a) and virtual absence (b) of secretory fraction no. 3 in backcross animals exhibiting repression of Balbiani ring 1. The reduced and alkylated secretion from single glands was separated by disc-electrophoresis as described for Figure 12. Lower part of gels is not shown; gel diameter 0.7 mm (From Grossbach, 1969)

chromosome IV and t/t homozygosity in both arms of chromosome I, frequently exhibit, at the site of the *pallidivittatus* Balbiani ring 1, a small puff or no puffing of chromomeres at all. The genetic conditions seem to be prerequisites of Balbiani ring 1 reduction which, however, is not their necessary consequence. In the limited amount of material that has been examined, larvae of this genetic constitution have been found which, nevertheless, apparently had normally developed Balbiani rings 1. It may be speculated that the manifestation of Balbiani ring 1 is under the influence of modifier genes in other chromosome arms as seems to be the case with Balbiani ring 4 (Beermann, 1961). For an ordinary manifestation of genes a certain degree of balances and checks may presumably be needed that may not always be maintained under conditions when genes from two different species are combined.

The repression of the *pallidivittatus*-Balbiani ring is a specific and selective one. The two homologous chromosomes IV are usually not somatically paired and attached to each other in the centromere regions, so that altogether 6 Balbiani rings are separately recognized. While Balbiani ring 1 in the *pallidivittatus*-homologue is reduced in these animals, the other Balbiani rings and, in particular, Balbiani ring 1 in the *tentans*-chromosome, are apparently normally developed.

A comparable selective change in the pattern of secretory proteins was observed in these animals. While most polypeptides were synthesized in amounts similar to those in controls, polypeptide no. 3 was found in much lower quantity, or more frequently, could not be detected at all (Fig. 29). This suggests a causal relation between the activity of *pallidivittatus*-Balbiani ring 1 and the production of polypeptide no. 3. It is, however, not easy to understand how a secretory component that is apparently identical in both species disappears from the secretion as a consequence of the repression of one haploid Balbiani ring.

A substantial shift in the activities of Balbiani rings 1 and 2 can be achieved by administering galactose to the larvae's culture medium (Beermann, 1973). Under natural and normal culture conditions, Balbiani ring 2 is largely expanded while Balbiani ring 1 varies somewhat in size but is always considerably smaller than the other. In the presence of 0.5% galactose, the activities of Balbiani rings 1 and 2 of the *pallidivittatus*-gland are remarkably increased and decreased, respectively, within 24 h, so that Balbiani ring 1 becomes maximally expanded and Balbiani ring 2 almost completely repressed. A parallel shift was observed with respect to the relative level of ^{3}H-uridine incorporation in the two sites (Beermann, 1973). It is important to note, as Beermann has pointed out, that the activities of other Balbiani rings or puffs are not noticeably influenced by the galactose treatment.

This situation provides a possibility to investigate cell-specific protein synthesis under conditions of selective changes in chromomere activity. Incorporation studies with ^{14}C-amino acids on salivary glands from animals that had been kept in galactose medium for some days, did not reveal shifts of ^{14}C-uptake in any of the secretory polypeptides as compared to controls. However, when the artificial proportion of Balbiani ring activities was maintained for two weeks or more, the relative synthesis rate of polypeptide no. 3 was found to become remarkably increased. Effects on other secretory polypeptides, particularly a decrease of the ^{14}C-uptake from amino-acids in polypeptide no. 2, were repeatedly observed but could not be reproduced in all experiments. It seems that a more stringent control of the experimental conditions must be achieved before the interesting features of the system can be fully exploited.

Concluding Remarks

The differentiated state of a cell may be defined in terms of specific protein synthesis. This state can be especially well characterized in cells being committed, like those of the *Camptochironomus* salivary gland, to the synthesis and secretion of a few predominant protein species. Obviously such cell types are very useful tools in developmental biology, as a satisfactory definition, in quantitative terms, of differentiated states must be a prerequisite to an investigation of the mechanisms that accomplish differentiation. The salivary glands of *Camptochironomus* with their beautiful giant chromosomes appear to be a system especially well suited for such studies. They comprise two closely related cell types which synthesize and secrete similar but not identical sets of polypeptides. The pattern of polypeptides can be experimentally changed under controlled conditions.

Moreover, the evidence available at the present time strongly suggests a correlation between some of the gland-specific giant puffs (Balbiani rings) and some of the quantitatively predominant secretory proteins (Fig. 30). The gene locus of polypeptide no. 6 has been localized in chromosome IV within a section of about 20 chromomeres comprising Balbiani ring 3. The gene locus of polypeptide no. 7 is

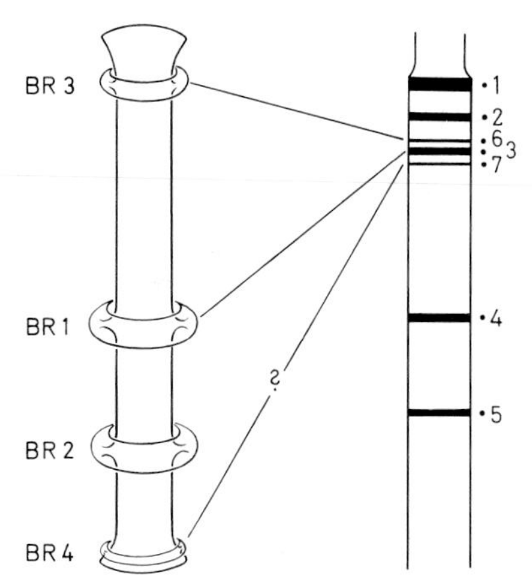

Fig. 30. Correlations between puffs (Balbiani rings) and secretory polypeptides in the salivary gland of *C. pallidivittatus*. Balbiani rings (BRs) 1–3 are developed in all gland cells whereas Balbiani ring 4 is restricted to the cells of the special lobe. (Redrawn from Grossbach, 1973)

situated in one of two regions of chromosome IV. One of these regions exhibits a Balbiani ring only in the cells that synthesize polypeptide no. 7. The production of polypeptide no. 3 appears to depend on the activity of Balbiani ring 1. Because of the limited number of secretory polypeptides, it seems possible to check whether a general correlation exists between the Balbiani rings and all the quantitatively predominant polypeptides of the gland, by experimentally modifying the activity of single Balbiani rings and by exploiting secretion mutants for a localization of additional genes. The recently isolated stocks of *C. tentans* carrying different alleles for polypeptides no. 3 and 5 will allow, by hybridization with *C. pallidivittatus*, a cytogenetic localization of the structural genes with a precision comparable to that which has been possible with polypeptides no. 6 and 7.

Korge (1975) has recently achieved the precise localization of a structural gene within a single chromomere that becomes puffed in the salivary gland of *Drosophila melanogaster* in cells where the gene is being expressed. This direct proof for a causal relationship between puff activity and synthesis of a cell-specific protein cannot be accomplished in *Camptochironomus* because of the lack of genetic markers and appropriate deficiencies. Even in the case of gene sz which determines the production of secretory granules, the identified chromosome section comprises more than 20 chromomeres (Beermann, 1961). By combining cytogenetic data with correlations between Balbiani ring activities and protein synthesis, it should, however, be possible to provide sufficient indirect evidence to prove or disprove the existence of causal relationships between the 3–4 Balbiani rings that altogether comprise 6 (and possibly a few more) chromomeres, and the 6–7 prominent secretory polypeptides of the gland cell.

Such causal relationships, should they turn out to exist for all Balbiani rings in the *Camptochironomus* gland, would imply that a level of puffing some orders of magnitude above average is generally required when proteins are to be synthesized at a continuously high level of about 10^7–10^8 molecules per genome per day. A general relationship of the kind which appears to exist at least in the case of certain

Balbiani rings (see also Baudisch, this vol., p. 204), would imply consequences for chromomere organization, as puffing is probably regulated on the level of the individual chromatid (Beermann, pers. comm.). Moreover, cells exhibiting a few extraordinarily puffed gene loci, from which RNA is transcribed at exceptionally high rates in order to maintain an outstanding synthesis level of certain proteins, are especially well-suited for the isolation and characterization of specific transcripts (see below). They may also lend themselves to an investigation of the mechanisms of gene activation on the chromomere level.

RNA species transcribed in the puffs of the *Camptochironomus* salivary gland have been studied by a number of authors (Edström and Daneholt, 1967; Daneholt et al., 1969; Pelling, 1970; Daneholt, 1972). The interest has been focused primarily on the transcript of Balbiani ring 2 which upon isolation from the chromosome exhibits a size of about 75 S and is transferred without noticeable reduction in size to the cytoplasm (Daneholt, 1972; Daneholt and Hosick, 1973a, b), where it can be identified unequivocally by its specifically hybridizing to Balbiani ring 2 in situ (Lambert, 1972, 1973; Lambert et al., 1972). 75 S RNA isolated from very large polysomes containing up to 100 or more ribosomes (Daneholt et al., 1976) has been shown to hybridize in situ to Balbiani rings 1 and 2 (Wieslander and Daneholt, in prep.), suggesting that at least some of the Balbiani ring transcripts actually enter polysomes. The presence of polyadenylic acid sequences also suggests a messenger function of the 75 S fraction (Edström and Tanguay, 1974).

The extraordinary size of the Balbiani ring 2 transcript has led to a discussion of the interesting possibility that the 75 S RNA may have a polycistronic organization (Daneholt et al., 1969; Daneholt and Hosick, 1973a; Rydlander and Edström, 1975). As the 75 S RNA comprises repetitive sequences (Daneholt, 1970; Lambert, 1973), this possibility is one of the alternatives to be considered especially if the salivary glands produce no polypeptides that require messengers of extraordinary size.

As has been mentioned above (Sect. IV.C), the glands produce a quantitatively predominant protein fraction (no. 1) of about 5×10^5 daltons. No cytogenetic evidence is as yet available concerning the location of the structural gene of this protein, but it may be speculated that the 75 S fraction, the preponderant nonribosomal RNA comprising 1.5% of the total cellular RNA (Daneholt and Hosick, 1973a) carries information coding for protein fraction no. 1, which accounts for about 50% of the secretory protein. If fraction no. 1 should turn out to represent a single polypeptide chain of about 5×10^5 daltons (for discussion, see Sect. IV.C.2), then the presence of giant transcripts in glandular polysomes may find a conventional explanation. The information for a polypeptide of this size would occupy a considerable part of the 75 S RNA though probably not its total length. Fibroin messenger RNA has also been reported to be larger than predicted from the molecular weight of fibroin (Lizardi et al., 1975). The reiteration of sequences within the 75 S RNA and the DNA of Balbiani ring 2 does not necessarily argue against its monocistronic nature, as the patterns of tryptic peptides and cyanogen bromide fragments of secretory fraction no. 1 seem to indicate an internally reiterated primary structure (Autio, Jacoby and Grossbach, unpubl.). As suggested by the results of Rydlander and Edström (1975) which could not, however, be confirmed in the author's laboratory, the large secretory fraction may turn out to be composed of subunits of a size of 30000–60 000 daltons. To explain the discrepancy between such translation products and the large size of the 75 S RNA in polysomes, Rydlander and Edström assume that the Balbiani ring transcripts are polycistronic messengers, a possibility which would render the Balbiani rings very untypical gene loci.

Acknowledgement. The original work was in part supported by the Deutsche Forschungsgemeinschaft.

References

Acton, A. B.: A study of the differences between widely separated populations of *Chironomus (= Tendipes) tentans* (Diptera). Proc. Roy. Soc. (London) Ser. B **151**, 277—296 (1959)

Andrews, P.: Estimation of the molecular weights of proteins on Sephadex G 100 and G 200. Biochem. J. **91**, 222–233 (1964)

Andrews, P.: The gel-filtration behaviour of proteins related to their molecular weights over a wide range. Biochem. J. **96**, 595–606 (1965)

Balbiani, E. G.: Sur la structure du noyau des cellules salivaires chez les larves de *Chironomus*. Zool. Anz. **4**, 637–641 (1881)

Baudisch, W.: Synthese von Oxyprolin in den Speicheldrüsen *Acricotopus lucidus*. Naturwissenschaften **48**, 56 (1961)

Baudisch, W.: Einbau von ^{14}C-Prolin in die Speicheldrüsen von *Acricotopus lucidus* (Chironomide). In: Proc. 2nd Intern. Congr. Histo. Cytochem., p. 236. Berlin-Heidelberg-New York: Springer 1964

Baudisch, W.: Spezifische Hydroxyprolinsynthese in den Speicheldrüsen von *Acricotopus lucidus*. Biol. Zbl. **86**, (Suppl.), 157—162 (1967)

Baudisch, W., Panitz, R.: Kontrolle eines biochemischen Merkmals in den Speicheldrüsen von *Acricotopus lucidus* durch einen Balbianiring. Exptl. Cell Res. **49**, 470–476 (1968)

Bauer, H.: Aufbau der Chromosomen aus den Speicheldrüsen von *Chironomus thummi* Kiefer. Z. Zellforsch. **23**, 280—313 (1935)

Bauer, H.: Beiträge zur vergleichenden Morphologie der Speicheldrüsenchromosomen (Untersuchungen an den Riesenchromosomen der Dipteren. II.) Zool. Jb. (Phys.) **56**, 239—276 (1936)

Bauer, H.: Chromosomenstruktur und -funktion. Jb. Max-Planck-Ges. 23—39 (1957)

Bauer, H., Beermann, W.: Die Polytänie der Riesenchromosomen. Chromosoma **4**, 630—648 (1952)

Becker, H.J.: Die Puffs der Speicheldrüsenchromosomen von *Drosophila melanogaster*. I. Mitt.: Beobachtungen zum Verhalten des Puffmusters im Normalstamm und bei zwei Mutanten, *giant* und *lethal-giant-larvae*. Chromosoma **10**, 654—678 (1959)

Beermann, W.: Chromomerenkonstanz und spezifische Modifikationen der Chromosomenstruktur in der Entwicklung und Organdifferenzierung von *Chironomus tentans*. Chromosoma **5**, 139–198 (1952a)

Beermann, W.: Chromosomenstruktur und Zelldifferenzierung in der Speicheldrüse von *Trichocladius vitripennis*. Z. Naturforsch. **7b**, 237—242 (1952b)

Beermann, W.: Cytologische Analyse eines *Camptochironomus*-Artbastards. I. Kreuzungsergebnisse und die Evolution des Karyotypus. Chromosoma **7**, 198—259 (1955)

Beermann, W.: Nuclear differentiation and functional morphology of chromosomes. Cold Spring Harbor Symp. Quant. Biol. **21**, 217—232 (1956a)

Beermann, W.: Inversions-Heterozygotie und Fertilität der Männchen von *Chironomus*. Chromosoma **8**, 1—11 (1956b)

Beermann, W.: Chromosomal differentiation in insects. In: Developmental Cytology. Rudnick, D. (ed.). New York: Ronald, 1959, pp. 83—103

Beermann, W.: Ein Balbiani-Ring als Locus einer Speicheldrüsenmutation. Chromosoma **12**, 1—25 (1961)

Beermann, W.: Riesenchromosomen. Protoplasmatologia. Handbuch der Protoplasmaforschung, Vol. IVd. Vienna: Springer, 1962

Beermann, W.: Chromomeres and Genes. In: Developmental Studies on Giant Chromosomes. Results and Problems in Cell Differentiation. Vol. IV, Beermann, W. (ed.), pp. 1—33. Berlin-Heidelberg-New York: Springer 1972

Beermann, W.: Directed changes in the pattern of Balbiani ring puffing in *Chironomus*: Effects of a sugar treatment. Chromosoma **41**, 297—326 (1973)

Beermann, W., Bahr, G.F.: The submicroscopic structure of the Balbiani-ring. Exptl. Cell Res. **6**, 195—201 (1954)

Berendes, H.D.: The hormone ecdysone as effector of specific changes in the pattern of gene activities of *Drosophila hydei*. Chromosoma **22**, 274—293 (1967)

Berendes, H.D.: The Control of Puffing in *Drosophila hydei*. In: Developmental Studies on Giant Chromosomes. Results and Problems in Cell Differentiation. Vol. IV, Beermann, W. (ed.), pp. 181—207. Berlin-Heidelberg-New York: Springer 1972

Berendes, H.D., Beermann, W.: Biochemical activity of interphase chromosomes (polytene chromosomes). In: Handbook of Molecular Cytology. Lima-de-Faria, A. (ed.), pp. 500—519. Amsterdam: North Holland 1969

Bishop, J.O., Pemberton, R., Baglioni, C.: Reiteration frequency of haemoglobin genes in the duck. Nature New Biol. **235**, 231—234 (1972)

Churney, L.: A contribution to the anatomy and physiology of the salivary gland system in the larva of *Chironomus* (Diptera). J. Morphol. **66**, 391—407 (1940)

Clever, U.: Genaktivitäten in den Riesenchromosomen von *Chironomus tentans* und die Beziehungen zur Entwicklung. I. Genaktivierung durch Ecdyson. Chromosoma **12**, 607—675 (1961)

Clever, U.: Genaktivitäten in den Riesenchromosomen von *Chironomus tentans* und ihre Beziehungen zur Entwicklung. II. Das Verhalten der Puffs während des letzten Larvenstadiums und der Puppenhäutung. Chromosoma **13**, 385—436 (1962)

Clever, U.: Genaktivitäten in den Riesenchromosomen von *Chironomus tentans* und ihre Beziehungen zur Entwicklung. IV. Das Verhalten der Puffs in der Larvenhäutung. Chromosoma **14**, 651—675 (1963a)

Clever, U.: Von der Ecdysonkonzentration abhängige Genaktivitätsmuster in den Speicheldrüsenchromosomen von *Chironomus tentans*. Develop. Biol. **6**, 73—98 (1963b)

Clever, U.: Chromosome activity and cell function in polytenic cells. II. The formation of secretion in the salivary glands of *Chironomus*. Exptl. Cell Res. **55**, 317—322 (1969)

Clever, U., Karlson, P.: Induktion von Puff-Veränderungen in den Speicheldrüsenchromosomen von *Chironomus tentans* durch Ecdyson. Exptl. Cell Res. **20**, 623—626 (1960)

Clever, U., Storbeck, I., Romball, C.G.: Chromosome activity and cell function in polytenic cells. I. Protein synthesis at various stages of larval development. Exptl. Cell Res. **55**, 306—316 (1969)

Daneholt, B.: Base ratios in RNA molecules of different sizes from a Balbiani ring. J. Mol. Biol. **49**, 381—391 (1970)

Daneholt, B.: Giant RNA transcript in a Balbiani ring. Nature New Biol. **240**, 229—232 (1972)

Daneholt, B.: Transfer of genetic information in polytene cells. Intern. Rev. Cytol. **39**, 417—462 (1974)

Daneholt, B.: Transcription in polytene chromosomes. Cell **4**, 1—9 (1975)

Daneholt, B., Andersson, K., Fagerlind, M.: J. Cell Biol. (in press) 1976

Daneholt, B., Case, S.T., Wieslander, L.: Gene expression in the salivary glands of *Chironomus tentans*. In: Progress in Differentiation Research. Proc. 2nd Intern. Conf. Differentiation. Müller-Bérat, N. (ed.), pp. 125—133. Amsterdam: North Holland Publ. 1976

Daneholt, B., Edström, J.-E.: The content of DNA in individual polytene chromosomes of *Chironomus tentans*. Cytogenetics **6**, 350—356 (1967)

Daneholt, B., Edström, J.-E., Egyházi, E., Lambert, B., Ringborg, U.: RNA synthesis in a Balbiani ring in *Chironomus tentans* salivary gland cells. Chromosoma **28**, 418—429 (1969)

Daneholt, B., Hosick, H.: Evidence for transport of 75S RNA from a discrete chromosome region via nuclear sap to cytoplasm in *Chironomus tentans*. Proc. Natl. Acad. Sci. **70**, 442—446 (1973a)

Daneholt, B., Hosick, H.: The transcription unit in Balbiani ring 2 of *Chironomus tentans*. Cold Spring Harbor Symp. Quant. Biol. **38**, 629—635 (1973b)

Deshmukh, A., Deshmukh, K., Nimmi, M. E.: Synthesis of aldehydes and their interactions during the in vitro aging of collagen. Biochemistry **10**, 2337—2342 (1971)

Doyle, D., Laufer, H.: Analysis of secretory processes in Dipteran salivary glands. In: Differentiation and Defense Mechanisms in Lower Organisms. In Vitro **3**, 93—103 (1968)

Doyle, D., Laufer, H.: Sources of larval salivary gland secretion in the Dipteran *Chironomus tentans*. J. Cell. Biol. **40**, 61—78 (1969a)

Doyle, D., Laufer, H.: Requirements of ribonucleic acid synthesis for the formation of salivary gland specific proteins in larval *Chironomus tentans*. Exptl. Cell Res. **57**, 205—210 (1969b)

Edström, J.-E., Daneholt, B.: Sedimentation properties of the newly synthesized RNA from isolated nuclear components of *Chironomus tentans* salivary gland cells. J. Mol. Biol. **28**, 331—343 (1967)

Edström, J.-E., Tanguay, R.: Cytoplasmic ribonucleic acids with messenger characteristics in salivary gland cells of *Chironomus tentans*. J. Mol. Biol. **84**, 569—583 (1974)

Elgin, S. C. R., Boyd, J. B.: The proteins of polytene chromosomes of *Drosophila hydei*. Chromosoma **51**, 135—145 (1975)

Engelmann, W., Shappirio, D. G.: Photoperiodic control of the maintenance and termination of larval diapause in *Chironomus tentans*. Nature (London) **207**, 548 (1965)

Fantoni, A., Lunadei, M., Ullu, E.: Control of gene expression during terminal differentiation of erythroid cells. In: Current Topics in Developmental Biology, Vol. IX, pp. 15—38. New York: Academic Press 1975

François, C., Marshall, R. D., Neuberger, A.: Carbohydrates in protein. 4. The determination of mannose in hen's egg albumin by radioisotope dilution. Biochem. J. **83**, 335—342 (1962)

Goetghebuer, M.: Tendipedinae (Chironominae). In: Die Fliegen der palaearktischen Region, Lindner, E. (ed.). Vol. XIIIC. Stuttgart: Schweizerbart 1937

Grossbach, U.: cf. Fig. 6 in: Beermann, W.: Gen-Regulation in Chromosomen höherer Organismen. Jb. Max-Planck-Ges. 1966, pp. 69—87

Grossbach, U.: Cell differentiation in the salivary glands of *Camptochironomus tentans* and *C. pallidivittatus*. Ann. Zool. Fennici **5**, 37—40 (1968)

Grossbach, U.: Chromosomen-Aktivität und biochemische Zelldifferenzierung in den Speicheldrüsen von *Camptochironomus*. Chromosoma **28**, 136—187 (1969)

Grossbach, U.: Chromomeren-Aktivität und zellspezifische Proteine bei *Camptochironomus*. Habilitationsschrift, Univ. Stuttgart-Hohenheim, 1971

Grossbach, U.: In: Tätigkeitsbericht 1970—1971 Max-Planck-Institut für Biologie, Abt. Beermann. Naturwissenschaften **59**, 551—553 (1972)

Grossbach, U.: Chromosome puffs and gene expression in polytene cells. Cold Spring Harbor Symp. Quant. Biol. **38**, 619—627 (1973)

Grossbach, U.: Korrelationen zwischen Chromosomen-Aktivität und Genexpression in differenzierten Zellen. Nachr. Akad. Wissensch. Göttingen, II. Math.-phys. Kl. pp. 176—179, 1975

Harrison, P. M.: The structure of apoferritin: Molecular size, shape, and symmetry from X-ray data. J. Mol. Biol. **6**, 404—422 (1963)

Heitz, E., Bauer, H.: Beweise für die Chromosomennatur der Kernschleifen in den Knäuelkernen von *Bibio hortulanus* L. Z. Zellforsch. **17**, 67—82 (1933)

Helmsing, P. J.: Induced accumulation of nonhistone proteins in polytene nuclei of Drosophila. II. Accumulation of proteins in polytene nuclei and chromatin of different larval tissues. Cell Diff. **1**, 19—24 (1972)

Helmsing, P. J., Berendes, H. D.: Induced accumulation of nonhistone proteins in polytene nuclei of *Drosophila hydei*. J. Cell Biol. **50**, 893—896 (1971)

Henrickson, P. A., Clever, U.: Protease activity and cell death during metamorphosis in the salivary gland of *Chironomus tentans*. J. Insect Physiol. **18**, 1981—2004 (1972)

Hochman, B.: Analysis of a whole chromosome in *Drosophila*. Cold Spring Harbor Symp. Quant. Biol. **38**, 581—589 (1973)

Hofmann, T., Harrison, P. M.: The structure of apoferritin: Degradation into and molecular weight of subunits. J. Mol. Biol. **6**, 256—267 (1963)

Holmgren, N.: Zur Morphologie des Insektenkopfes. II. Einiges über die Reduktion des Kopfes der Dipterenlarven. Zool. Anz. **27**, 343—355 (1904)

Holtzer, H.: Induction of chondrogenesis, a concept in quest of mechanisms. In: Epithelial Mesenchymal Interactions. Billingham, R. (ed.). Baltimore: Williams and Wilkins 1968

Holtzer, H., Weintraub, H., Mayne, R., Mochan, B.: The cell cycle, cell lineages, and cell differentiation. In: Current Topics in Developmental Biology, Vol. VII, pp. 229—256. New York: Academic Press 1972

Judd, B. H., Shen, M. W., Kaufman, T. C.: The anatomy and function of a segment of the X chromosome of *Drosophila melanogaster*. Genetics **71**, 139—156 (1972)

Judd, B. H., Young, M. W.: An examination of the one cistron: one chromomere concept. Cold Spring Harbor Symp. Quant. Biol. **38**, 573—579 (1973)

Kafatos, F. C.: Cocoonase synthesis: Cellular differentiation in developing silk moths. In: Problems in Biology: RNA in Development, Hanly, E. W. (ed.), pp. 111—144. Salt Lake City: Univ. Utah Press 1969

Kafatos, F. C.: The cocoonase zymogen cells of silkmoths: A model of terminal cell differentiation for specific protein synthesis. In: Current Topics in Developmental Biology, Vol. VII, pp. 125—191. New York: Academic Press 1972

Keyl, H.-G.: Verdopplung des DNS-Gehalts kleiner Chromosomenabschnitte als Faktor der Evolution. Naturwissenschaften **51**, 46—47 (1964)

Keyl, H.-G.: A demonstrable local and geometric increase in the chromosomal DNA of *Chironomus*. Experientia **21**, 191—193 (1965)

Kloetzel, J. A., Laufer, H.: Fine structure analysis of larval salivary gland function in *Chironomus thummi*. J. Cell Biol. **39**, 74a (1968)

Kloetzel, J. A., Laufer, H.: Developmental changes in fine structure associated with secretion in larval salivary glands of *Chironomus*. Exptl. Cell Res. **60**, 327—337 (1970)

Klotz, I. M., Darnall, D. W.: Protein subunits: A Table (2nd ed.). Science **166**, 126—128 (1969)

Korge, G.: Chromosome puff activity and protein synthesis in larval salivary glands of *Drosophila melanogaster*. Proc. Natl. Acad. Sci. **72**, 4550—4554 (1975)

Laemmli, U. K.: Cleavage of structural proteins during the assembly of the head of bacteriophage T4. Nature (London) **227**, 680—685 (1970)

Lambert, B.: Repeated DNA sequences in a Balbiani ring. J. Mol. Biol. **72**, 65—75 (1972)

Lambert, B.: Repeated nucleotide sequences in a single puff of *Chironomus tentans* polytene chromosomes. Cold Spring Harbor Symp. Quant. Biol. **38**, 637—644 (1973)

Lambert, B., Beermann, W.: Homology of Balbiani ring DNA in two closely related *Chironomus* species. Chromosoma **51**, 41—47 (1975)

Lambert, B., Wieslander, L., Daneholt, B., Egyházi, E., Ringborg, U.: In situ demonstration of DNA hybridizing in *Chironomus tentans*. J. Cell Biol. **53**, 407—418 (1972)

Laufer, H.: Developmental studies of the Dipteran salivary gland. III. Relationships between chromosomal puffing and cellular function during development. In: Developmental and Metabolic Control Mechanisms and Neoplasia, pp. 237—250. Baltimore: Williams and Wilkins 1965

Laufer, H.: Developmental interactions in the dipteran salivary gland. Am. Zoologist **8**, 257—271 (1968)

Laufer, H., Nakase, Y.: Salivary gland secretion and its relation to chromosomal puffing in the Dipteran, *Chironomus thummi*. Proc. Natl. Acad. Sci. **53**, 511—516 (1965)

Laufer, H., Nakase, Y., Vanderberg, J.: Developmental studies of the dipteran salivary gland. I. The effects of actinomycin D on larval development, enzyme activity, and chromosomal differentiation in *Chironomus thummi*. Develop. Biol. **9**, 367—384 (1964)

Layman, D. L., Narayanan, A. S., Martin, G. R.: The production of lysyl oxidase by human fibroblasts in culture. Arch. Biochem. Biophys. **149**, 97—101 (1972)

Lefevre, G.: The one band—one gene hypothesis: Evidence from a cytogenetic analysis of mutant and nonmutant rearrangement breakpoints in *Drosophila melanogaster*. Cold Spring Harbor Symp. Quant. Biol. **38**, 591—599 (1973)

Lellák, J.: Positive Phototaxis der Chironomiden-Larvulae als regulierender Faktor ihrer Verteilung in stehenden Gewässern. Ann. Zool. Fennici **5**, 84—87 (1968)

Lizardi,P.M., Williamson,R., Brown,D.D.: The size of fibroin messenger RNA and its polyadenylic acid content. Cell **4**, 199—205 (1975)
Lorand,L., Jacobsen,A.: Specific inhibitors and the chemistry of fibrin polymerization. Biochemistry **3**, 1939—1943 (1964)
Lucas,F., Shaw,J.T.B., Smith,S.G.: The silk fibroins. Advan. Protein Chem. **13**, 107—242 (1958)
Lucas,F., Shaw,J.T.B., Smith,S.G.: Comparative studies of fibroins. I. The amino acid composition of various fibroins and its significance in relation to their crystal structure and taxonomy. J. Mol. Biol. **2**, 339—349 (1960)
Mechelke,F.: Reversible Strukturmodifikationen der Speicheldrüsenchromosomen von *Acricotopus lucidus.* Chromosoma **5**, 511—543 (1953)
Mechelke,F.: The timetable of physiological activity of several loci in the salivary gland chromosomes of *Acricotopus lucidus.* Proc. 10th Intern. Congr. Genetics., Vol. II, p. 185, Toronto: Univ. Toronto Press 1958
Mechelke,F.: Beziehungen zwischen der Menge der DNS und dem Ausmaß der potentiellen Oberflächenentfaltung von Riesenchromosomenloci. Naturwissenschaften **46**, 609 (1959)
Mechelke,F.: Spezielle Funktionszustände des genetischen Materials. Wiss. Konf. Ges. Dtsch. Naturf. u. Ärzte. Rottach-Egern 1962, pp. 15—29. Berlin-Göttingen-Heidelberg: Springer 1963
Metz,C.W., Lawrence,E.G.: Preliminary observations on *Sciara* hybrids. J. Heredity **29**, 179—186 (1938)
Miall,L., Hammond,A.: The structure and life-history of the Harlequin fly *(Chironomus).* Oxford: Clarendon Press 1900
Painter,T.S.: A new method for the study of chromosome rearrangements and the plotting of chromosome maps. Science **78**, 585—586 (1933)
Palmén,E., Aho,L.: Studies on the ecology and phenology of the Chironomidae (Dipt.) of the Northern Baltic. 2. Ann. Zool. Fennici **3**, 217—244 (1966)
Palmiter,R.D., Wrenn,J.T.: Interaction of estrogen and progesterone in chick oviduct development. III. Tubular Gland Cell Differentiation. J. Cell Biol. **50**, 598—615 (1971)
Panitz,R.: Innersekretorische Wirkung auf Strukturmodifikationen der Speicheldrüsenchromosomen von *Acricotopus lucidus.* (Chironomide). Naturwissenschaften **47**, 383 (1960)
Panitz,R.: Hormonkontrollierte Genaktivitäten in den Riesenchromosomen von *Acricotopus lucidus.* Biol. Zbl. **83**, 197—230 (1964)
Pankow,W.: Entwicklungsspezifische Balbianiringaktivität und Sekretproteinsynthese in Speicheldrüsen von *Chironomus tentans.* Thesis 5166, Fed. Tech. Univ. Zürich, 1973
Panyim,S., Chalkley,R.: The molecular weights of vertebrate histones exploiting a modified sodium dodecyl sulfate electrophoretic method. J. Biol. Chem. **246**, 7557—7560 (1971)
Parekh,A.C., Glick,D.: Studies in histochemistry LXV. Heparin and hexosamine in isolated mast cells: determination, intracellular distribution, and effects of biological state. J. Biol. Chem. **237**, 280—286 (1962)
Pelling,C.: Chromosomal synthesis of ribonucleic acid as shown by incorporation of uridine labeled with tritium. Nature (London) **184**, 655—656 (1959)
Pelling,C.: Ribonukleinsäure-Synthese der Riesenchromosomen. Autoradiographische Untersuchungen an *Chironomus tentans.* Chromosoma **15**, 71—122 (1964)
Pelling,C.: Puff RNA in polytene chromosomes. Cold Spring Harbor Symp. Quant. Biol. **35**, 521—531 (1970)
Pichon,Y.: The pharmacology of the insect nervous system. In: The Physiology of Insecta. Rockstein,M. (ed.), Vol. IV, pp. 102—174. New York: Academic Press 1974
Piez,K.A.: Cross-linking of collagen and elastin. An. Rev. Biochem. **37**, 547—570 (1968)
Pisano,J.J., Finlayson,J.S., Peyton,M.P.: Chemical and enzymic detection of protein cross-links. Measurement of ε-(γ-glutamyl) lysine in fibrin polymerized by factor XIII. Biochemistry **8**, 871—876 (1969)
Poulson,D.F., Metz,C.W.: Studies on the structure of nucleolus forming regions and related structures in the giant salivary gland chromosomes of Diptera. J. Morph. **63**, 363—395 (1938)
Pozdniakoff,D. de: L'appareil séricigène des larves de *Chironomus.* Ann. Soc. Sci. Bruxelles,T. **52**, Ser. B, 209—214 (1932)

Rambousek,F.: Sitzungsber. königl. böhm. Ges. Wiss., math.-naturw. Klasse 1912. Cited after Beermann (1962)
Reynolds,J.A., Herbert,S., Polet,H., Steinhardt,J.: The binding of divers detergent anions to bovine serum albumin. Biochemistry **6**, 937—947 (1967)
Reynolds,J.A., Tanford,C.: Binding of dodecyl sulfate to proteins at high binding ratios. Possible implications for the state of proteins in biological membranes. Proc. Natl. Acad. Sci. U.S. **66**, 1002—1007 (1970a)
Reynolds,J.A., Tanford,C.: The gross conformation of protein-sodium dodecyl sulfate complexes. J. Biol. Chem. **245**, 5161—5165 (1970b)
Rodems,A.E., Henrikson,P.A., Clever,U.: Proteolytic enzymes in the salivary gland of *Chironomus tentans*. Experientia **25**, 686—687 (1969)
Rutter,W.J., Kemp,J.D., Bradshaw,W.S., Clark,W.R., Ronzio,R.A., Sanders,T.G.: Regulation of specific protein synthesis in cytodifferentiation. J. Cell. Physiol. **72**, Suppl. 1, 1—18 (1968)
Rydlander,L., Edström,J.-E.: Expression of Balbiani ring genes. In: Proc. 10th FEBS Meeting, pp. 149—156. Amsterdam: North Holland Publ. 1975
Schin,K.S., Clever,U.: Lysosomal and free acid phosphatase in salivary glands of *Chironomus tentans*. Science **150**, 1053—1055 (1965)
Schin,K.S., Clever,U.: Ultrastructural and cytochemical studies of salivary gland regression in *Chironomus tentans*. Z. Zellforsch. **86**, 262—279 (1968)
Schin,K.S., Laufer,H.: Studies of programmed salivary gland regression during larval-pupal transformation in *Chironomus thummi*. I. Acid hydrolase activity. Exptl. Cell. Res. **82**, 335—340 (1973)
Schin,K.S., Laufer,H.: Uptake of homologous hemolymph protein by salivary glands of *Chironomus thummi*. J. Insect Physiol. **20**, 405—411 (1974)
Seifter,S., Gallop,P.M.: The structure proteins. In: The Proteins, Neurath,H. (ed.), Vol. IV, pp. 155—458. New York: Academic Press 1966
Shannon,M.P., Kaufman,T.C., Shen,M.W., Judd,B.H.: Lethality patterns and morphology of selected lethal and semi-lethal mutations in the zeste-white region of *Drosophila melanogaster*. Genetics **72**, 615—638 (1972)
Shapiro,A.L., Vinuela,E., Maizel,I.V.: Molecular weight estimation of polypeptide chains by electrophoresis in SDS-polyacrylamide gels. Biochem. Biophys. Res. Commun. **28**, 815—820 (1967)
Siegel,R.C., Pinnell,S.R., Martin,G.R.: Cross-linking of collagen and elastin. Properties of lysyl oxidase. Biochemistry **9**, 4486—4492 (1970)
Sorsa,V., Green,M.M., Beermann,W.: Cytogenetic fine structure and the chromosomal localization of the white gene in *Drosophila melanogaster*. Nature New Biol. **245**, 34—37 (1973)
Suzuki,Y., Brown,D.D.: Isolation and identification of the messenger RNA for silk fibroin from *Bombyx mori*. J. Mol. Biol. **63**, 409—429 (1972)
Suzuki,Y., Gage,L.P., Brown,D.D.: The genes for fibroin in *Bombyx mori*. J. Mol. Biol. **70**, 637—649 (1972)
Tashiro,Y., Morimoto,T., Matsuura,S., Nagata,S.: Studies on the posterior silk gland of the silkworm, *Bombyx mori*. I. Growth of posterior silk gland cells and biosynthesis of fibroin during the fifth larval instar. J. Cell Biol. **38**, 574—588 (1968)
Tashiro,Y., Otsuki,E., Shimadzu,T.: Sedimentation analyses of native silk fibroin in urea and guanidine HCl. Biochim. Biophys. Acta **257**, 198—209 (1972)
Walshe,B.M.: Feeding mechanisms of *Chironomus* larvae. Nature (London) **160**, 474 (1947)
Weber,K., Osborn,M.: The reliability of molecular weight determinations by dodecyl sulfate-polyacrylamide gel electrophoresis. J. Biol. Chem. **244**, 4406—4412 (1969)
Weber,K., Osborn,M.: Proteins and sodium dodecyl sulfate: Molecular weight determination on polyacrylamide gels and related procedures. In: The Proteins, 3rd ed., Neurath,H. (ed.), Vol. VII, pp. 179—223. New York: Academic Press 1975
Wieslander,L., Daneholt,B.: in preparation
Wobus,U., Popp,S., Serfling,E., Panitz,R.: Protein synthesis in the *Chironomus thummi* salivary gland. Mol. G. Genet. **116**, 309—321 (1972)

Balbiani Ring Pattern and Biochemical Activities in the Salivary Gland of *Acricotopus lucidus* (Chironomidae)

W. BAUDISCH

Zentralinstitut für Genetik und Kulturpflanzenforschung der Akademie der Wissenschaften der Deutschen Demokratischen Republik, 4325 Gatersleben, GDR

I. Introduction

The phenomenon of puffing in polytene chromosomes has been interpreted as an expression of differential gene activity in relation to cell function and cell differentiation (Beermann, 1952). A series of observations and experiments demonstrating that the development of a puffing pattern is tissue and phase-specific (Beermann, 1952; Mechelke, 1953; Becker, 1959; Clever, 1961; Gabrusewycz-Garcia, 1964; Berendes, 1965; Ashburner, 1967; Ribbert, 1967) strongly support this hypothesis. The puffed regions of giant chromosomes are sites of RNA synthesis (Pelling, 1959; Sirlin, 1960), the rate of synthetic activity being roughly proportional to the state of expansion of the individual puff (Pelling, 1964; Edström and Daneholt, 1967; Daneholt et al., 1969).

Cytological (Mechelke, 1953, 1958) and biochemical (Baudisch, 1961, 1963a, 1964) investigations on the larval salivary gland of Acricotopus lucidus have demonstrated that the gland is composed of three cell-types which exhibit clear functional as well as chromosomal puffing differences. Therefore, the glands seem especially suited for the investigation of gene activation as related to cell differentiation.

II. The Salivary Gland and Its Balbiani Rings

The larvae of *Acricotopus lucidus* (Diptera, Chironomidae subfamily Orthocladiinae) possess a pair of salivary glands which consist of three morphologically distinct lobes (Fig. 1). The lumina of these lobes are interconnected and have a common duct. According to Mechelke (1953) they are designated main lobe, side lobe, and anterior lobe. The average dry weight of a gland from a larva in the last

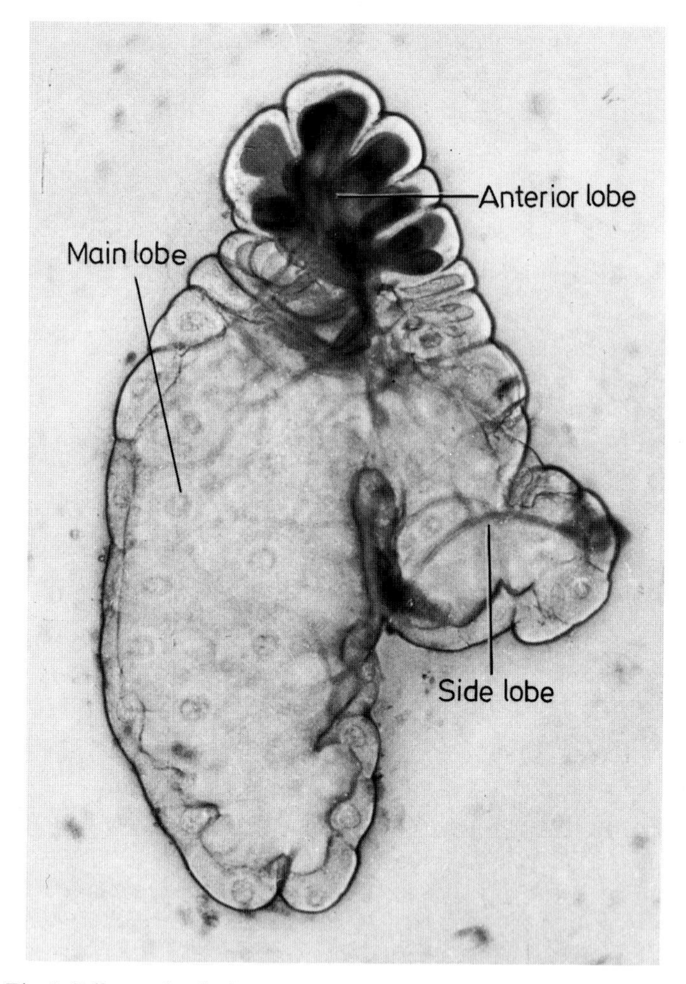

Fig. 1. Salivary gland of *Acricotopus lucidus* (By courtesy F. Mechelke)

stage is 12 µg (Baudisch, 1963b). The main lobe is approximately as large as the two other lobes together. The anterior lobe is the smallest one; the puffing pattern (Mechelke, 1963) as well as its secretory activity differ greatly from those in the other parts of the gland.

A. lucidus has a haploid set of three chromosomes which in the salivary gland develop into large polytene chromosomes. Each lobe of the gland exhibits a specific pattern of puffs in which the Balbiani rings are most conspicuous (Mechelke, 1953, 1963). In the cells of the main lobe, two Balbiani rings (BR1 and BR2) are developed in chromosomes I and II, respectively (Fig. 2). These Balbiani rings are also present in the side lobe; in addition, a third cell-specific Balbiani ring (BR6) is developed in chromosome II (Fig. 3).

An entirely different puffing pattern is found in the anterior lobe (Fig. 3). While the sites of BR1, BR2, and BR6 are not puffed in these cells, different Balbiani rings

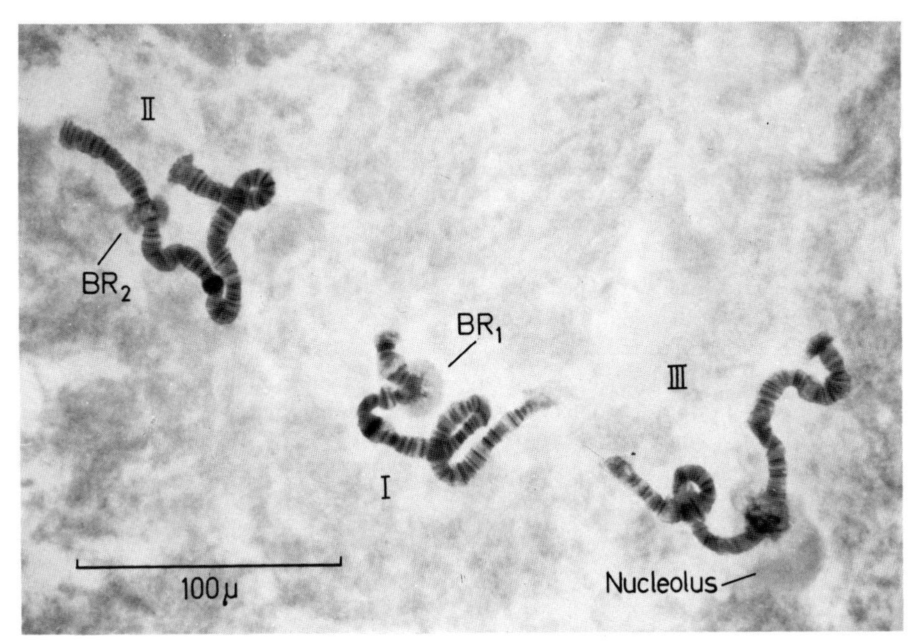

Fig. 2. The three giant chromosomes in the main lobe of the salivary gland from *Acricotopus lucidus*. (By courtesy F. Mechelke)

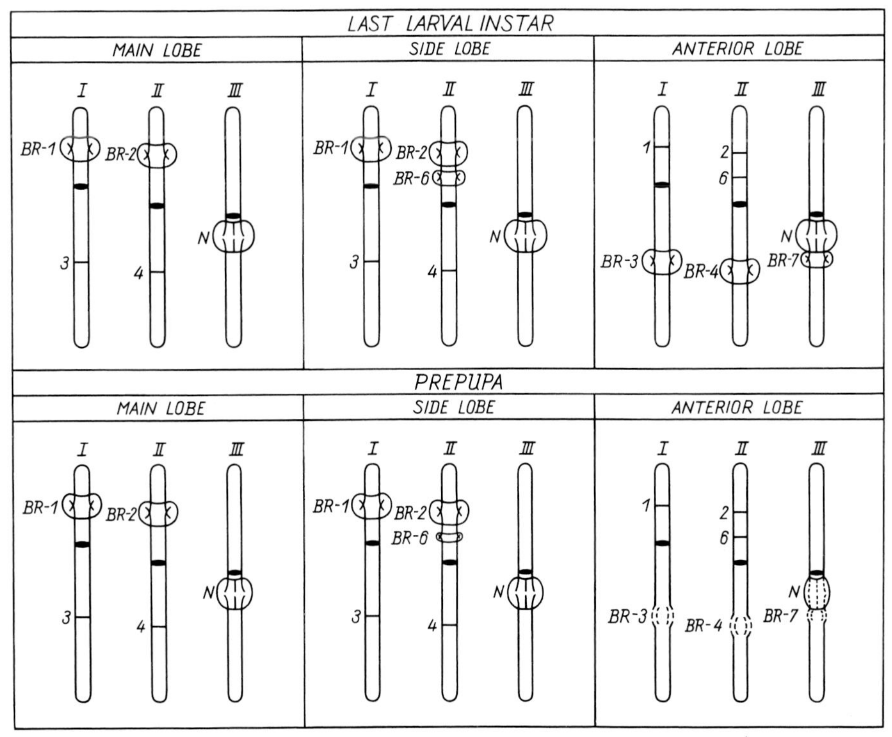

Fig. 3. Balbiani ring pattern in the chromosomes of the three different cell types in larval and prepupal *Acricotopus* salivary glands. BR1–BR7: Balbiani rings 1–7. N: nucleolus. (Adapted from Mechelke, 1967)

are formed in other regions of chromosomes I and II which have been designated Balbiani ring 3 (BR3) and Balbiani ring 4 (BR4), respectively. Furthermore, an additional Balbiani ring (BR7) is found in a site close to the nucleolus in chromosome III (Mechelke, 1953, 1958; Panitz, 1972).

In contrast to the main and side lobes of the gland, where the physiological activity and the puffing pattern remain unchanged during the prepupal stage, characteristic hormone-induced changes (Panitz, 1960; Panitz et al., 1972) occur in the anterior lobe at the beginning of the prepupal stage: The Balbiani rings 3 and 4 as well as the composite structure formed by BR7 and the nucleolus regress and stop synthesizing RNA. The physiological activity in the anterior lobe declines concomitantly and the endoplasmatic reticulum which is characteristic of the gland cells is broken down (Döbel, 1968). The fine-structure of the gland cells finally dissolves and a yellow-colored secretory lumen is formed which contains mainly lysed cell proteins.

The striking yellow color acquired by the secretion in the anterior lobe at the beginning of the prepupal stage is due to the presence of carotinoids (Baudisch, 1963b), the main carotinoid constituent being β-carotine. The composition of carotinoids in the secretion shows great similarity of the carotinoid content of the algae (Spirogyra) upon which *Acricotopus* larvae feed. It appears unlikely that there is a causal relation between the regression of the three Balbiani rings and the appearance of carotinoids in the anterior lobe, at least not in the sense of a direct effect of differential gene suppression. As shown by Panitz (1972), there occurs no yellow coloration of the anterior lobe upon experimental repression of the Balbiani rings by ecdyson treatment.

III. Protein Differentiation of the Salivary Gland and Its Secretion

The functional character of a cell is primarily expressed by the synthesis of specific proteins. For the interpretation of chromosome activity it is, therefore, crucial to know whether the synthesis of proteins in the salivary gland can be correlated with the activity of specific gene loci. The salivary gland of *Acricotopus*, as we have seen, comprises three cell types with different puffing patterns (Mechelke, 1953) and functions (Baudisch, 1964). In analogy to the case of *Chironomus* (Grossbach, 1973), the main function of the salivary gland consists in producing large quantities of structural proteins that are used throughout the entire larval- and prepupal stage to construct typical housing tubes (Baudisch, 1963a). It was, therefore, of interest to find out whether the three lobes of the *Acricotopus* gland contribute different protein components to the saliva as suggested by the differences observed in the Balbiani ring puffing patterns.

In an early attempt to study functional differentiation within the salivary gland of *Acricotopus* the total protein of each of the three lobes was separately hydrolyzed and the hydrolysates separates by two-dimensional chromatography on paper (Baudisch, 1960). It was discovered that, while all regions of the gland contain the amino acids usually found in proteins, hydroxyproline is present in the hy-

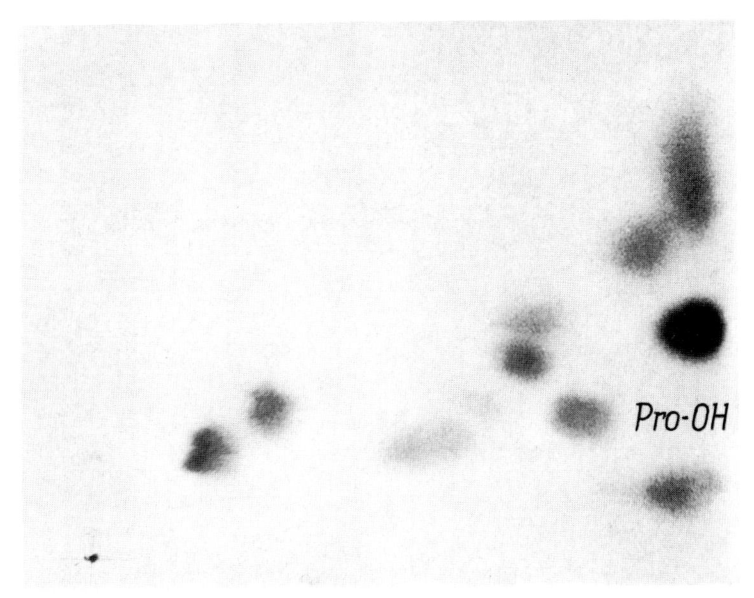

Fig. 4. Paper chromatogram of the amino acids from an acid hydrolysate of main lobes of the salivary gland. Pro-OH = Hydroxyproline

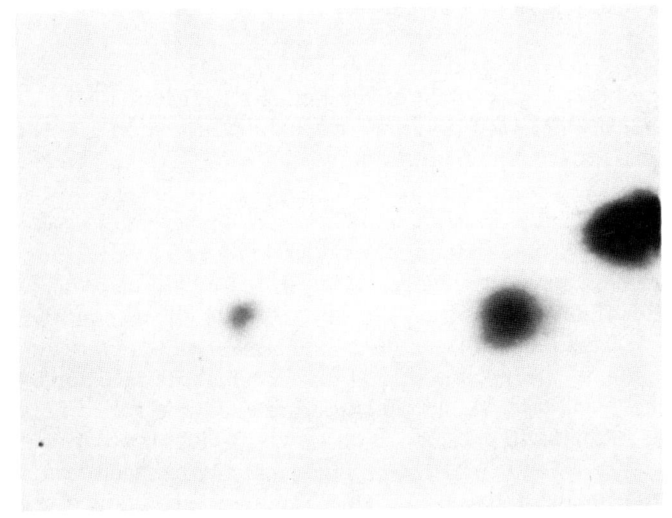

Fig. 5. ^{14}C-labeled amino acids in an acid hydrolysate of *Acricotopus* salivary glands which had been incubated in ^{14}C-proline. Autoradiograph of a paper chromatogram. *Left* to *right*: glutamic acid, hydroxyproline, proline

drolysates from the main and side lobes, but not in those from the anterior lobes (Fig. 4). Aside from the unexpected presence of hydroxyproline, it is of interest to note that the amino acid composition of *Acricotopus* salivary glands, while basically similar to that of the secretion of Drosophila salivary glands (Kodani, 1948), differs markedly from that of the Chironomus salivary gland secretion which

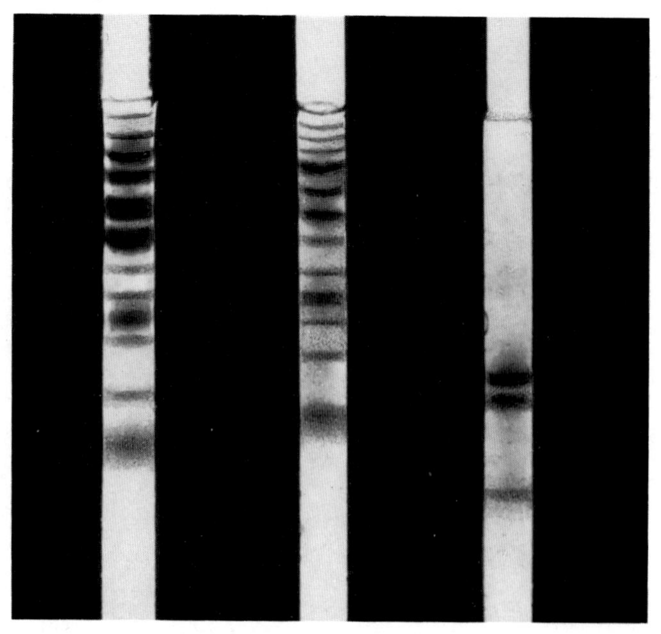

Fig. 6. Electrophoretic separation of the secretory proteins from the three gland lobes of *Acricotopus* in 15% acrylamide. *Left:* main lobe secretion; *middle:* side lobe secretion; *right:* anterior lobe secretion

is characterized by a high content of basic amino-acids (Grossbach, 1969). Nor is there any similarity with the Lepidopteran silk thread secretion, extensive studies of which have proved the dominating amino acids to be glycine, alanine, and serine (Lucas et al., 1958; Suzuki and Brown, 1972).

The hydroxyproline as found in the secretion from the main and side lobes of the *Acricotopus* salivary gland (Baudisch, 1963a) is known to be a component of some structural proteins, especially collagen. It is synthesized during protein synthesis by hydroxylation of proline residues already integrated in the polypeptide chain (Stetten, 1949; Gianetto and Bouthillier, 1954). Incubation of salivary glands in a medium containing ^{14}C-proline (Fig. 5) has demonstrated the same pathway of hydroxyproline synthesis (Baudisch, 1967). The exclusive occurrence of hydroxyproline in the main and side lobes points to qualitative differences in the protein synthesis capacities of the three lobes. Technical difficulties arise due to the small amount of material that is available. The main lobe of the gland, e.g., yields approximately 2 µg of secretory protein. The method used for the analysis of these small amounts of proteins was based on the technique of microelectrophoresis on acrylamide gel as described by Grossbach (1965). The secretory proteins of *Acricotopus* could be dissolved only within a narrow pH-range in the presence of urea and detergents. The electrophoretic separation shows that each lobe of the gland produces a characteristic pattern of proteins.

The main and the side lobe exhibit very similar complements of secretory proteins. This situation may be compared to the similarity between the two cell types in the patterns of major puffs (Balbiani rings). No difference in the proteins of

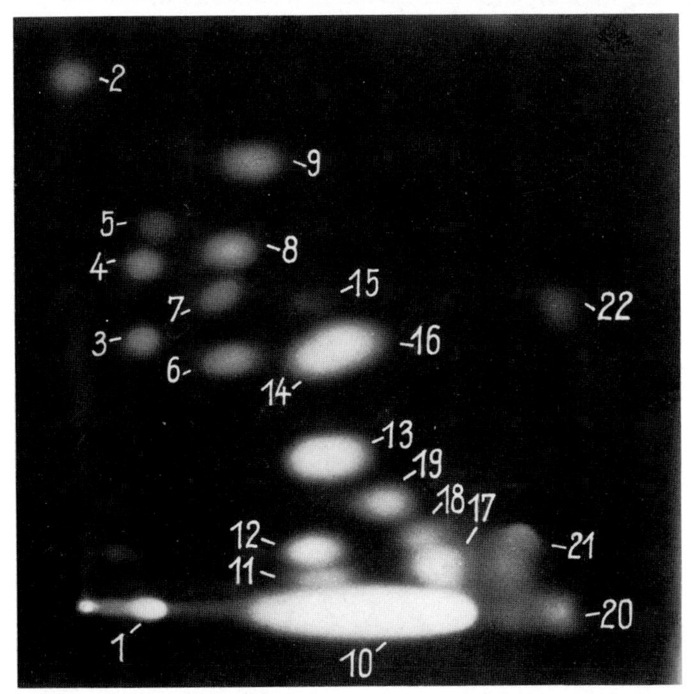

Fig. 7. Amino acid composition of those secretory proteins which are not precipitated by the fixing medium. After electrophoresis of the secretion, the proteins were fixed in the gels. Proteins soluble in the fixing medium were obtained by evaporation of the medium and hydrolysed in acid. The amino acids were dansylated and separated by chromatography on thin layers. *1*: cysteic acid; *2*: unknown; *3*: Phe; *4*: Leu; *5*: Ile; *6*: unknown; *7*: unknown; *8*: Val; *9*: Pro; *10*: dansyl-chloride; *11*: Asp; *12*: Glu; *13*: Gly; *14*: Ala; *15*: dansyl-NH_3; *16*: reference value indentified as β-alanine; *17*: Ser; *18*: Thr; *19*: Pro-OH; *20*: His; *21*: Arg; *22*: Tyr

the last larval and the prepupal stages could be detected in these two gland lobes (Fig. 6).

On the other hand, the cells of the anterior lobe produce a totally different pattern of secretory proteins. None of the three fractions separated can be homologized to any of the secretory constituents from the other parts of the gland. This obviously corresponds to the diversity of the puffing pattern: Balbiani rings 3, 4, and 7 are developed only in the anterior lobe while all the Balbiani rings characteristic of the other lobes are lacking here.

The pattern of proteins from the anterior lobe was found in the fourth larval instar only. With the onset of lysis in the prepupal stage, defined secretory fractions were no longer detectable.

It is interesting to note that the number of protein fractions in the main and side lobes does not correspond to the number of Balbiani rings. The main lobe of the gland, e.g., exhibits two Balbiani rings but 12 fractions of secretory proteins. In the anterior lobe, on the other hand, the number of Balbiani rings equals the number of secretory fractions. This must be said with reservations, however, since the anterior lobe, being the smallest, yields only submicrogram quantities of secretion for analysis and could conceivably contain smaller protein fractions in undetectable quantities.

The presence of hydroxyproline in the gland secretions from the main and side lobes was established early in the biochemical work on *Acricotopus* (Baudisch, 1960). After the electrophoretic separation of the secretion it was surprisingly found that none of the stained proteins contained hydroxyproline (Baudisch and Herrmann, 1972). It became apparent that the protein or proteins containing hydroxyproline were not precipitated by the medium used for fixation in the gels (acetone:water:acetic acid 70:23:7, trichloracetic acid 20%, sulfosalicilic acid 20%). As shown in Figure 7, by a micromethod according to (Neuhoff et al., 1969), hydroxyproline was detected in the fixation medium by chromatography (Woods and Wang, 1967) of the samples after evaporation, acid hydrolysis of the residue, and dansylation (Gray and Hartley, 1963). During electrophoresis the nonfixable proteins moved approximately as far as the middle protein fractions. According to this migratory distance and the acrylamide concentration of the gels, their molecular weight should be around 35000 Daltons.

IV. Balbiani Ring Patterns and Protein Synthesis

As pointed out in the foregoing chapter, the observations on the parallels as well as the discrepancies between Balbiani ring and protein patterns in the various portions of the Acricotopus gland raised the question of a possible interpretation in terms of gene activation pattern. This required the demonstration of a more direct relationship between the activity of a given Balbiani ring locus and the presence of one, or more, individual protein fractions, either using cytogenetic techniques (cf. Beermann, 1961), or by means of an experimental repression or induction of specific Balbiani rings. Both lines of attack have been tried.

It has been stated that chromosomes I and II in the main and side lobes of the salivary gland develop Balbiani ring 1 and 2 respectively. The sizes of these two Balbiani rings are subject to fluctuations that are frequently correlated to an inversion in the left, shorter arm of the first chromosome in the region of Balbiani ring 1 (Mechelke, 1953; Wobus et al., 1971a; Panitz, 1972). This inversion can appear either heterozygous (type B) or homozygous (type A) with a subterminal Balbiani ring 1, while in the normal situation this Balbiani ring is more or less medially situated (type C; Mechelke, 1953, 1960). In *Acricotopus* strains which had been bred several years in the laboratory, two groups of type C animals occurred: one with a comparatively small Balbiani ring 1 (BR1 < BR2) and one with a comparatively large Balbiani ring 1 (BR1 > BR2). In strains containing a chromosome I of type A, most animals had developed a small Balbiani ring 1. If it is assumed that at least a part of the RNA synthesized in the Balbiani rings functions as a messenger for the synthesis of cell-specific proteins (Edström and Tanguay, 1974), it can be argued that size differences of the two Balbiani rings with parallel differences in RNA-synthesis rates should result in quantitative differences in the protein composition of the gland secretion. This question was investigated in experiments in which the secretion from one gland was separated electrophoretically, while the other gland from the same animal was used as cytological control. To rule out position-effects, all experiments were carried out on animals of one structure-type (Figs. 8 and 9). The preliminary results of these investigations confirm earlier observations that the proteins from the main lobe can be separated into two groups according to solubility. Band 1 to 4 are relatively easy to dissolve, while bands 7 to 12 are more difficult to dissolve. Bands 5 and 6 are the most prominent and contain approximately half of the total protein of the secretion. According to the migratory distance and the acrylamide concentration (Shapiro et al., 1967; Ingram et al., 1967), the molecular weight of fractions 5 and 6 is calculated to be approximately 30000 Daltons.

After staining the electropherograms with Coomassie Brilliant Blue, the relative quantitative amounts of the 12 protein fractions from the main lobe were compared by densitometry. It was found that the amounts of the different protein fractions were distinctly dependent upon the sizes of the Balbiani rings 1 and 2. Fractions 1 to 4 appeared especially intense in animals where Balbiani ring 1 was large (Fig. 10). On the other hand, fractions 7–12 were most prominent when Balbiani ring 2 was large (Fig. 11). It cannot yet be decided whether the intensities of fractions 5 and 6 change with different sizes of the Balbiani rings in the main lobe.

Several substances, among them the insect molting hormone ecdysone (Clever, 1963; Panitz et al., 1972) are capable of significantly changing the puffing pattern of giant chromosomes. However, a direct influence on the puffing behavior of Balbiani rings has in no case been demonstrated. Therefore, it is the more surprising that a group of plant growth substances, the gibberellins, have been shown to cause the specific regression of Balbiani ring 2 in a salivary gland of Acricotopus (Panitz, 1967). Autoradiographs following administration of ^{3}H-uridine under these conditions show a complete block of RNA-synthesis (Panitz, 1972) even in nuclei where the morphological structure of the Balbiani ring is maintained almost normally. This inhibition is completely reversible.

A short-time exposure to gibberellins was found to inhibit exclusively the RNA-synthesis activity of Balbiani ring 2. A slight regression of Balbiani ring 1 was observed only after a longer exposure to gibberellins.

Balbiani ring 2 is characteristic of the main and side lobes in the salivary gland. The cells of these lobes specifically synthesize a protein which can be identified by its content of hydroxyproline. Therefore, the influence of the gibberellin-treatment on the cell-specific protein synthesis could easily be tested in this case (Baudisch and Panitz, 1968). After 30 min incubation with gibberellin, a distinct inhibition of hydroxy-proline-synthesis from ^{14}C-proline was found (Fig. 12). After one hour of incubation, hydroxyproline-synthesis was no longer detectable at all (Fig. 13).

The selective blocking of RNA-synthesis in a large Balbiani ring made it possible to ascertain the influence of this gene-locus on the composition of the gland secretion. The results can be explained by the mechanism of hydroxyproline-synthesis during protein synthesis and justify the assumption that, following the inhibition of RNA-synthesis in Balbiani ring 2, the production of hydroxyproline-containing protein is suspended. The inactivation of Balbiani ring 2 and the subsequent inhibition of hydroxyproline synthesis strongly suggest a causal relationship between the function of this gene locus and cell-specific activity.

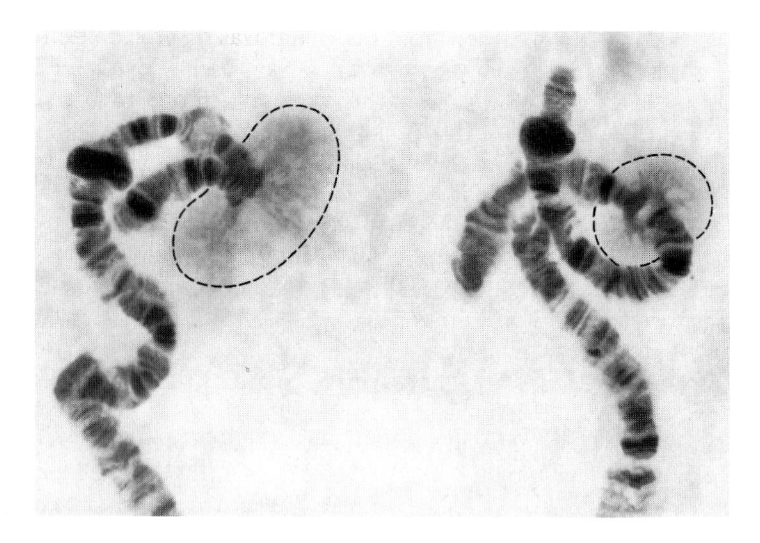

Fig. 8. Chromosomes I and II with Balbiani rings BR1 *(left)* and BR2 *(right)*. Structure-type C (BR1 > BR2)

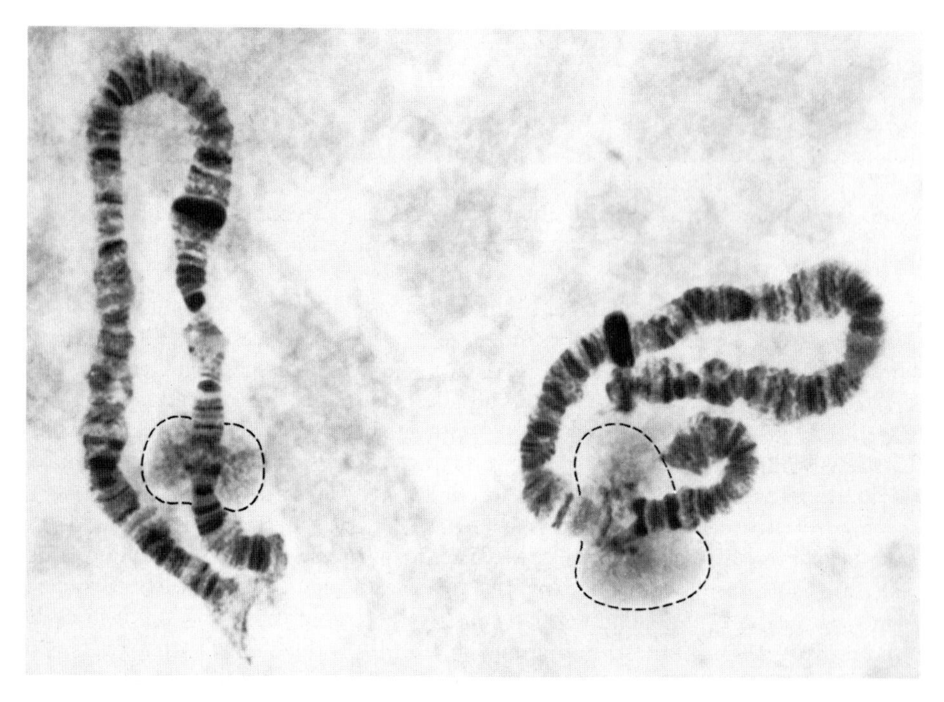

Fig. 9. Chromosomes I and II with Balbiani rings BR1 *(left)* and BR2 *(right)*. Structure-type C (BR1 < BR2)

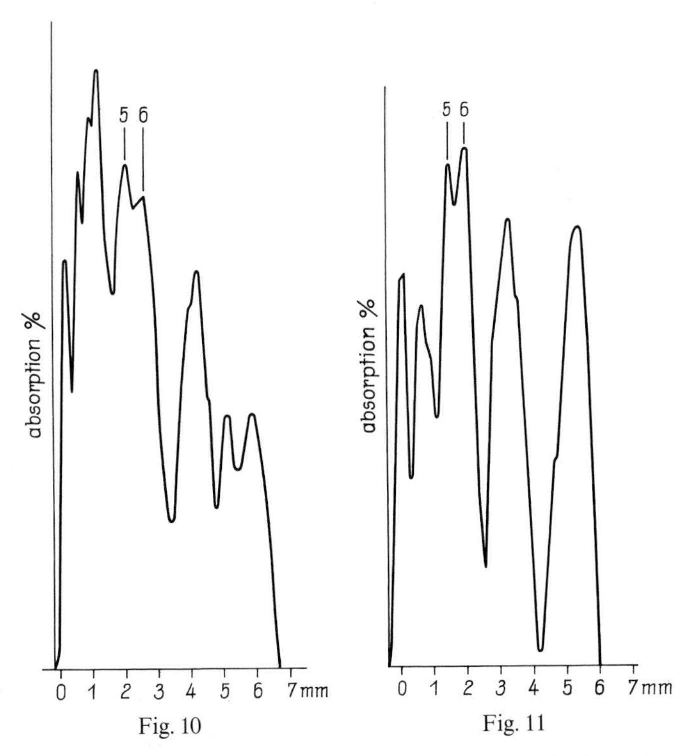

Fig. 10 Fig. 11

Figs. 10 and 11. Electrophoretic separation of the secretory proteins from the main lobe of the salivary gland. Densitometer tracks after staining the proteins with Coomassie Brilliant Blue

Fig. 10. Relative dominance of fractions 1–4 in glands which had large Balbiani ring 1 (BR1 > BR2)

Fig. 11. Prominence of fractions 7–12 in glands which had large Balbiani ring 2 (BR2 > BR1)

Although the mechanism of gibberellin action in the preparations used is not known, it is striking that such a short treatment, a maximum of one hour, was sufficient to fully block hydroxyproline synthesis. For this reason further experiments with actinomycin D were carried out. Actinomycin D inhibits RNA-synthesis in all chromosome loci of the salivary gland in vitro (Clever, 1964, 1966; Kiknadze, 1965; Serfling, 1968). Upon this treatment, a rapid decrease of hydroxyproline synthesis was observed. One hour after administration of the drug, hydroxyproline synthesis was totally inhibited (Baudisch, in prep.). The similar effect of two very different substances, gibberellin and actinomycin D, on the synthesis of a specific protein suggests a rather short life time of the messenger-RNA for this protein. On the other hand, highly stable m-RNA molecules may be present in Chironomus salivary glands (Clever, 1969; Clever et al., 1969; Doyle and Laufer, 1969; Wobus et al., 1972). Under the influence of actinomycin D, the incorporation of radioactively labeled amino acids in the gland-specific secretory proteins was only 20–40% below that in untreated salivary glands (Clever et al., 1969).

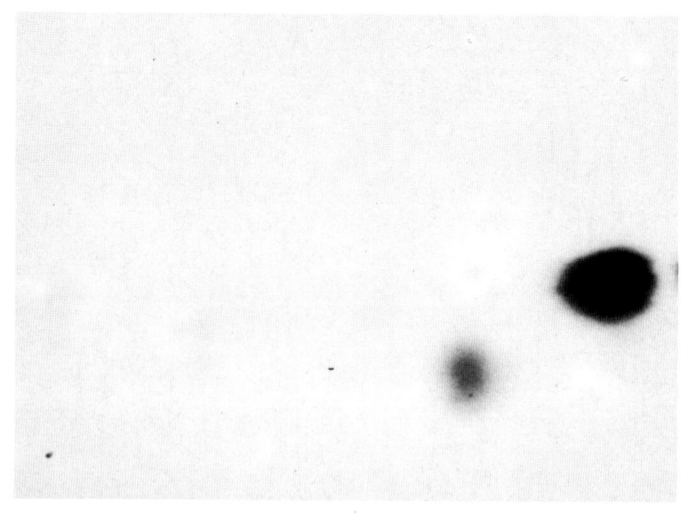

Fig. 12

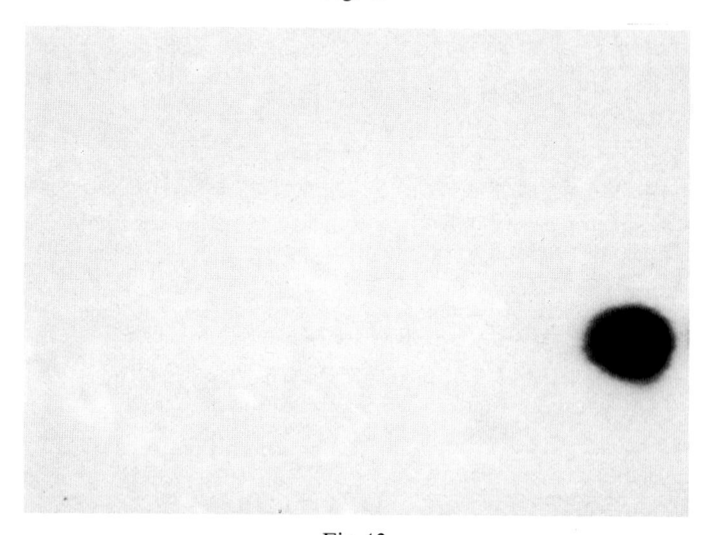

Fig. 13

Figs. 12 and 13. Influence of gibberelline on hydroxyproline synthesis in the gland. ^{14}C-labeled amino acids in an acid hydrolysate of glands which had been incubated in ^{14}C-proline. Autoradiograph of a paper chromatogram

Fig. 12. 30-min treatment with gibberelline

Fig. 13. 60-min treatment with gibberelline

V. Conclusions

Correlations of active gene loci and gene products have been investigated primarily in polytene cell types. Such correlations, which bear importance on the hypothesis of differential gene activity (Beermann, 1959), have been demonstrated in a number of cases. With cytogenetic methods, Beermann (1961) has shown a morphologically visible constituent of the salivary gland secretion to depend on a cell specific Balbiani ring. Grossbach (1968, 1969, 1973) has isolated and separated the secretory proteins and has in this and other cases demonstrated correlations between Balbiani rings and gland specific proteins. From these and other investigations (Wobus et al., 1970; Wobus et al., 1971b) it became clear that there is no simple numerical relation between active chromosome loci and proteins. Instead, Balbiani rings may contain the genetic information for several proteins.

The biochemical analysis of the three lobes of the salivary gland of *Acricotopus* showed a clear physiological differentiation of the cytologically distinct cell types (Baudisch, 1963a, b, 1967). This differentiation must be based on a differential activation of genes in the gland lobes. The production of a hydroxyproline-containing protein, which shows an extraordinary precipitation behaviour (Baudisch and Herrmann, 1972), only in two of the three cell types, is the most obvious character of the intraglandular differentiation. Selective inhibition of Balbiani ring 2 (Panitz, 1967) demonstrated the dependency of hydroxyproline synthesis on the activity of this locus (Baudisch and Panitz, 1968) and thus rendered likely that the genetic information for a hydroxyproline-containing protein (or proteins) is present in this Balbiani ring. From the short time interval between Balbiani ring regression and cessation of hydroxyproline synthesis, it follows that the half life of the messenger-RNA involved must in this instance be rather short.

Electrophoretic analysis of the secretory proteins from the three gland lobes substantiated the physiological differences of the cell types. The main and side lobes which differ only in one Balbiani ring produce a very similar set of proteins, while the cytologically distinct anterior lobe exhibits a completely different pattern of secretory components. Furthermore, the relative quantities of several proteins in the main lobe seem to mirror size differences of Balbiani rings 1 and 2 and concomitant variations in the RNA synthesis rates of these loci. In animals with large Balbiani rings 1, a predominance of protein fractions 1–4 was regulary observed, while the development of large Balbiani ring 2 was found to be correlated with a preponderance in the pattern of fractions 7–12.

The results obtained so far from experimental investigations of *Acricotopus* salivary glands and of other Chironomids indicate that the structural proteins of the salivary secretion are coded in those chromosomal loci which show a characteristically high puffing activity in these glands.

References

Ashburner, M.: Patterns of puffing activity in the salivary gland chromosomes of *Drosophila*. I. Autosomal puffing patterns in a laboratory stock of D. melanogaster. Chromosoma **21**, 398–428 (1967)

Ashburner, M.: Puffing patterns in *Drosophila melanogaster* and related species. In: Results and Problems in Cell Differentiation. Vol. IV, pp. 101–151. Berlin-Heidelberg-New York: Springer 1972

Baudisch, W.: Spezifisches Vorkommen von Carotinoiden und Oxyprolin in den Speicheldrüsen von *Acricotopus lucidus*. Naturwissenschaften **47**, 21 (1960)

Baudisch, W.: Synthese von Oxyprolin in den Speicheldrüsen von *Acricotopus lucidus*. Naturwissenschaften **47**, 2 (1961)

Baudisch, W.: Aminosäurezusammensetzung der Speicheldrüsen von *Acricotopus lucidus*. Biol. Zbl. **82**, 351–361 (1963a)

Baudisch, W.: Chemisch-physiologische Untersuchungen an den Speicheldrüsen von *Acricotopus lucidus*. 100 Jahre Landwirtsch. Inst. Univ. Halle, pp. 152–159. Halle 1963b

Baudisch, W.: Untersuchungen zur physiologischen Charakterisierung der einzelnen Speicheldrüsenlappen von *Acricotopus lucidus*. Struktur und Funktion des genetischen Materials. Erwin-Baur-Gedächtnisvorlesungen 3, pp. 231–234. Berlin: Akademie-Verlag 1964

Baudisch, W.: Spezifische Hydroxyprolinsynthese in den Speicheldrüsen von *Acricotopus lucidus*. Biol. Zbl. **86**, (Suppl.) 157–162 (1967)

Baudisch, W., Herrmann, F.: Über Proteine mit ungewöhnlichem Löslichkeitsverhalten im Speicheldrüsensekret von *Acricotopus lucidus* (Chironomidae). Acta Biol. Med. Germ. **29**, 521–525 (1972)

Baudisch, W., Panitz, R.: Kontrolle eines biochemischen Merkmals in den Speicheldrüsen von *Acricotopus lucidus* durch einen Balbiani-Ring. Exptl. Cell Res. **49**, 470–476 (1968)

Bauer, H.: Der Aufbau der Chromosomen aus den Speicheldrüsen von *Chironomus thummi* Kiefer Z. Zellforschung **23**, 280–313 (1935)

Becker, H.-J.: Die Puffs der Speicheldrüsenchromosomen von *Drosophila melanogaster*. I. Beobachtungen zum Verhalten des Puffmusters im Normalstamm und in zwei Mutanten, giant und lethal-giant-larvae. Chromosoma **10**, 654–678 (1959)

Beermann, W.: Chromosomenkonstanz und spezifische Modifikationen der Chromosomenstruktur in der Entwicklung von Chironomus tentans. Chromosoma **5**, 139—198 (1952)

Beermann, W.: Chromosomal differentiation in insects. In: Developmental Cytology, Rudnick, D. (ed.), pp. 83–103. New York: Ronald 1959

Beermann, W.: Ein Balbianiring als Locus einer Speicheldrüsenmutation. Chromosoma **12**, 1–25 (1961)

Berendes, H. D.: Salivary gland function and chromosomal puffing patterns in *Drosophila hydei*. Chromosoma **17**, 35–77 (1965)

Clever, U.: Genaktivitäten in den Riesenchromosomen von *Chironomus tentans* und ihre Beziehung zur Entwicklung. I. Genaktivierung durch Ecdyson. Chromosoma **12**, 607–675 (1961)

Clever, U.: Von der Ecdysonkonzentration abhängige Genaktivitätsmuster in den Speicheldrüsen von *Chironomus tentans*. Develop. Biol. **6**, 73–98 (1963)

Clever, U.: Actinomycin and puromycin: Effect on sequential gene activation by ecdysone. Science **146**, 794–795 (1964)

Clever, U.: Induction and repression of a puff in *Chironomus tentans*. Develop. Biol. **14**, 421–438 (1966)

Clever, U.: Chromosome activity and cell function in polytenic cells. II Exptl. Cell Res. **55**, 317–322 (1969)

Clever, U., Storbeck, I., Romball, C. G.: Chromosome activity and cell function in polytenic cells. I. Protein synthesis at various stages of larval development. Exptl. Cell Res. **55**, 306–316 (1969)

Daneholt, B., Edström, J.-E., Egyhazi, E., Lambert, B., Ringborg, U.: Chromosomal RNA synthesis in polytene chromosomes of *Chironomus tentans*. Chromosoma **28**, 399–417 (1969)

Döbel, P.: Über die plasmatischen Zellstrukturen der Speicheldrüse von *Acricotopus lucidus*. Kulturpflanze **16**, 203–214 (1968)

Doyle, D., Laufer, H.: Requirements of ribonucleic acid synthesis for the formation of salivary gland specific proteins in larval *Chironomus tentans*. Exptl. Cell Res. **57**, 205–210 (1969)

Edström, J.-E., Daneholt, B.: Sedimentation properties of the newly synthesized RNA from isolated nuclear components of *Chironomus tentans* salivary gland cells. J. Mol. Biol. **28**, 331–343 (1967)

Edström, J.-E., Tanguay, R.: Cytoplasmic ribonucleic acids with messenger characteristics in salivary gland cells of *Chironomus tentans*. J. Mol. Biol. **84**, 569–583 (1974)

Gabrusewycz-Garcia, N.: Cytological and autoradiographic studies in *Sciara coprophila* salivary gland chromosomes. Chromosoma **15**, 312–344 (1964)

Gianetto, R., Bouthillier, L.P.: Some aspects of the prolin metabolism. Can. J. Biochem. Physiol. **32**, 154–160 (1954)

Gray, W.R., Hartley, B.S.: A Fluorescent End-Group Reagent for Proteins and Peptides. Biochem. J. **89**, P_{59} (1963)

Grossbach, U.: Acrylamide gel electrophoresis in capillary columns. Biochem. Biophys. Acta **107**, 180–182 (1965)

Grossbach, U.: Cell differentiation in the salivary glands of *Camptochironomus tentans* and *C. pallidivittatus*. Ann. Zool. Fenn. **5**, 37–40 (1968)

Grossbach, U.: Chromosomen-Aktivität und biochemische Zelldifferenzierung in den Speicheldrüsen von *Camptochironomus*. Chromosoma **28**, 136–187 (1969)

Grossbach, U.: Chromosome puffs and gene expression in polytene cells. Cold Spring Harbor. Symp. Quant. Biol. **38**, 619–627 (1973)

Ingram, L., Tombs, M.P., Hurst, A.: Mobility-molecular weight relationships of small proteins and peptides in acrylamide-gel electrophoresis. Analyt. Biochem. **20**, 24–29 (1967)

Kiknadze, I.I.: Functional changes of giant chromosomes under conditions of inhibited RNA synthesis. Citologija **7**, 311–318 (1965)

Kodani, M.: The protein of the salivary secretion in *Drosophila*. Proc. Nat. Acad. Sci. **34**, 131—135 (1948)

Lucas, F., Shaw, J.T.B., Smith, M.S.G.: The silk fibroins. Advan. Prot. Chem. **12**, 107–242 (1958)

Mechelke, F.: Reversible Strukturmodifikationen der Speicheldrüsenchromosomen von *Acricotopus lucidus*. Chromosoma **5**, 511–543 (1953)

Mechelke, F.: The timetable of physiological activity of several loci in the salivary gland chromosomes of *Acricotopus lucidus*. In: Proc. Tenth Intern. Congr. Gen., Vol. II, p. 185. Toronto: Univ. Toronto, 1958

Mechelke, F.: Beziehungen zwischen der Menge der DNS und dem Ausmaß der potentiellen Oberflächenentfaltung von Riesenchromosomen-Loci. Naturwissenschaften **47**, 334–335 (1960)

Mechelke, F.: Spezielle Funktionszustände des genetischen Materials. Wiss. Konf. Ges. deut. Naturf. u. Ärzte, Rottach-Egern, pp. 15–29. Berlin-Göttingen-Heidelberg: Springer 1963

Mechelke, F.: Biologische Grundfragen aus der Sicht der Genetik. Landwirtschaftl. Hochschule Hohenheim. Reden und Abhandlungen. Stuttgart: Ulmer 1967

Neuhoff, V., Haar, F., Schlimme, E., Weise, M.: Zweidimensionale Chromatographie von Dansyl-Aminosäuren im pico-Mol-Bereich, angewandt zur direkten Charakterisierung von Transfer-Ribonucleinsäuren. Hoppe-Seylers, Z. Physiol. Chem. **350**, 121–128 (1969)

Panitz, R.: Innersekretorische Wirkung auf Strukturmodifikationen der Speicheldrüsenchromosomen von *Acricotopus lucidus*. Naturwissenschaften **47**, 383 (1960).

Panitz, R.: Hormonkontrollierte Genaktivitäten in den Riesenchromosomen von *Acricotopus lucidus*. Biol. Zbl. **83**, 197–230 (1964)

Panitz, R.: Funktionelle Veränderungen an Riesenchromosomen nach Behandlung mit Gibberellinen. Biol. Zbl. **86**, (Suppl.), 147–156 (1967)
Panitz, R.: Balbiani ring activities in Acricotopus lucidus. In: Results and Problems in Cell Differentiation, Vol. IV, pp. 209–227. Berlin-Heidelberg-New York: Springer 1972
Panitz, R., Wobus, U., Serfling, E.: Effect of Ecdysone and analogues on two Balbiani rings of *Acricotopus lucidus*. Exptl. Cell Res. **70**, 154–160 (1972)
Pelling, C.: Chromosomal synthesis of ribonucleic acid as shown by the incorperation of uridine labelled with tritium. Nature (London) **184**, 655–656 (1959)
Pelling, C.: Ribonukleinsäuresynthese der Riesenchromosomen. Autoradiographische Untersuchungen an *Chironomus tentans*. Chromosoma **15**, 71–122 (1964)
Pozdniakoff, D.: L'appareil séricigène des larves de *Chironomus* Ann. Soc. Sci. Bruxelles **52**, 214–219 (1932)
Ribbert, D.: Die Polytänchromosomen der Borstenbildungszellen von Calliphora erythrocephala unter besonderer Berücksichtigung der geschlechtsgebundenen Heterozygotie und des Puffmusters während der Metamorphose. Chromosoma **21**, 296–344 (1967)
Ritossa, F., Pulitzer, J. P.: Aspects of structure of polytene chromosome puffs of *Drosophila busckii* derived from experiments with antibiotics. J. Cell Biol. **19**, 60A (1963)
Serfling, E.: Die Induktion funktioneller Veränderungen an Riesenchromosomen. Diplomarbeit, Martin-Luther-Univ. Halle-Wittenberg, 1968
Shapiro, A. L., Vinuela, E., Maizel, J. V.: Molecular weight estimation of polypeptide chains by electrophoresis in SDS-polyacrylamide gels. Biochem. Biophys. Res. Commun. **28**, 815–820 (1967)
Sirlin, J. L.: Cell sites of RNA and protein synthesis in the salivary gland of Smittia (Chironomidae). Exptl. Cell Res. **19**, 177–180 (1960)
Stetten, M. R.: Some aspects of the metabolism of hydroxyproline, studied with the acid of isotopic nitrogen. J. Biol. Chem. **181**, 31–37 (1949)
Suzuki, Y., Brown, D. D.: Isolation and identification of the messenger RNA for silk fibroin from *Bombyx mori*. J. Mol. Biol. **63**, 409–429 (1972)
Wobus, U., Panitz, R., Serfling, E.: Tissue specific gene activities and proteins in the *Chironomus* salivary gland. Mol. Gen. Genetics **107**, 215–223 (1970)
Wobus, U., Popp, S., Serfling, E., Panitz, R.: Protein synthesis in the *Chironomus thummi* salivary gland. Mol. Gen. Genetics **116**, 309–321 (1972)
Wobus, U., Serfling, E., Baudisch, W., Panitz, R.: Chromosomale Strukturumbauten bei *Acricotopus lucidus* korreliert mit Änderungen im Proteinmuster. Biol. Zbl. **90**, 433–441 (1971a)
Wobus, U., Serfling, E., Panitz, R.: The salivary gland proteins of a *Chironomus thummi* strain with an additional Balbiani ring. Exptl. Cell Res. **65**, 240–245 (1971b)
Woods, K. R., Wang, K. T.: Separation of dansyl-amino acids by polyamide layer chromatography. Biochim. Biophys. Acta **133**, 369–370 (1967)

Subject Index

Results and Problems in Cell Differentiation

A Series of Topical Volumes in Developmental Biology

Editors: W. Beermann, W. Gehring, J. B. Gurdon, F. C. Kafatos, J. Reinert

Springer-Verlag
Berlin
Heidelberg
New York

Volume 1
The Stability of the Differentiated State
Editor: H. Ursprung
56 figures. XI, 144 pages. 1968
ISBN 3-540-04315-2

Volume 2
Origin and Continuity of Cell Organelles
Editors: J. Reinert and H. Ursprung
135 figures. XIII, 342 pages. 1971
ISBN 3-540-05239-9

Volume 3
Nucleic Acid Hybridization in the Study of Cell Differentiation
Editor: H. Ursprung
29 figures. XI, 76 pages. 1972
ISBN 3-540-05742-0

Volume 4
Developmental Studies on Giant Chromosomes
Editor: W. Beermann
110 figures. XV, 227 pages. 1972
ISBN 3-540-05748-X

Volume 5
The Biology of Imaginal Disks
Editors: H. Ursprung and R. Nöthiger
56 figures. XVII, 172 pages. 1972
ISBN 3-540-05785-4

Volume 6
W. J. Dickinson, D. T. Sullivan
Gene-Enzyme Systems in Drosophila
32 figures. XI, 163 pages. 1975
ISBN 3-540-06977-1

Volume 7
Cell Cycle and Cell Differentiation
Editors: J. Reinert and H. Holtzer
92 figures. XI, 331 pages. 1975
ISBN 3-540-07069-9

N. J. STRAUSFELD

Atlas of an Insect Brain

81 figures, partly coloured, 71 plates.
XII, 214 pages. 1976
ISBN 3-540-07343-4

This atlas is the first presentation of the main regions and pathways of an arthropod brain to combine both mass-staining of fibres and selective impregnation of neurons. It displays in detail the basic structures of the neuropils and schematizes them into a comprehensive and simple plan of sensory compartments and core neuropil. The main section of the book illustrates serial sections through the brain of the fly *Musca domestica* with reference to a coordinate system that relates covert structures to the head capsule. There follow detailed drawings of the forms and locations of Golgi-stained elements in the brain. The introductory chapters summarize the history of insect neuroanatomy, sketch the cellular constituents of neuropils, and outline the neuropil's basic organization. There is an appendix on histological methods applicable to insects and a multilingual glossary of terms relating to brain structures. The atlas is richly illustrated with 160 carefully prepared photographs and many beautiful drawings.

Contents:

Springer-Verlag
Berlin
Heidelberg
New York